An Overcrowded World ?

Population, Resources and the Environment

edited by
Philip Sarre and John Blunden

The five volumes of the series form part of the second-level Open University course D215 *The Shape of the World*. If you wish to study this or any other Open University course, details can be obtained from the Central Enquiry Service, PO Box 200, The Open University, Milton Keynes, MK7 6YZ.

For availability of the video- and audiocassette materials, contact Open University Educational Enterprises Ltd (OUEE), 12 Cofferidge Close, Stony Stratford, Milton Keynes, MK11 1BY.

Oxford University Press, Walton Street, Oxford OX2 6DP
Oxford New York
Athens Auckland Bangkok Bombay
Calcutta Cape Town Dar es Salaam Delhi
Florence Hong Kong Istanbul Karachi
Kuala Lumpur Madras Madrid Melbourne
Mexico City Nairobi Paris Singapore
Taipei Tokyo Toronto
and associated companies in
Berlin Ibadan

Oxford is a trade mark of Oxford University Press

Published in the United States
by Oxford University Press Inc., New York

Published in association with The Open University

First published 1995

Edited, designed and typeset by The Open University

Printed in the United Kingdom by
The Alden Press Limited

A catalogue record for this book is available from the British Library

Library of Congress Cataloguing in Publication Data applied for

ISBN 0 19 874189 8 (paper)

ISBN 0 19 874188 X (cloth)

9789C/d215v3pi1.1

An Overcrowded World?

Population, Resources and the Environment

Contents

Preface

An Overcrowded World? Population, Resources and the Environment is the third of five volumes in a new series of human geography teaching texts. The series, entitled *The Shape of the World: Explorations in Human Geography* is designed as an introduction to the principal themes of geographical thought: namely, those of space, place and the environment. The five volumes form the central part of an Open University course, with the same title as that of the series. Each volume, however, is free-standing and can be studied on its own or as part of a wide range of social science courses in universities and colleges.

The series is built around an exploration of many of the key issues which are shaping our world as we move into the twenty-first century and which, above all else, are geographical in character. Each volume in various ways engages with taken-for-granted notions such as those of nature, distance, movement, sustainability, the identity of places and local cultures to put together what may be referred to as the building blocks of our geographical imagination.

In fact, our understanding of the nature of the geographical imagination is one of three shared features which distinguish the five volumes as a series. In developing the contribution that geography can make to our understanding of a changing world and our place within it, each volume has something distinct to offer. A second feature of the volumes is that the majority of chapters include a number of selected readings – extracts drawn from books and articles – which relate closely to the line of argument and which are integral to the discussion as it develops. The relevant readings can be found at the end of the chapter to which they relate and are printed in two columns to distinguish them from the main teaching text. The third shared feature of the volumes is the student-orientated nature of the teaching materials. Each volume is intended as part of an interactive form of study, with activities and summaries built into the flow of the text. These features are intended to help readers to grasp, consider and retain the main ideas and arguments of each chapter. The wide margins – in which you will find highlighted the concepts that are key to the teaching – are also intended for student use, such as making notes and points for reflection.

While each book is self-contained, there are a number of references back (and a small number of references forward) to the other books in the series. For those readers who wish to use the books as an exploration in human geography, you will find the references to chapters in the other volumes printed in bold type. This is particularly relevant to the final chapters of Volumes 2–5 as they form a sequence of chapters designed to highlight the uneven character of global development today. On a related teaching point, we have sometimes referred to the group of less developed and developing countries by the term 'third world', in inverted commas to convey the difficulty of continuing to include the diverse range of countries – which embraces some rapidly industrializing nations – under this conventional category. The 'disappearance' of a second world, with the demise of the Communist bloc, also questions the usefulness of the category and, in one way, simply reaffirms the significance of the world's changing geography.

Finally, it remains only to thank those who have helped to shape this Open University course. The names of those responsible for the production of this course are given in the list of Course Team members on page *ii*. Of those, we would like to extend our thanks to a number in particular. It is fair to say that the course would not have had the shape that it does were it not for the breadth of intellectual scholarship provided by our external assessor, Professor Nigel Thrift. Over a two-year period, Nigel, among other activities, commented and offered constructive advice on every draft chapter discussed by the Course Team – in all, some eighty-plus drafts! The Course Team owe him a major debt. We also owe a special debt to our Tutor Panel – Rick Ball, Jenny Meegan and Phil Pinch – for their ceaseless concern that the teaching materials were precisely that: materials which actually do teach. Our editors at the Open University, Melanie Bayley and Fiona Harris, not only raised the professional standard of the series as a whole with their meticulous editing, they also became involved at an early stage in the Course's life and thus were able to smooth the path of its development. Thanks also to Ray Munns for his cartographic zeal and to Paul Smith, our media librarian, who, as ever, translated our vague descriptions of this or that image into an impressive array of illustrations. The typographic design and initial cover idea were developed by Diane Mole who then relinquished the course to Debbie Crouch; their expertise has transformed our typescripts into this handsome series of volumes. The speed and accuracy with which the multiple drafts were turned round by Margaret Charters and Doreen Warwick also deserves our special thanks. Without their excellent secretarial support, the course would not be in any shape at all.

Lastly, in the collaborative style of work by which Open University courses are produced, the awesome task of co-ordinating the efforts of the Course Team and ensuring that the materials are produced to schedule, falls to the course manager. It is still a source of amazement to us how our course manager, Christina Janoszka, managed this task as if it were no task at all. We owe her probably more than she is aware.

John Allen
on behalf of
The Open University Course Team

Introduction

This book asks whether, at the turn of the twentieth to the twenty-first century, the world is becoming overcrowded. Many people feel that it is, whether they think of developers moving into their areas to build new homes, the traffic jams they experience on journeys to work or shop, the increasing number of people they meet in their favourite recreation areas, or the huge crowds they see on the television or encounter when they visit exotic places. Fears of overcrowding underpin policies to control immigration and to promote 'family planning' at home and abroad. But are these fears real or imaginary? Do we fear overcrowding out of dislike of any change, or because it affects our lifestyle choices, or because it threatens the natural basis of human existence?

Even at the common-sense level, a little reflection suggests that we tend to describe an area or event as 'overcrowded' not solely on the basis of the number of people. We may be inspired or elated by being with many other people at a carnival, festival, sporting event or demonstration. 'Many people' tends to become 'too many' when they behave, or threaten to behave, in ways we do not like: they destroy something we want to use, prevent our access to it or just make us feel uncomfortable using it. Overcrowding is not just a matter of the number of people in an area; it relates to what those people are doing, to what facilities exist to cater for them, and to cultural expectations about what is appropriate.

Different interest groups may have very different views about overcrowding – for example, in nineteenth-century Britain the residents of overcrowded industrial cities began to seek access to the open spaces of the Pennines, only to find the land closed to them and guarded by armed gamekeepers. The resulting mass trespasses undoubtedly seemed a very threatening result of overcrowding to the wealthy landowners who feared for the quality of the grouse shooting. Today, one of the major problems of the Peak District National Park is the large numbers of visitors and the consequent overcrowding of well-known 'beauty spots'.

The most influential statement on overcrowding was the Reverend Thomas Malthus' famous *Essay on the Principle of Population.* Malthus argued that overcrowding was inevitable because population tends to grow geometrically (that is, it can double and double again over given intervals of time), while 'the means of subsistence' can grow only arithmetically (that is, extra production can only add a fixed amount in each time period). Later economists made the situation worse by proposing that productive activities faced a 'law of diminishing returns', where extra inputs produced successively smaller additions to productivity. Malthus' argument made it seem inevitable that overpopulation would occur, first in Britain and then across the whole world, and that population growth could only realistically be held back by famine, disease and war.

His argument has been much criticized, both on principle as an argument against provision of financial relief to the poor and in practice as false. In the two centuries since his essay was written, the growth in food supply has been as fast as that of world population, and world economic output has been much faster still, so that many more people exist, most with much higher living standards. Over the period, the supposed difference in arithmetic and geometric growth rates has not appeared. Optimists conclude that the argument was false and can be ignored. Pessimists, or neo-Malthusians, argue that special circumstances have blurred the difference, but that the current rapid growth of population means that overpopulation is imminent.

If the concept of overcrowding is so uncertain in everyday use, how can we focus it for academic use? Some clarification can be obtained by looking at geographical ideas about globalization (discussed, for example, by **Allen, 1995***). It seems that these do begin to explain ways in which people might feel increasingly overcrowded. First, the notion that space is shrinking (due to more efficient transport and communication) clearly puts us into more immediate contact with other people. Second, the processes of unequal interdependence may involve us in more threatening contacts: the firm we work for may face strong competition from overseas rivals or be threatened with takeover by a large multinational firm. Third, the mixing of people from different ethnic or social backgrounds in particular areas as a result of international migration may increase our contacts with people whose cultural ways we do not share or even understand and whose interests seem to be in competition with ours. However, these processes do not unequivocally demonstrate an increase of overcrowding in the world: it may simply be that we live in one of the more densely populated and highly developed parts of the unevenly developed world and have an *illusion* of overcrowding. So to understand the issue of overcrowding, this book will develop a more environmentally aware concept of globalization.

Since 1970, environmentalists have put forward more or less critical views of economic growth and globalization. A recurrent theme has been that there is a limit to growth: the earth has been seen as finite, with a limited capacity to supply resources for human need – or, as many environmentalists would put it, human greed. Whereas the social worlds of the economy, the multinationals and the world cities are apparently able to grow and speed up indefinitely, the natural world is finite in size, in the availability of materials and in biological variety (over the short term). As the natural world provides society's life-support system and resource base, its finiteness could be, or could become, a problem. If overcrowding were to have a damaging effect on natural productivity, so that increasing demand caused a decrease in supply, the problem would be compounded. It is this prospect of damage to natural systems and exhaustion of natural resources that underpins environmentalist fears of overcrowding.

As at the common-sense level, therefore, overcrowding of the world involves more than merely numbers of people. Growing numbers do have an impact, but human impacts on each other and on the earth are dramatically altered by the way we live: affluent, consumer-oriented societies have many times greater impact per head than do subsistence or spiritually oriented societies. Numbers and lifestyles may have multiplied effects where societies are experiencing both population growth and economic growth. It is the spectre of a rapidly growing world population with increasingly materialist lifestyles, making ever greater demands on the earth for resources and having increasing impacts through waste and pollution, which prompts environmentalist fears of global overcrowding. In this sense, the question 'Is the world overcrowded?' problematizes every aspect of the relations between society and nature. Population numbers and growth rates are obvious issues,

* A reference in emboldened type denotes a chapter in another volume of the series.

but the kinds of lifestyles people adopt, and the nature of the production systems that support these, are increasingly questioned by environmentalists. In effect, this means that human values and social divisions are also called into question: are materialism and economic growth sufficient goals for human existence and economic policy, especially when they seem to generate social inequality and environmental damage? Many environmentalists argue that they are not and that new directions are needed which accord higher priority to the environment and to those people who suffer economic deprivation. Effectively, they are arguing for new forms of globalization.

Activity 1 Read the extract 'From one earth to one world' in Box 1 below.

How many representations of globalization are implied by this short reading?

Box 1 From one earth to one world: an overview by the World Commission on Environment and Development

'In the middle of the 20th century, we saw our planet from space for the first time. Historians may eventually find that this vision had a greater impact on thought than did the Copernican revolution of the 16th century, which upset the human self-image by revealing that the Earth is not the centre of the universe. From space, we see a small and fragile ball dominated not by human activity and edifice but by a pattern of clouds, oceans, greenery, and soils. Humanity's inability to fit its doings into that pattern is changing planetary systems, fundamentally. Many such changes are accompanied by life-threatening hazards. This new reality, from which there is no escape, must be recognized – and managed.

Fortunately, this new reality coincides with more positive developments new to this century. We can move information and goods faster around the globe than ever before; we can produce more food and more goods with less investment of resources; our technology and science gives us at least the potential to look deeper into and better understand natural systems. From space, we can see and study the Earth as an organism whose health depends on the health of all its parts. We have the power to reconcile human affairs with natural laws and to thrive in the process. In this our cultural and spiritual heritages can reinforce our economic interests and survival imperatives.

This Commission believes that people can build a future that is more prosperous, more just, and more secure. Our report, *Our Common Future,* is not a prediction of ever increasing environmental decay, poverty, and hardship in an ever more polluted world among ever decreasing resources. We see instead the possibility for a new era of economic growth, one that must be based on policies that sustain and expand the environmental resource base. And we believe such growth to be absolutely essential to relieve the great poverty that is deepening in much of the developing world.

But the Commission's hope for the future is conditional on decisive political action now to begin managing environmental resources to ensure both sustainable human progress and human survival. We are not forecasting a future; we are serving a notice – an urgent notice based on the latest and best scientific evidence – that the time has come to take the decisions needed to secure the resources to sustain this and coming generations. We do not offer a detailed blueprint for action, but instead a pathway by which the peoples of the world may enlarge their spheres of co-operation.'

Source: World Commission on Environment and Development, 1987, pp. 1–2 (The Brundtland Report: after the chairperson, Gro Harlem Brundtland)

The image of the earth from space, a favourite one with environmentalists, clearly represents the earth as a single natural system – indeed, a vulnerable natural system, 'a small and fragile ball'. This image emphasizes the unity of the 'one earth' (see Plate 1).

It is immediately succeeded by a view of economic/technical globalization, which can 'move information and goods faster around the globe than ever before' and so shrink space. Later in the Brundtland Report, from which the extract in Box 1 is taken, it is pointed out that, until recently, human activities were 'neatly compartmentalized within nations', but that these compartments have begun to dissolve.

Even this short extract contains hints of other representations of globalization: scientific evidence and understanding is growing; there is a vision of a future which is more just, relieves poverty and is based on enlarged spheres of collaboration: cultural and spiritual heritage are in some way relevant. The title itself is a call for a unified human response to sustaining prosperous lifestyles and the 'one earth': it is calling for globalization of environmentalism, welfare and political action.

This book adopts a more evaluative posture than the Brundtland Report, as befits a teaching text rather than a political prospectus. The Report advocates that environment and development, society and nature, can be reconciled through the concept and policies of 'sustainable development'. This book asks: *'Are current lifestyles sustainable?'* In doing so, it emphasizes the conservation of natural systems, but maintains a concern (inevitable in the light of Brundtland's inclusion of equity as part of sustainability) with uneven development. It focuses on topics where 'the natural' is most centrally concerned: conservation of wilderness, population growth and resource availability – three areas where environmentalists fear the negative consequences of overcrowding.

The starting points of the book – fear of overcrowding and doubts about sustainability – make it inevitable that it must also deal with values. The lack of a clear border between crowding and overcrowding, and the inclusion of greater equity within sustainability, mean that judgements must be value laden. The extract in Box 1 demonstrates that aspects of the globalization of human–nature interaction can be represented in a variety of different ways and that the shrinking of space improves our ability to deal with problems as well as making the problems seem worse. A central concern with issues of representation links this book to other volumes in the series.

Chapter 1 relates to the idea of overcrowding in a direct way. It is a perspective which is calculated to highlight both the globalization of society and issues of representation. It is the idea of *wilderness.* This has been an important idea through much of recorded history, especially from Ancient Greece to the USA in the nineteenth century, and it continues to inspire some environmentalists today. The chapter both extends the analysis of uneven development, asking whether development is now truly global or whether any substantial areas can be regarded as undeveloped, and identifies a need to rethink, and ideally to redefine in practice, the relationship between human society and nature, in order to make our values more consistent with a globalized environment and economy.

Chapter 1 argues that, in the current era, with settlement and resource extraction covering almost the whole land surface of the world, we need to replace the idea of wilderness or natural environment as a 'safety valve' *beyond* developed areas and to avoid the conceptual separation of society and environment, or humanity and nature, which underlies it. Instead, we need to recognize that the relationships are much more integral: the human species is a product of nature through the process of evolution and is dependent on natural processes as well as being uniquely able to divert or disrupt them through social organization and use of technology. We need to recognize that nature is in society and society is in nature. This is the basis for the selection of the two major topics of the book.

Chapters 2 and 3 analyse the topic which pre-eminently embodies the natural in the social: *the growth of the human population.* Of course, this is of more than conceptual significance: the size and rate of growth of the human population are unprecedented and are regarded by some environmentalists as two of the major threats to sustainability. If the population of the world is to double in the next generation and reach ten billion or more, it is hard to see how current lifestyles could be unaffected. If that doubling is compounded by an increase in the demand each person places on the environment, we face a period of dramatic environmental change, much of it regarded as detrimental from today's viewpoint. But rapid policy response to this perceived problem faces great difficulties – demographers are still trying to explain why population grows more or less quickly in different places at different times, and any population policy faces cultural, political and ethical obstacles, not the least of which is the refusal of many people and institutions to accept that population growth is a problem at all. The interplay of the population question with environment, uneven development and ethics is well illustrated by the newspaper extract reproduced below.

Adrian Hamilton

Population is not the agenda

There can have been few international conferences quite so misconceived as the current United Nations-sponsored debate on population in Cairo. Hailed as the successor to the 1992 Earth Summit in Rio, it derives from precisely the same mistaken impetus: not concern with the Third World, but fear of it.

In the case of the environment it was alarm by the rich that the developing Third World would eat up resources and pollute the atmosphere as fast as the West did when it industrialised generations ago. The Third World went along in the fond hope (largely disappointed as it turned out) that the rich would then pay them to cleanse their smokestacks and stop the logging.

In the case of population, what is driving Western concern is the nightmare of a Third World exploding with people who would then sweep in great migratory waves into the comfortable industrialised nations. Forget the calls for development aid, education and all the measures that normally bring down birth rates as people become richer. How much easier (and cheaper) if you intervene directly to control the population among the poor within their own borders.

Well it won't do, and thank heavens there have been some powerful enough political figures to blow the whistle. Inevitably, and rightly, they have been women. For if population is not to be about development and education, then it is about society, families and, above all, mothers.

After all the consensual waffle of the opening speeches by the UN Secretary General, Boutros Boutros-Ghali, the US Vice-President Al Gore and the Egyptian President Hosni Mubarak (all men – but that may be coincidental), it was the two women Prime Ministers who set out the real argument.

On the liberal left, Gro Harlem Brundtland, Prime Minister of Norway – a country with a population of 4.3 million, a population growth rate of 0.4 per cent and one of the highest per capita incomes in the world, thanks to oil. Births, she argued, were about the rights of a woman to do what she wished with her own body.

From the Islamic world came Benazir Bhutto, Prime Minister of Pakistan, a country with a population of 125 million, a population growth rate of 2.87 per cent and one of the lowest per capita incomes in the world. Empowerment for women, she argued, was a key to the problem, but not at the expense of Islam's belief in the sacredness of life and the core of the family.

It's a real debate about Western and Eastern values that is not just about religious divides. Indeed the Pope with his opposition to birth control is an irrelevance to much of it. Islam does not have those objections, but it does oppose what it sees as secularist influences that promote promiscuity and the undermining of social relationships.

Unfortunately, it's a debate that cannot be held, let alone produce effective results, in such a political forum, as this. Ever on the look-out for causes to play on anti-Western sentiments, Islamic fundamentalists, and even moderate leaders, have all too easily portrayed the conference as an attempt to force Western values on their societies.

As it is, Benazir Bhutto only just made the conference at all against the political clamour from religious leaders to stop her. Saudi Arabia gave in to its mullahs and declined to attend. So did several others.

This is the worst of all possible worlds. Relations between the West and Muslims are bad enough as it is. The world cannot afford constantly to sharpen the divides by politicising issues of genuine social concern.

That is as true inside as outside. If change is to come within the Islamic world, it will almost certainly come from educated women. Paint reform in West-versus-East terms, and their cause will be forced backwards.

'This conference', declared Benazir Bhutto, 'must not be viewed by the teeming masses of the world as a universal social charter seeking to impose adultery, abortions, sex education and other such matters on individuals, societies and religions, which have their own social ethos'.

But that is exactly how it is being seen, and used. If only the UN had kept to development. But that would have involved money and the West doesn't want to give that away.

Source: *The Observer*, 11 September 1994, p. 25

Chapters 2 and 3 deal with the extremes of the population question. Chapter 2 focuses on Africa, often represented as the extreme case of runaway population growth and environmental degradation. Chapter 3 considers the experience of Europe, conventionally seen as the continent which first experienced rapid population growth and which has now overcome this by reducing fertility to replacement levels or below. Detailed analysis shows that quite different representations are possible. In some respects, Africa may be underpopulated and may therefore require more people to manage its environment more successfully. Malnutrition in Africa may not be the result of local incompetence, but reflect the use of much African land to grow crops for export to the developed world, especially Europe. Europe may indeed have solved the problem of runaway population growth, but in doing so it was able to draw on land and resources from the whole world during the colonial era. Its high standard of living means that, after the USA, it is the world's second largest user of resources, and thus its demands on resources arguably have a much larger impact than do those of Africa.

The chapters show that the experiences of both continents are complex, that different interpretations can be brought to bear and that all are problematic. They contribute to a shift of attention from population growth in isolation to the balance between population and development. With this they lead to a change of emphasis in the book: Europe is already a net importer of resources and Africa's resource use is likely to grow dramatically through a combination of population growth and economic development. In this respect, both continents point to resource use, rather than population *per se*, as being a greater problem for sustainability.

Resources are natural objects or processes which society has defined as useful and which are identified and exploited for social – usually economic or political – purposes. They are parts of the natural world which become part of the social world. In the course of exploitation, natural systems are altered, sometimes with catastrophic effects on habitats, species and local societies. Again, this is a major issue for environmentalists who oppose the impact on natural systems, fear the increasing extent of pollution and argue that rapidly growing demand, caused by population growth and materialist lifestyles, will lead to resource exhaustion and economic collapse. As with population growth, other analysts dispute this view and argue that use of renewable resources, recycling and substitution will solve all problems of resource provision and ensure sustainability into the foreseeable future.

Chapter 4 identifies the conventional distinction between stock resources, notably metal ores, and flow or renewable resources, such as solar radiation. Environmentalists often argue that depletion of stock resources will inevitably slow economic development unless a rapid move is made towards renewables. The chapter shows that the real world is more complex. Shortages of stock resources may not result from natural limits but from social limits, such as fear of political instability or the cost of pollution control. Most stock resources can be made sustainable into at least the medium term through recycling and substitution. Fossil fuels are a conspicuous exception, but here the option of renewable energy supply has considerable potential. The example of water supply, however, shows that renewable resources are not infinite or immutable: they may be locally scarce or politically disputed and their supply can be disrupted. Reliance on renewable resources does not automatically guarantee sustainability.

Chapter 5 argues that the most serious impact of resource use results from unequal interdependence. Resource extraction is dominated by a small number of giant corporations who service markets in the developed world with insufficient regard to negative local impacts on society and environment near to mines and smelters, much of which occurs in remote or undeveloped areas. Available methods of pollution control may not be applied because of cost or political pressures. Resource problems are not global in the sense that they affect us all; they affect some localities and regions while the benefits for producers and consumers are concentrated elsewhere. In other words, geographically uneven development is crucial both in generating environmental and social problems and in inhibiting efforts to reduce or eliminate them.

The crucial effects of uneven development in relation to both population and resources are emphasized in the concluding Chapter 6. This chapter also argues that, although resources do not seem to impose global limits at present or in the near future, neither stock nor flow resources can sustain infinite expansion of populations and economies: there will be a need to contain, or even reduce, consumption. This can be achieved in some measure through greater efficiency in use of materials and energy, but at some point it will be necessary to impose limits on our material and energy consumption. This requires global political responses, which are also made difficult by uneven development.

It should now be clear why this book is called *An Overcrowded World?* The prospect is for rapid population increase and economic growth, the implications of which are an urgent need for more resources of all kinds and a greater production of pollutants. The consequence could be overcrowding and degradation of settled areas and the elimination of existing areas of wilderness. However, none of this scenario is undisputed: pessimists suggest that crisis is imminent while more optimistic commentators argue that economic growth can solve the problem of population growth and provide the means to use resources more efficiently and with less pollution. There are debates about explanations and policy responses, in which issues are represented very differently by different interest groups. These debates are compounded by social divisions within societies and uneven development between them. At present, the practical outcome seems to be that the silent majority of the human population are too committed to the materialistic lifestyles that they have achieved, or to which they aspire, to support political moves towards more sustainable forms of development. That may be one of the main threats to sustainability. At a deeper level, the commitment to material consumption may be precluding pursuit of different goals, from tourism to exotic places and the conservation of biodiversity through human solidarity to aesthetic satisfaction and spiritual salvation. More practically, shortages of food, space, water and vital resources help to provoke armed conflict. There could be a danger, presaged in the urban ghettos, in Iraq, Somalia, Rwanda and in Bosnia, that the world will become a wilderness in a new sense.

Philip Sarre and John Blunden

References

ALLEN, J. (1995) 'Global worlds', in Allen, J. and Massey, D. (eds) *Geographical Worlds*, Oxford, Oxford University Press in association with The Open University.

MALTHUS, T. (1798) *An Essay on the Principle of Population.*

WORLD COMMISSION ON ENVIRONMENT AND DEVELOPMENT (1987) *Our Common Future*, Oxford and New York, Oxford University Press.

Paradise lost, or the conquest of the wilderness

Chapter 1

by Philip Sarre

1.1 Introduction: why wilderness?

wilderness

The introduction to this book makes clear that *wilderness* was chosen as the topic for this first chapter for two reasons. It spotlights the historic process whereby human society has pushed back the boundaries around settled, developed and 'humanized' areas until only residual areas of wild land are left. It also raises deeper issues of definition, representation and values.

Activity 1 Because the concept of wilderness is value laden, there are no right answers to some of the questions raised in this chapter. Reflect for a moment on what the concept of wilderness means to you. Is it the antithesis of settlement or civilization? Is it desirable or undesirable? Is it objective reality or state of mind? Can you identify where wilderness occurs today?

The process of expansion of settled areas has long had important implications for both society and the environment. The globalization of the economy has created a new situation: we seem to be very close to a closure of the frontier between the settled world and the wilderness, leaving only Antarctica and a few fragments of wilderness elsewhere. This has made environmentalists increasingly concerned about wilderness preservation. If the whole surface of the earth is becoming humanized, there will be little, if any, remaining untouched wilderness left as a 'safety net' for individuals, society or natural processes. The possibility of a world without wilderness raises stark questions about how we should relate to nature in the humanized world.

Even in this last paragraph, issues of representation and values have arisen. It is true that wilderness has been seen in a positive light by environmentalists during the last century. However, for most of human history the dominant attitude to wilderness has been one of fear: it was seen as threatening – actively so as the habitat of dangerous animals, more passively in its composition of forests, mountains and deserts which were regarded as inhospitable terrain. Usually, wilderness was avoided and fenced off, unless its wildness was needed as a refuge by criminals, as a challenge to 'turn boys into men' or as a path towards the creator used by prophets. Over time, former wilderness has been enclosed, cultivated and exploited for its resources as populations have grown and technology has advanced. Only since the balance has shifted towards human dominance over most of the world's wild places, has it been argued that wilderness should be valued and preserved.

A central argument of this chapter is that, as human societies have developed progressively more powerful technologies of production, distribution and consumption, there has been a parallel evolution of representations of the relationship between society and nature and of the values which influence change. It will be argued that societies with different modes of production have tended to have different attitudes and values towards the natural world.

This is not to say that there is a simple correlation: concepts dealing with the actual or proper relationships between individual, society and nature may have an influence for millenia. Many of the ideas with which this chapter is concerned are articulated in religions, though the chapter is not about religion *per se*: it merely reflects the fact that most human societies have explained their relationships to nature as part of their religious beliefs. Only in the last century has this task been given in part to science. The argument here is that science can represent what is and has been, but that it cannot dictate what ought to be. To do that, societies need value systems, and one of the environmentalist criticisms of contemporary societies is that the emphasis on growth and consumption pre-empts important questions of value.

The chapter will show that contemporary pressures on the last areas of wilderness are just the final phase of a process which has been accelerating for ten millennia. This process has brought us to a position where unprecedented numbers of people live affluent lifestyles and place unprecedented demands on natural systems. The chapter will argue that the dominant values of the last decades, some of which have very ancient origins, will worsen rather than solve environmental problems. It will use the closure of the global frontier between society and nature as a test and a symbol. As a test, preservation of wilderness is an issue which determines whether everything is to be decided in human terms or whether nature's interests are to be considered. As a symbol, globalization identifies a wholly new era in which we can no longer assume that nature will take care of parts of the earth's surface. In this situation, we seem to need new values to cope with a new reality.

The chapter is seeking to answer two key questions:

1 Is global society crowding out the last wilderness?

2 What is wilderness and how has wilderness been represented?

It does so through five sections:

- This first section introduces the concept of wilderness.
- Section 1.2 considers two case studies of current pressures on wilderness.
- Section 1.3 explores the past spread of human influence.
- Section 1.4 shows how different past and present societies have related their productive technologies and their attitudes to wilderness and nature.
- Section 1.5 considers how appropriate different kinds of attitude are to current circumstances, and in particular to a globalizing society which could eliminate the last vestiges of wilderness in a very short time.

1.1.1 The significance of wilderness

Some environmentalists have attributed the proliferation of environmental problems to the attitudes spelt out in the Bible and carried through into industrialization in the Christian (and initially the Protestant) world. The Book of Genesis gives humans an explicit mandate: 'And God blessed them, and God said unto them, Be fruitful, and multiply, and replenish the earth, and subdue it: and have dominion over the fish of the sea, and over the fowl of the air, and over every living thing that moveth upon the earth' (Genesis 1: 28, Authorized Version). Although this instruction is seen as justification of human supremacy over *nature,* other aspects of Genesis are more equivocal: Adam was made *in the image* of God, but made *of* earth. After eating the fruit of the tree of knowledge and becoming self-conscious, God fears that Adam and Eve are becoming too God-like and expels them to struggle for a living cultivating more marginal land. Eden, too, is somewhat equivocal: on the one hand, it can be seen as a reference back to the idyllic life of the hunter-gatherer, able to harvest fruit without work (see Plate 2), but, on the other, it could also be seen as the highly productive irrigated agriculture of the city states of Sumer, from which the Israelites

nature

Map showing the location of the Garden of Eden. This nineteenth-century map demonstrates the eagerness of European Christians to locate Biblical events in the known world. This location for Eden puts it at the head of the Tigris–Euphrates delta, a prime site for early irrigated agriculture

moved to more marginal land. Indeed, the writers of the Old Testament seem to be more sympathetic to pastoralism than cultivation: the origin of the dispute between Cain and Abel is that God had respect for Abel's offerings of young animals but not for Cain's offerings of crops. This positive view of nomadic pastoralism is linked to a disenchantment with urban, and hence agricultural, civilization, as illustrated by the destruction of Sodom and Gomorrah, and is accompanied by a positive attitude to wilderness. Old Testament leaders repeatedly led their tribes into the wilderness to distance them from human wickedness and bring them nearer to God. Given this ambiguity, some Christians have read Genesis as giving humans domination over nature, and this domination of nature has been emphasized throughout most of Christian history.

The theme of the domination of nature was powerfully stated in a celebrated thesis put forward by Frederick Jackson Turner in the USA and first published in 1893. The fact that this thesis was argued for the USA is doubly significant. First, the USA became the world's dominant power and played a key role in establishing competitive capitalism as the ideology and practice of the globalizing economy. Second, the USA more recently has been a major centre for debate about environmentalism and originated the two international pressure groups Greenpeace (with Canadian participation) and Friends of the Earth. The thesis argued a century ago has uncanny resonance with current pressures on the last global wildernesses.

Frederick Jackson Turner's paper offered a radical new interpretation of American history. He argued that the USA had been decisively shaped by the existence of large areas of wilderness, an open frontier and free land. The frontier both affected individuals and institutions:

The result is that to the frontier the American intellect owes its striking characteristics. That coarseness and strength combined with acuteness and inquisitiveness; that practical, inventive turn of mind, quick to find expedients; that masterful grasp of material things, lacking in the artistic but powerful to effect great ends; that restless, nervous energy; that dominant individualism, working for good and for evil, and withal that buoyancy and exuberance which comes with freedom – these are traits of the frontier, or traits called out elsewhere because of the existence of the frontier. Since the days when the fleet of Columbus sailed into the waters of the New World, America has been another name for opportunity, and the people of the United States have taken their tone from the incessant expansion which has not only been open but has even been forced upon them. He would be a rash prophet who should assert that the expansive character of American life has now entirely ceased.

(Turner, 1947, p. 37)

Thomas Cole, Landscape with Figures and a Mill, *1825. The Minneapolis Institute of Arts. This image of the American landscape contains classic pioneer themes: the axe and felled tree symbolize the clearance of virgin forest to create pasture for cattle and building materials. The mill also involves transformation of natural energy into forms usable by industry*

The peculiarity of American institutions is, the fact that they have been compelled to adapt themselves to the change of an expanding people – to the changes involved in crossing a continent, in winning a wilderness, and in developing at each area of this progress out of the primitive economic and political conditions of the frontier into the complexity of city life.

(Turner, 1947, p. 2)

Across the Continent: 'Westward the Course of Empire Takes its Way', *drawn by E.F. Palmer. The Museum of the City of New York. This is a classic representation of the westward expansion of the USA. More than rugged individualists are involved: industry, in the form of the railway and the telegraph, is on an irresistable course, overtaking the wagon trains. The line separates civilization and culture from wild nature and the smoke aggressively indicates the imminent fate of the native peoples – although they have already adopted horses and rifles from the European invaders*

The peculiarity which Turner stresses is the democratic basis of US politics, limited by individualism and the impossibility of regulating the frontier. The dangers from 'Indians' helped to bring the early states together, while the establishment of new states helped to consolidate federal power. The limited power of government not only allowed individual citizens to flout the law or to take it into their own hands, but affected standards of public life generally:

Individualism in America has allowed a laxity in regard to governmental affairs which has rendered possible the spoils system and all the manifest evils which follow from a lack of a highly developed civic spirit. In this connection may be noted also the influence of frontier conditions in permitting lax business honor, inflated paper currency and wild-cat banking.

(Turner, 1947, p. 32)

His later writing both fills in local detail within his broad thesis and considers what differences came about after the official closure of the frontier in 1890. In an essay written in 1910 and included in the 1947 collection, he marvels at the rate of industrial and economic growth, noting, for example, that pig iron production in the USA in 1907 exceeded that of Britain, France and Germany combined. However, he notes that agricultural production was much less buoyant: 'Already population is pressing on food supply while capital consolidates in billion dollar organisations' (Turner, 1947, p. 314). Not only was capital consolidating, but the expansion had not stopped with the closing of the frontier:

Having colonized the Far West, having mastered its internal resources, the nation turned at the conclusion of the nineteenth and the beginning of the twentieth century to deal with the Far East, to engage in the world-politics of the Pacific Ocean. Having continued its historic expansion into the land of the old Spanish empire by the successful outcome of the recent war, the United States became the mistress of the Philippines at the same time that it came into possession of the Hawaiian Islands, and the controlling influence in the Gulf of Mexico. It provided early in the present decade for connecting its Atlantic and Pacific coasts by the Isthmian Canal, and became an imperial republic with dependencies and protectorates – admittedly a new world-power, with a potential voice in the problems of Europe, Asia, and Africa.

(Turner, 1947, p. 315)

Turner's thesis has been much debated by historians, many arguing that he overstressed the role of the frontier and neglected the roles of slavery, industrialization and the labour movement. However, it is now widely accepted that the frontier was a major influence. *Its importance from our point of view is that the USA articulated, in practice and in theory, the dominant view of wilderness in the era of industrialism: wilderness is to be conquered, settled and civilized with all possible speed.* This is a view which is rarely expressed explicitly but which continues to be expressed in practice.

There were always different views. In the mid-nineteenth century, for example, the American naturalist and philosopher Henry David Thoreau was writing and lecturing about the spiritual qualities of the New England landscape, but his views were largely ignored for a century or more. It was only in the last quarter of the century as the frontier was closing that the positive view of wilderness as something to be preserved came to be influential – due, especially, to the writings of John Muir – and a system of national parks and forests was established. Today, Americans remain the strongest advocates of the desirability of wilderness, but at least as many stress the need to hunt, fish and struggle against wild nature as stress the need to 'walk softly on the earth'. Both sets of wilderness advocates find themselves opposed to the continuing and powerful development lobby wishing to cut old forests, dam rivers and extract minerals wherever economically viable, even when they do so in the name of conservation, which in the view of the turn-of-the-century 'Conservation Movement' meant 'wise use'. (The contrast between conservation and preservation is further explored in Reading C by Robyn Eckersley.)

This American conflict between exploiters and preservers of nature is now being played out at an international scale. Indeed, Frederick Jackson Turner's emphasis on the closing of the American frontier in 1890 now looks a century premature. The continental territory may have been settled by then, but American capital continued to spread out into Canada and Latin America, then into Europe and Asia. Although US territory spread to Alaska and into the islands of the Pacific and Caribbean, it was an increasingly international capitalism which spread out across the world. In

John Muir, pioneer environmentalist and founder of the Sierra Club, in Yosemite

doing so, it was increasingly involved in exploiting resources in areas that were already settled. Today, both global capital and growing indigenous populations are making inroads into the last large wild areas of the globe – the rainforest, the high mountains of Asia and the Arctic tundra. This incipient closure of the global frontier has galvanized international environmental groups (including Friends of the Earth, whose roots go back to John Muir and the preservationists) to defend these last wildernesses. However, just as the pioneer environmentalists contributed to the dispossession of the native Americans by excluding them from national parks, so today's debates are confounded by the presence of indigenous peoples whose ideas of wilderness are quite different from those held by international business. As well as the overt struggle to control resources, there is an implicit struggle between different ideas about nature and its relationship to society.

Activity 2 Can you think of a British equivalent of the nineteenth-century push to close the US frontier? If so, what were the key similarities and differences?

Although it started much earlier, British colonial expansion gained pace during the nineteenth century and into the twentieth. Part of this expansion, especially into Australia, was portrayed as filling empty lands, but most colonial situations clearly dealt with inhabited territories. As a result, colonialism was more focused on the social dimensions than on the land itself. Like the American frontier, colonialism was presented as character-forming for colonists as well as for colonized peoples. The withdrawals from the colonies, which have taken place since 1945, would have seemed as unthinkable then as withdrawals from wilderness seem now.

1.2 Pressures on wilderness

In his recent study of the ways in which human society has changed the face of the earth, Simmons (1989) concludes that most of the earth is dominated by human management. Five kinds of area remain close to their wild state: the polar regions, hot deserts, tropical forests, the high mountain chains and some small remote islands. Each kind of wilderness area has its own experience, but this section will focus on two in particular – the tropical forests and the Arctic region – because these are experiencing rapid change over which there is intense debate. Both rapid population growth and outside commercial interests inflict pressure on the tropical forests, and the Arctic region has been opened up for resource exploitation by the USA, Canada and Russia, causing conflict with its indigenous peoples.

1.2.1 Deforestation and the Philippines

As recently as 1945, tropical forests covered much of the Equatorial zone: most extensively in Amazonia, but also in West Africa and on the islands of South East Asia. Though each area has different species of plants and animals, they share a similar ecological structure. In these areas heavy rain and high temperatures throughout the year encourage a dense and multi-layered forest with large numbers of different species of trees, climbers, epiphytes, animals and insects. So impenetrable is the vegetation that only a small percentage of light reaches the forest floor and most animal life occurs in the higher layers. Early commentators, from Alfred Wallace onward, stressed the amazing fertility of these forests. During the nineteenth century there was some exploitation of accessible parts of the forest for special timbers such as teak, as well as for forest products such as ratten, latex and gums. Nevertheless, deforestation was not a cause of concern at least until the 1960s and did not become a global issue until the 1980s.

A plethora of studies has established that deforestation is taking place in almost every tropical forest and that the rate of clearance is fast. However, although there is concern that all rainforests could be cleared within a few decades, the author of a recent study points out that we simply do not know how much forest has been cleared, how much remains or what the rate of clearance is (though estimates range from 78,000 to 200,000 square kilometres per year) (Kummer, 1991). The aim of this section is not to grapple with inadequate statistics, though, but rather to consider the causes, consequences and implications of deforestation. In doing so, it follows Kummer when he argues:

... on a global scale, the factors that appear to be most important in explaining tropical deforestation are increasing populations and the spread of agriculture. Population in particular seems to stand out in this regard. However, at a deeper level, some authors feel that the real issue is access to the forest resource and social conditions which produce widespread poverty. In short, at the global level, it is difficult to articulate the specific causes of deforestation for a single country.

(Kummer, 1991, p. 16)

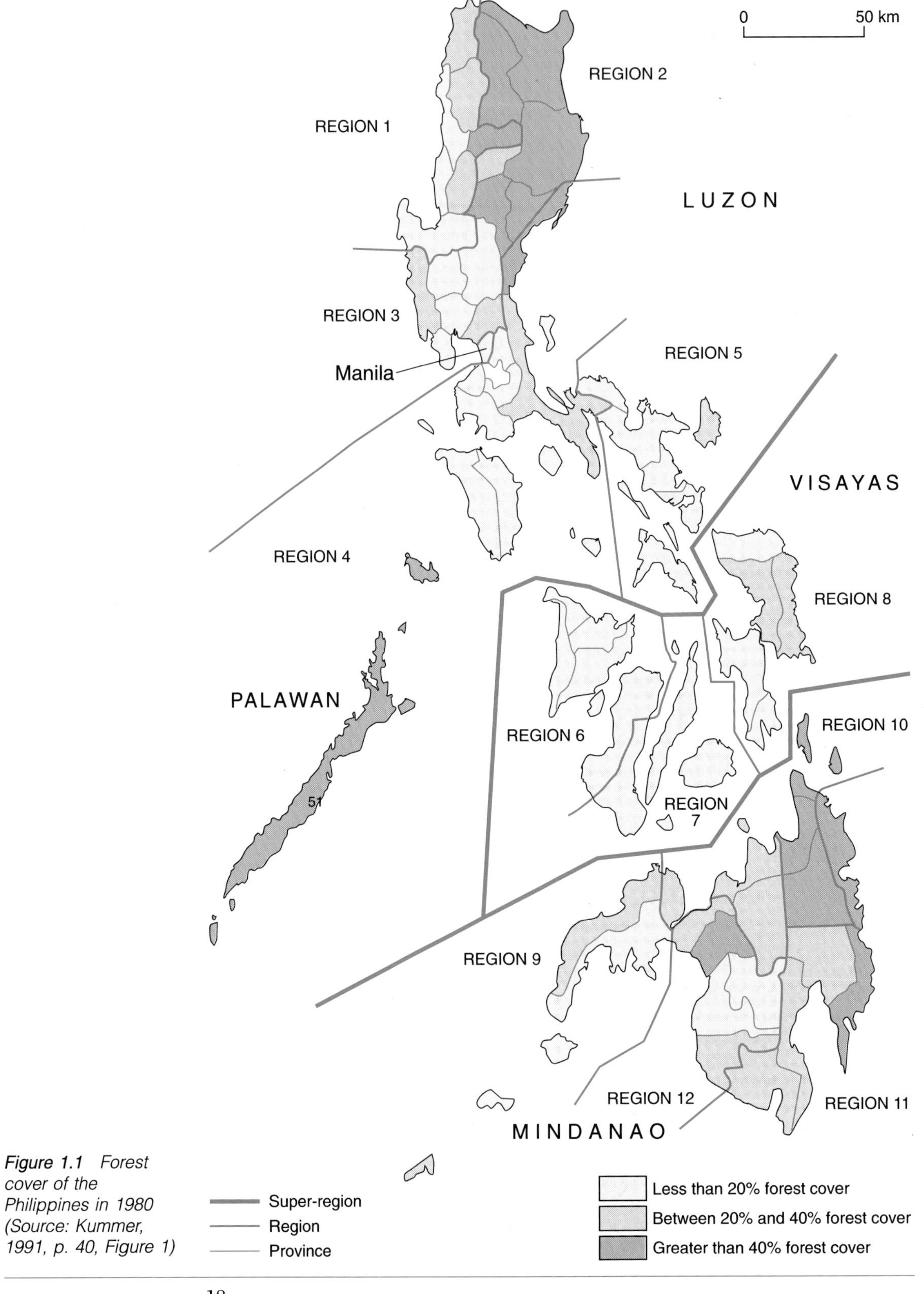

Figure 1.1 *Forest cover of the Philippines in 1980 (Source: Kummer, 1991, p. 40, Figure 1)*

The studies do suggest that certain factors are more relevant to some areas: for example, ranching mainly affects Latin American rainforests, while the more arid forests in Africa and South Asia suffer greater damage from the collection of firewood. However, the meaning of deforestation can be better understood through a detailed case study of the Philippines.

The Philippines comprise an archipelago with a total area of 300,000 square kilometres. When first visited by Europeans in 1521, about 90 per cent of the area was covered with a forest consisting of a high proportion of dipterocarps, the most prized timber trees. By 1900 the area covered by forest had fallen to about 70 per cent; in 1950 it was less than 50 per cent and by 1987 less than 25 per cent (see Figure 1.1). This was the earliest and worst case of deforestation in South East Asia. As in other parts of the world, deforestation in the Philippines has produced many problems, directly in depriving former forest-dwellers of their livelihoods and indirectly through soil erosion, with loss of topsoil, clogging of stream channels, accelerated runoff and hence disastrous flooding followed by depleted water levels. Increased sediment in streams has polluted the sea and choked coral reefs. Perhaps the ultimate problem is that by about the year 2000 the country will face a shortage of wood for domestic use.

Many studies of the causes of deforestation have been carried out in the Philippines by organizations such as the Food and Agriculture Organization (FAO) (of the United Nations) and the World Bank. The most frequently cited culprits are shifting cultivators, settled agriculturists and commercial loggers. Critics argue that shifting cultivators damage the forest by burning, especially if they return to burn and cultivate the same plot in less than 10 to 15 years. However, Kummer's recent four-year study argues that traditional shifting cultivators are not a serious problem and that the process of deforestation is actually one which combines loggers and small-scale cultivators. In his view a major factor in deforestation has been corruption. Officially, logging can take place only with a permit from the Forestry Department; however, there is strong evidence that the controls are ineffective. Data on exports of logs from the Philippines show their numbers to have reduced from the late 1970s and into the 1980s. Data on import of logs from the Philippines into their major markets, especially Japan, on the other hand, show the level remaining constant and suggest that illegally exported logs exceeded legal exports in about 1983 and were worth five times as much (US $1 billion) by 1988. Illegal, as opposed to legal, export has two major advantages from the point of view of the loggers: first, it is tax free and, second, it allows logging organizations to build up large overseas deposits of hard currency. These features are the cause of serious disadvantages for the economy of the Philippines.

The second stage of deforestation, in Kummer's view, occurs when land which has been logged is settled by poor migrants and used for subsistence farming. This prevents regeneration of trees which might otherwise allow the forest to re-establish itself. The origin of these poor migrants would appear to be obvious since the Philippines has the highest population density and fastest population growth rate of any South East Asian country – almost doubling between 1969 and 1980 to 48 million. In 1990 the population had increased to 66 million and population density measured 220 per square kilometre, much higher than Indonesia (100), Malaysia (54) or Brazil (18) – all of which started to experience deforestation later, though they now have higher rates.

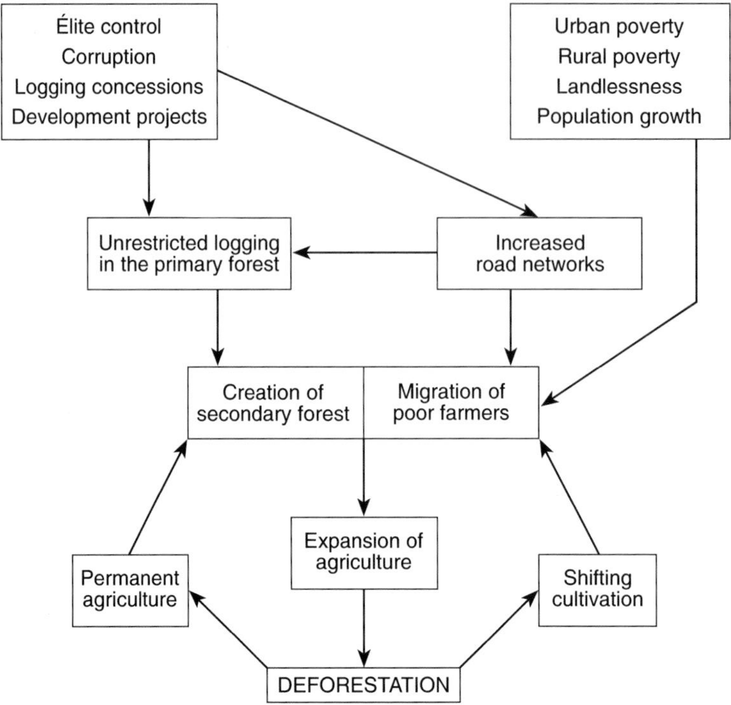

Figure 1.2 The deforestation process in the Philippines (Source: Kummer, 1991, p. 96, Figure 2)

However, Kummer's analysis (see Figure 1.2) challenges any suggestion that population pressure is the root cause of deforestation. Rather, he points to control of the Philippines economy and society by the ruling élite as the cause of both strands of deforestation. Logging concessions have been given to the élite, who also have the means to use influence and bribes to export logs illegally. The financial gains have not been reinvested in the Philippines economy which has grown more slowly than those of its neighbours, with a serious decline in the early 1980s. Even when economic growth was occurring at around 5 per cent in the period from the 1950s to the 1970s, living standards among the poor majority did not rise. Birth rates remained high and there was substantial migration to the capital, Manila, in search of work, and into Mindanao and the highlands in search of land. Indeed, it has been suggested that the élite encouraged migration into logged land to defuse political protest about poverty or logging.

Finally, it is worth noting that the physical geography of the Philippines has also played a part in deforestation: it is clearly easier to export logs and escape legal control in a country consisting of hundreds of islands than it is in a country like Thailand where most exports pass through the capital city.

The experience of deforestation in the Philippines is distinctive; in other countries different factors have been influential. For example, Brazil has been less affected by log exports, but greater interest in developing the forest has been shown by large landowners. Malaysia has been a major exporter of timber, but has established commercial plantations on logged areas. Indonesia is a more recent parallel to the Philippines, partly because logging organizations from the Philippines are active in Indonesia. In Papua New

Guinea forest-dwellers have legal ownership of forest, but they have been persuaded to grant concessions to loggers at very low cost. Underlying these national differences, however, are some broader factors: (i) the international market for tropical timber makes logs a very valuable commodity; (ii) the existence of large areas of mature forest makes destructive cutting the cheapest form of extraction; (iii) forest concessions are usually short term so no incentive exists for longer-term management; and (iv) the balance of power favours the urban élite over the rural poor, and forest-dwellers are usually economically and politically powerless. Within this complex of factors, population growth undoubtedly plays a role, but it is increasingly seen as a symptom rather than the root cause.

Although the pressure from logging and from cultivators remains intense, there are indications that the future need not be as exploitative as the past has been. Ecological studies now suggest that tropical forest has considerable capacity to recover from catastrophe, so logging need not be the end. Foresters are developing techniques to give sustainable yield. Economic analyses in Amazonia show that the income from collection of forest products can exceed that from logging or ranching. The potential value of tropical plant species in medicinal use is being increasingly recognized (though this could lead to a new form of exploitation in which multinational drug companies exploit and patent tropical plant products without recompensing the countries from which the product is collected). Well-managed plantations can avoid the soil erosion and flooding caused by indiscriminate deforestation, though they may be faced with problems of chemical pollution and social inequality. Finally, in a number of countries, forests have been designated as national parks with both local and international support and with the intention of preserving some examples of primary forest for ecological study and recreation.

Tropical rainforests are the earth's most complex and varied ecosystems and it is unlikely that any universal panacea can save them. Their extent will certainly reduce in the near future, though there are grounds for hoping that some parts, at least, can be protected from destruction. However, if they are to be saved, change must also be implemented outside the forest areas – the regulation of the international trade in timber and the provision of jobs or agricultural land for poorer people in the less developed world must be improved, and people everywhere must adopt a more responsible attitude to nature. The American tradition of conservation as 'wise use' may provide a prototype: this was formulated at the turn of the last century to provide a sound basis for the use of the new lands of the western United States.

1.2.2 The last frontier: Alaska

When it became the 49th state of the United States in 1959, Alaska increased the nation's area by nearly 20 per cent. The new area included vast stretches of unexplored land and untapped resources. But when Secretary of State William H. Seward negotiated its purchase from Russia in 1867, it was known as 'Seward's Folly.' Since its acquisition, its settlement and exploitation have been hindered by its distance from the rest of the nation, the climate and terrain, and the slowness of communications. Many problems still stand in the way of immigration and economic development, and Alaska continues to be the country's last frontier.

(The New Encyclopædia Britannica, 1985, Vol. 29, p. 438)

This quotation shows that, in some respects, Alaska was seen as a continuation of the western frontier of the USA, for, like the western frontier, it was new, vast, remote, 'unexplored' and 'untapped' and the assumed response was 'immigration and economic development'. In terms of climate and terrain, it was even more formidable than the western frontier, having the highest mountains, the largest glaciers and the only areas of permafrost in the USA. It was widely accepted as the last frontier, though more frequently through the fiction of Jack London (*White Fang, The Call of the Wild*) than from direct experience of actually going there. But, as with other areas of wilderness, the acquisition of Alaska gave rise to questions and doubt. How had it become a possession of Russia? Were there no indigenous inhabitants? Was it calculation or folly that had led to its acquisition by the USA?

The 'new' area was not, in fact, very new, even in terms of its discovery in 1741 by Vitus Bering, a Dane working for the Tsar of Russia. Neither was it exactly 'untapped': nearly a century of exploitation for the fur trade had reduced an abundant sea otter population to near extinction. In broader terms, this was in fact the first part of North America to be settled by humans, who crossed the Bering Strait when sea levels were lowered by the Ice Age. After at least 15,000 years of human occupation, Alaska had four distinct cultures: the Tlingit on the south-east coast, the Aleuts on the Alaska Peninsula, the Inuit on the western and northern coasts and the Athabaskan in the interior. All lived as hunter-gatherers, and, as a result of the abundance of salmon, the Tlingit were among the most affluent pre-agricultural cultures known.

Early exploitation by Russians, British and Americans relied on the coastal resource-base of fish and seals, but finds of gold began to attract prospectors inland and the Gold Rush to the Klondyke in 1897 brought the first substantial settlement. This was hardly a usual settlement, however: before 1914 there were five times as many men as there were women, and all vegetable food had to be imported until agriculture was established in the Matanuska Valley near Anchorage in 1934. Nevertheless, gold, copper, coal, fish and timber were all being exploited before the Second World War. During the 1940s and 1950s the population tripled to 226,000, of whom 80,000 were military personnel and their families. The first influx of military personnel was related to the war against Japan; later the proximity of the Soviet Union was seen to be the threat against which the territory needed to be protected. The military presence resulted in the establishment of a transport network, including the Alaska Highway, which opened up the whole territory to exploration and exploitation. In 1959 Alaska became the 49th state, although it had less than half the population of Rhode Island spread across an area over twice the size of Texas.

Soon after Alaska gained statehood, oil was discovered on the Kenai Peninsula and in Cook Inlet, but the state remained a drain on the national economy (especially after the major earthquake and tsunami of 1964) until the discovery of oil on the North Slope. This triggered a major debate between, on the one hand, those whose chief interests were economic, and, on the other, conservationists: a debate which came to focus on the proposal to build a 789-mile pipeline to link the oilfield on the North Slope to the ice-free port of Valdez. In spite of conservationist fears of damage to land and wildlife, the pipeline was approved; it was opened in 1977. Twelve years later the fears of the conservationists were realized when the *Exxon Valdez* spilled 10 million gallons of crude oil into Prince William Sound, with devastating effects on marine life, including the sea otter population recovering from the earlier effects of hunting. Less spectacularly, but more threateningly, production from the Prudhoe Bay field is now in decline and interests are focusing on the possible exploitation of oil in the Alaska Wildlife Reserve.

There have always been interests other than those of resource exploitation. Pioneer environmentalist John Muir visited Alaska several times in the 1870s and 1880s and his book *Travels in Alaska* celebrates it as even more rugged than his beloved Sierra Nevada and less threatened by sheep herders and loggers. The state has remained a magnet for wilderness enthusiasts ever since, and this has been recognized by the Federal Government: Denali National Park (1917) and Katmai (1918) and the Glacier Bay (1925) National Monuments together total over seven million acres, to say nothing of the 19 million acres of wildlife reserves maintained by the US Fish and Wildlife Service, or the 104 million acres of federal land transferred to parks, refuges and wilderness areas in 1980. Wilderness recreational activities, including fishing and hunting, remain major attractions to tourists and to settlers in Alaska (see Figure 1.3).

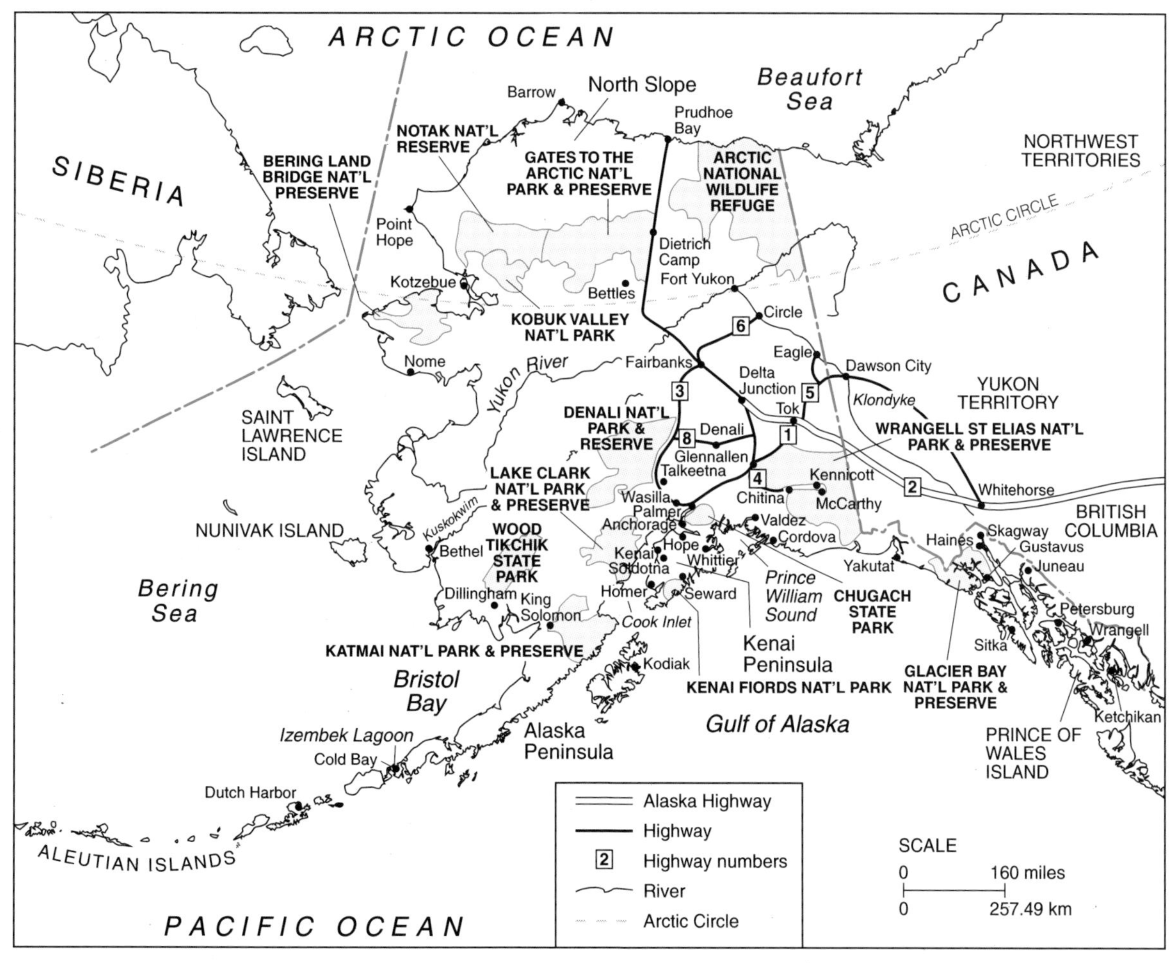

Figure 1.3 Alaska (Source: adapted from Dalby, 1994, p. xiv)

The least well-integrated set of interests in Alaska for the last century have been those of its native peoples who comprise approximately a sixth of the population. As in other parts of the Arctic, contact with Europeans brought devastating epidemics, change in cultural values and economic dependency. Reaction to this finally came in 1967 with the formation of the Alaska Federation of Natives to press land claims. Four years later, in 1971, the Alaskan Native Claims Settlement Act was passed by the US Congress, establishing twelve regional Native Corporations with control of one ninth of the state (an area larger than England) and a settlement of US $962.9 million to extinguish all claims outside these areas. Although this settlement seems generous in relation to previous treatment of native Americans, it actually contradicts their cultural values in crucial ways.

Activity 3 Turn now to Reading A by Mark Nuttall, entitled 'Environmental policy and indigenous values in the Arctic: Inuit conservation strategies', which you will find at the end of this chapter. As you read, note the contrasts between the way Alaska is represented by US interests and linkage with other Inuit communities outside Alaska.

Two striking features are readily apparent from this reading. First, the Inuit regard Arctic environments, not as wilderness, but as a 'system of shared relations', involving both persons and animals, rights and responsibilities. This brings them into conflict both with resource exploiters, who do not recognize many of these relationships, and environmentalists who wish to preserve animals from hunting. Second is the internationalization of Inuit political relationships through the Inuit Circumpolar Conference which has led to contacts with indigenous groups elsewhere, including those from the rainforest regions. This stresses that non-European and non-capitalist views of nature are not just a vanishing residue of the past, but a distinctive and tested approach which could well be relevant to the future. This makes it all the more ironic that the US Congress should have conceded land rights to native Alaskans in the form of shareholdings in corporations owning land as defined by US law and in defiance of the Alaskans' own cultural relations to the land. In contrast, the government of Greenland is evolving a balance between conservation and exploitation which is more sensitive to Inuit culture (and which, in being so, offends international environmentalists in relation to whaling and hunting).

Summary of section 1.2

The case studies of the Philippines and of Alaska, plus references out to similar environments, show that the Western perception of wilderness is largely illusory:

- Both Arctic and rainforest have a long history of occupation by hunter-gatherer groups.
- The apparent 'naturalness' of these areas results from low population densities, nomadism, use of simple technologies and, crucially, cultural practices which sustain ecosystems over a long period.
- These sustainable lifestyles are under pressure both from growing populations of cultivators and from mineral and timber extraction companies, most of which are multinationals.
- The operations of both cultivators and extraction companies cause ecological and cultural disruption and the indigenous peoples gain little economic advantage to compensate them.
- As the major consumers of oil, metals, timber and hamburgers, the citizens of the developed world are the unwitting motivators of these environmental changes.

In some cases, the role of the developed world citizens is more direct. **Cater (1995)*** shows how trekkers have destabilized remote areas of the Himalayas both socially and ecologically. However, like the Inuit Circumpolar Conference and the Greenland government, the government of Nepal has begun to develop a response which seeks to use tourism and improve the living standards of local people without damaging the landscape and ecosystems which bring in tourists from the other side of the world. Their response rejects the notion of a national park in favour of a conservation area where local people are active participants. This challenges the 'exploit or preserve' dualism, which has become central in US thinking, in favour of a more integrated view of society and nature. In so doing it also challenges views which have been central to globalizing capital and seeks alternatives, whether older or newer, as a basis to sustain future lifestyles.

In sections 1.3 and 1.4, I outline the attitudes to nature which have underlain past and present lifestyles of different societies at different times.

Activity 4 If these wilderness areas are in fact occupied, how can the concept of wilderness have such currency? Make a brief note of your response before you continue.

There are a number of possible strands to an answer to this question. First, low-impact nomadic lifestyles may not be perceived by incomers (after all, the thoroughly humanized agricultural landscapes of Britain's national parks are seen as 'natural' by many visitors). Second, and more insidiously, indigenous

* A reference in emboldened type denotes a chapter in another volume of the series.

inhabitants may not be recognized as 'fully human' by more 'advanced' cultures: from the Spanish conquest of Mexico, through the Wild West, to twentieth-century colonies, clearance and conquest have been justified on the grounds that, as perceived by the colonizers, there was no already existing human society or culture. Third, there may be a need to experience, or at least imagine, a wilderness beyond the everyday humanized environments: after all, the presence of wilderness has been the common experience of succeeding generations of humanity up to this century. We may still carry with us attitudes that were formed in very different conditions in the past.

1.3 The transformation of nature

Earlier parts of this chapter have given some examples of contemporary pressures on wild nature and some indications of a range of ideas about wilderness, nature and conservation. This section takes a broader perspective, showing how impacts of human societies on environments have developed over space and time. This is followed in section 1.4 by an account of how different kinds of societies have used different concepts to relate themselves to nature. The account given here relies particularly on Glacken (1967), Simmons (1989) and Oelschlaeger (1991).

1.3.1 The growth and spread of human impacts on environments

The extent and severity of human impacts on environments depend on the way societies organize production. Three broad methods have existed. For most of human history, human groups were small and mobile and lived as hunter-gatherers. From origins in East Africa, human hunter-gatherers spread over all parts of the world, except Antarctica. From about 10,000 BC, agricultural societies began to appear, at first in the Near East from where they spread into Asia and Europe, especially in temperate latitudes. Over the last few centuries, industrialized societies have appeared, starting in the UK and spreading into Europe, the USA and East Asia. Industrialization has not only transformed relations with the environment in industrial areas or countries, it has done so even in surviving agricultural and hunting societies. This 'rippling out' of the effects of industrialization was carried by trade and imperialism; now it covers the whole globe, though the level and nature of development remains extremely uneven. Where early impacts on environment were local and often reversible, today's impacts are global and perhaps irreversible.

However, we should not underestimate the effects of past societies. Hunting and gathering could only support very sparse populations, but they were practised for tens of thousands of years in environments ranging from tropical forests to hot deserts and to the high Arctic, and some remnants exist even today in remote and extreme locations. Many hunter-gatherer groups may have existed in balance with their environments for long periods, taking only a proportion of the yield of fruits, vegetables and animals. But there is ample evidence that some groups had significant impacts. The first major impact, evident in the fossil record, is the extinction of large animals by human groups. Such extinctions were numerous in North America but can also be traced in Australia, New Zealand, Madagascar and Indonesia. Large animals, mainly mammals but also flightless birds and marsupials, often had long life-spans and few natural enemies, and, consequently, low rates of reproduction. As such they would have been quickly reduced in numbers by hunters. The loss of major species was itself a significant change, but it would also have led to change in vegetation cover as a result of reduced demand for grazing. Other groups have changed vegetation cover directly and deliberately by the use of fire: this was done to favour certain species over others and even to ensure that seeds or nuts were all produced at a given time. Fire was extensively used by Australian Aborigines and was a crucial factor in moulding the distinctive vegetation cover which European migrants thought of as natural. Regular burning has also been used in the Mediterranean lands, at first by hunters and later by shepherds, to produce the distinctive thorny scrub called *maquis* or *garrigue* in preference to the natural cover of oak. There are some indications of hunter-gatherers using other techniques, such as diversion of stream channels and transplanting seedlings, but change to vegetation cover by fire and modification of animal numbers are the most extensive effects. Surviving hunter-gatherer societies show complex adaptations to ensure long-term survival, in terms both of detailed knowledge of potential food resources, and of beliefs and customs which prevent them from degrading the systems they depend on. Recent cases of negative effects, such as the over-hunting of beaver in nineteenth-century Canada or elephant-poaching in contemporary East Africa, result from international markets which provide the demand the hunters respond to. The result of the long history of occupation by hunter-gatherers is that virtually all parts of the world have ecosystems modified by humans. Unless we regard humans as part of nature, natural environments are extremely rare.

Settled agriculture, which began to emerge around 9000 BC, and pastoralism, which emerged somewhat later, have involved much more systematic environment change than has hunting. The early spread of crop plants, animal husbandry and associated techniques showed that there were early long-distance links between settled areas. Perhaps the greatest single change was deforestation: the temperate woodlands of Europe, North America, the Mediterranean lands and China were all decimated by agriculturalists. Nowhere was this process more dramatic than in the UK where a largely forest-covered island was transformed into the least forested part of Europe. In turn, deforestation exposed the soil to erosion and much soil was lost, especially in drier and hillier areas. The loss of the cedar forests of Lebanon and the erosion of hillsides in Greece were already being commented upon 2,000 years ago. In long-settled areas some riddles persist: for example, the Romans grew wheat in areas of North Africa which are now desert – but is this desert a result of climate change, or of mismanagement of the land? A very early environmental change was brought about by irrigation – indeed, irrigation was crucial to some of the earliest civilizations. Where some of those civilizations collapsed, as they did in Mesopotamia, it is difficult to establish whether the breakdown was brought about by environmental change, the collapse of the necessary social organization, or some combination of the two. But irrigation has transformed areas as diverse as Egypt and the mountains and plains of South and East Asia, where tier after tier of terraces are a testament to generations of sustained effort by agriculturalists.

Agriculturalists have also transformed landscapes by drainage, from simple field ditches to the massive dykes and channels used, especially by the Dutch, to reclaim land from the sea. The use of windmills to pump away surplus water is a reminder that early agriculturalists could call not only on the energy of domesticated draft animals but on sources of energy like wind and water. Wind power was also crucial to the sailing ships that drew the world together in the fifteenth and sixteenth centuries. Trade in precious metals, agricultural products and handicrafts contributed to larger-scale production of luxury goods from gold to spices and silk. Crop plants were spread to new areas, most famously in the establishment of Brazilian rubber trees in Malaya. Natural vegetation was increasingly replaced by plantations of tea, coffee, cocoa, rubber and sugar. In turn, the profits from trade created the institutions and markets which could finance and sell new industrial products and generate new kinds of environmental impact (see Plate 3).

It was no coincidence that Britain, the country which had come to dominate world trade in the eighteenth century, became the cradle of the industrial revolution in the nineteenth. First the textile industry was mechanized and powered by water or by steam engines. Then iron and steel production was revolutionized by the use of coke in blast furnaces, initially in the Ironbridge Gorge and later in industrial districts in South Wales, Cumbria, Central Scotland, the North-East, Yorkshire and the Black Country. The last name is a clear indication that this early industrialization caused massive pollution of land, air and water. Coal smoke was perhaps the most widespread pollutant, but in areas like Cheshire there were even more devastating pollutants as the infant chemical industry released hydrogen chloride gas into the air, causing acid clouds which killed vegetation for miles around.

Industrialization did more than revolutionize production. It prompted rapid population growth and even more rapid migration. In 1801, Britain had a population of 10 million, of which 10 per cent lived in towns and cities. By 1901, the population had grown to 25 million, of which 77 per cent lived in urban areas. The rapid growth of large industrial towns and cities was one of the major phenomena of the age, and towns like Birmingham and Manchester grew from small beginnings to centres of a million people or more. It is one of the major paradoxes of the time that mass movement of people into cities, where overcrowding and squalor were widespread, was actually part of a process in which population grew rapidly. This is explained by the fact that overall mortality rates fell well below fertility rates. Only well into the twentieth century did European fertility rates fall towards mortality rates and population growth reduce. Exactly how and why this occurred remains a mystery because it involved new attitudes and lifestyles rather than any technological innovation (Chapter 3 discusses this issue further). From this it can be seen that industrialization involves the transformation of whole societies, not just the building of a few factories.

One of the crucial ramifications of industrialization has been the consequent developments in agriculture, forestry and fishing. The application of machines and chemicals to agriculture has both transformed productivity and increased the capacity to convert woodland, wetlands and heathlands to arable use. But it has also dramatically increased adverse impacts on wild plants and animals and begun to contaminate water with nitrates. The increased capacity of modern fishing boats and whalers has severely reduced fish and whale populations, in some cases effectively destroying fisheries and perhaps threatening extinction of some whale species. In partially enclosed seas, the deleterious effects of over-fishing are compounded by pollution from urban and industrial waste.

One threatening impact of industrialization has been the development of technologies that control rates of mortality: living standards and attitudes and practices of fertility control have not been able to keep pace. As a result, world population has grown at an ever-accelerating rate (Figure 1.4). For most of human history there have been only a few million hunter-gathers on earth. Agricultural societies supported a few hundred million, probably reaching about 600 million by 1500 AD. The century of industrial revolution doubled world population to 1.7 billion in 1890, and growth has accelerated since then, giving a total population of over 5 billion by 1990. To make matters worse, the current growth rate is 2 per cent a year, implying a further doubling in only 35 years. So a world population of 10 billion is now regarded as probable in the immediate future. Such a growth implies even further extension of agricultural land, more intensive food production and additional environmental impacts. If economic growth is fast enough to continue to increase average living standards, as it has done in the past, growing population multiplied by growing consumption will produce rapidly multiplying impacts on the environment. The prospects must be bleak unless we can work out, and implement, new ways of organizing production which have reduced effects on the environment. In doing so, we will have to reconsider our attitudes to nature. As a starting point, it will be helpful to know what kinds of attitude existed in the past.

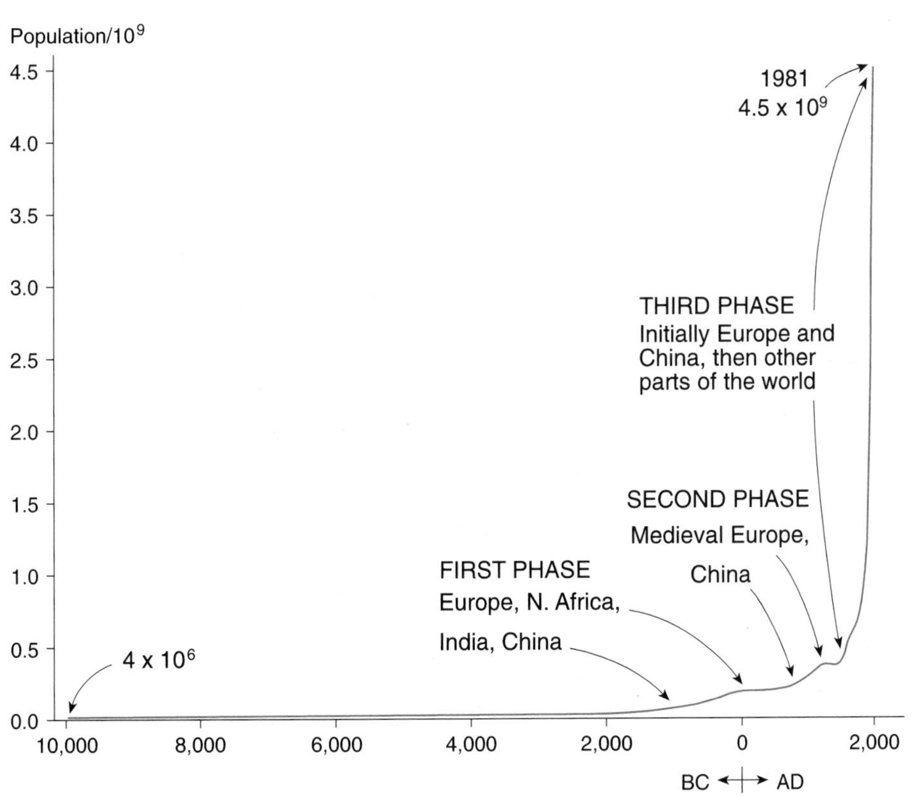

Figure 1.4 Global population growth from 10,000 BC to the present (Source: Silvertown and Sarre, 1990, p. 149, Figure 4.19)

1.4 Concepts of nature and wilderness

1.4.1 Hunter-gatherers

Local variations aside (tools, geography, game) our prehistoric ancestors lived well by hunting and foraging; they buried human remains and were religious; they had an understanding of nature's ways that reflects an intelligence equal to our own; and their art reveals a rich imaginative life. In short, Paleolithic people were not the ignorant, fierce brutes that civilized humans imagine, a fact that places the onus on us – especially on those who would grapple with the idea of wilderness – to re-assess our concept. We come from that green world of the hunter-gatherers.

(Oelschlaeger, 1991, p. 6)

In spite of the definitive tones of the quotation above, it is not possible to be certain about the customs, let alone the concepts, of Stone Age hunter-gatherers. The sources available are archaeological evidence (especially from caves: of living arrangements, food remains and art forms), descriptions of first contacts with aborigines by explorers, and study of contemporary hunter-gatherers' lifestyle, memories and myths. Unfortunately, archaeological evidence is often enigmatic, explorers' accounts are often coloured by their own world views, and contemporary groups have been affected by more modern lifestyles. So accounts of hunter-gatherers must be 'imaginative reconstructions' – and as such are probably also shaped by the author's particular concerns. Early biologists' accounts of primate-hunting, for example, were used to argue for the decisive role of aggression and male domination; later critiques stressed co-operation among females. Any attempt to identify hunter-gatherers' concepts about their environment must therefore be modest in its claims.

What is clear is that hunter-gatherers were typically nomadic, with a small kin group moving around a very large area in search of a varied, and usually abundant, food supply. Nomadism dictates a very limited weight of material possessions, and it may also have been that births of children were spaced so that there were not too many infants to carry at any one time. Population densities were very low by today's standards. Technology and social co-operation allowed these groups successfully to hunt large animals and exploit large numbers of plants for specialized uses.

Box 1.1

Anthropologists have recognized a range of belief systems about the relation of observable natural objects or events to supernatural causes:

- Animism describes beliefs that individual animals, trees, rocks, springs, etc. have their own spirits.
- Pantheism asserts that god is everything and everything is god.
- Polytheism recognizes many gods, each of which has a particular domain of nature or society. These gods are located partially outside the world but intervene in it actively.
- Monotheism believes that there is only one God.

Cave paintings and studies of recent groups suggest that hunter-gatherers were animistic – recognizing spirits in animals, trees and rock formations – and that particular groups often had a special relationship with totemic animals (see Box 1.1). In Europe, bears and bulls seem to have had a special significance, while to Australian Aborigines kangaroos, emus and even honey ants are totems. The need to hunt sacred animals required special rituals or procedures, so food gathering could not become exploitative.

As groups were dependent on their territory, concepts of use rights must have existed, but there seems to be no evidence of ownership of land. On the contrary, Australian Aborigines see themselves as belonging to the land. Many hunter-gatherer societies have myths which account for their origins and reasons for moving to their present area. For many, their history is charted in the landscape.

Oelschlaeger emphasizes the importance in many hunter-gatherer societies of myths of the Great Hunt and of the central role of the Magna Mater or Great Mother. His analysis of these societies is summarized in Table 1.1 – but note the key word 'likely'. His overall conviction is that hunter-gatherer (or hunter-forager) groups do not separate wild nature from culture, but see themselves as organically related to other living things. His argument is that the conceptual separation of society and nature came with agriculture (Oelschlaeger, 1991, p. 32).

Table 1.1 *Conjectures on a palaeolithic idea of wilderness*

Palaeolithic hunter-foragers likely:
– believed that, irrespective of place, nature was home
– regarded nature as intrinsically feminine
– thought of nature as alive
– assumed that the entire world of plants and animals, even the land itself, was sacred
– surmised that divinity could take many natural forms and that metaphor was the mode of divine access
– believed that time was synchronous, folded into an eternal mythical present
– supposed that ritual was essential to maintaining the natural and cyclical order of life and death

Source: Oelschlaeger, 1991, p. 12, Table 1

1.4.2 Agricultural society

Information about agricultural societies is much more abundant: they are more recent, produced larger and permanent settlements and, above all, left written records. Domestication of crops and animals opened up a separation between society and nature in which natural species were devalued – becoming 'weeds' and 'pests'. Harmony with nature was displaced by a concern with the fertility of domestic species. Humans became more self-conscious as agents of environmental change. These concerns became much more significant around 4000 BC as large-scale civilizations emerged for the first time in Mesopotamia and Egypt. The use of irrigation resulted in food surpluses, thus allowing increases in population, which in turn gave rise to complex social hierarchies and substantial cities, and to public works such as pyramids. These cultures were polytheistic rather than animistic and their writings suggest both a growing conviction that the earth had been designed for humanity, as well as pride in building on and progressing beyond the natural foundations.

Many societies express their ideas about the relationship between people and wild nature in their myths of origin. In this way, though such ideas may be implicit and even contradictory, they nevertheless have lasting influence. One of the earliest examples of such a myth is the Epic of Gilgamesh, dating from the Sumerian civilization of Mesopotamia which flourished 5,000 years ago.

As the myth goes, Gilgamesh and a friend entered the forest and experienced the thrill of confrontation with the wilderness: 'They stood in awe at the foot of the green mountain. Pleasure seemed to grow from fear for Gilgamesh. As when one comes upon a path in woods unvisited by men, one is drawn near the lost and undiscovered in himself: He was revitalized by danger' (Mason, 1970, p. 35).

However, Gilgamesh went on to kill the forest god Humbaba, symbolizing the deforestation needed to establish irrigated agriculture between the Tigris and Euphrates rivers. This agriculture allowed the growth of the population and specialized urban and priestly occupations, but the environment was difficult and necessitated substantial planning to sustain the irrigation systems and resist threats from invaders, leading to an autocratic society later analysed as 'oriental despotism'. Indeed, cycles of floods and droughts, plus salination of irrigated soils, ultimately led to a decline in yields and cultivated areas by about 2000 BC, the first large-scale ecological catastrophe in human history.

The difficulty of the environment required active divine intervention:

A pantheon of gods, numbering in the hundreds, supervised the universe. They differed greatly in function and importance. The four most important duties were the heaven-god (An), the air-god (Enlil), the water-god (Euki) and the great mother goddess (Ninhursag). They usually headed the god lists and were often portrayed as acting together in a group. The Mesopotamian, unlike the Egyptian, did not take the cosmic order as given; it was something that had to be constantly maintained and administered like a state by a council of the gods. Mesopotamian nature, unlike Egyptian nature, was unruly.

(Tuan, 1974, pp. 89–90)

Two significant developments occurred in the eastern Mediterranean up to two millenia ago: the emergence of Judaism and, later, Christianity as monotheistic religions, and the pursuit of rationalism in ancient Greece. The Old Testament antipathy to cities, and the positive role of wilderness, were reversed by Christianity's encounter with Greek rationalism, initially brought about by St Paul. It has been powerfully argued by Plumwood (1993) that Plato's philosophy signified a decisive step in the alienation of human society from nature, as well as of women from men, masters from slaves. She argues that Plato's view, which separates the 'world of abstract forms' (or 'ideals') and reason from the visible 'world of changes', and his logic, which stresses identity through negation, provided the basis for a 'master rationale' of

dualism

dualisms. These dualisms put in opposition to one another human reason and the chaos of nature; male reason and female weakness; human and animal; master and slave. The dualisms were more than dichotomies, however: they identified superior and inferior and subordinated those categories regarded as inferior – nature, woman, animal, slave – to the rationality of the 'master'. They implied that the subordinated, negative sides of the dualisms somehow had an identity, so woman was natural, animal, slave. Plato's logic gave earthly superiority to selected men but identified the purpose of existence as lying in the 'world of abstract forms'.

This logic was also applied to Christianity, identifying the purpose of existence as lying in the future of the human spirit in heaven. Since Plato's philosophy and the Christian religion were the dominant influences on European society for a millenium and a half, Plato's definitions of superiority and inferiority have had enduring consequences for exploitation and oppression of the 'inferior categories'. This is particularly ironic as Plumwood identifies the driving force behind Plato's philosophy as the need for a belief system which would motivate warrior heroes not only to live for the state but to die for it in the inevitable and frequent wars of the time.

Plato's influence created a hierarchy of attitudes to land as well as to life forms. Wilderness was devalued to meaningless chaos, while settled areas became the earthly optimum through the rationalizing effects of men (as the 'masters'). However, even the rational landscape of city and agriculture remained far inferior to the abstract realm of the 'world of forms', or, as defined by Christianity, the world of heaven.

Both religion and rationality contributed to a further separation between human society and nature: worship of a single God located *outside* nature devalued nature as a reference point, while stress on self-consciousness and rationality emphasized the difference between the human and the natural (see Plate 4). Of course, this separation was far from absolute, and Glacken (1967), in his magisterial analysis of European writing about society and nature from Classical Greece to 1800, recognized three major themes: (i) that the earth had been designed by God as an abode for mankind (*sic*); (ii) that human cultures were shaped by the natural features of the places in which people lived; and (iii) that human society was an agent of environmental change. Most of the emphasis was put on the perfection of the creation, but the ancient Greeks were already aware of deforestation and soil erosion as result of human activity.

Asian agricultural societies, on the other hand, had different attitudes to nature from those current in Europe, mainly because they were less affected by monotheistic religion and shared basic assumptions about the unity of humankind, nature and divinity. India's complex religious framework is partly explained in Gadgil and Guha's (1992) account of the ecological history of India as the result of the encapsulation of animistic groups of hunter-gatherers by monotheistically inclined cultivators, who had a need to clear at least part of the forest.

Hinduism is ambiguous in relation to environmental attitudes: the animistic elements promote a reverence for nature for its own sake, but the monotheism can lead to a devaluation of nature as merely the product of and pointer towards the underlying deity (Brahman). At best, however, as outlined by Balasubramian (1991), Hinduism seems to have the potential to balance the spiritual and the organic, and self-improvement and responsible behaviour towards nature.

Buddhism, a religion without a named god, originated as a protest against Hinduism and initially appealed to lower-caste groups, many of which were descended from animistic hunter-gatherers. It is humanistic and conscious of nature and argued to be an ideal basis for environmental ethics (de Silva,

1991). In Japan it has been modified both by contact with Confucianism in China and with Shintoism in Japan to produce the version known as Zen Buddhism. Shintoism was another polytheistic village religion with a high regard for nature and a belief in the sacred character of natural features such as mountains, forests and promontories. It was characterized particularly by the link it made between the sun, the gods and people, since the royal family was seen as divine and ancestors took their place as gods. As a result, Zen Buddhism encourages people to live simply and in harmony with nature in their locality and to submit to imperial authority. However, the emphasis is on a tamed nature, and, perhaps encouraged by the Japanese distinction between in-groups and out-groups (Hashimoto, 1993), these positive attitudes to nature are applied selectively. The result, for Japan, has been a contradiction between high standards of energy efficiency and environmental regulation at home and a reputation as an industrial polluter and international eco-predator abroad, especially in relation to exploitation of the rainforest and of marine mammals (Maull, 1992; Tsumi and Weidner, 1989).

The indigenous attitudes to nature of the Asian agricultural societies are in principle less alienated from nature than those of Europe. They seem to indicate a rather socially conformist but environmentally low-impact lifestyle. However, in practice they have been largely displaced by attitudes of European origin – British or Dutch in South and South East Asia, Marxism-Leninism in China and American in Japan and the Philippines – which began to emerge between the sixteenth and eighteenth centuries as a result of imperialism and the globalizing economy. Nevertheless, Asian religions have had a significant impact on environmentalists from Thoreau to the present day.

1.4.3 Europe: modernity and beyond

In Europe the most extreme conceptual opposition between humanity and nature was achieved during the Enlightenment and embodied in the vision of *modernity* and the industrialized societies to which this gave rise. Modernity involved an anthropocentric view: a view centred on people that was secular, democratic, rational and materialistic. Trade and science combined to launch a new industrial society in which economic growth was increasingly geared to mass consumption. Nature was to serve as a source of raw materials, made available as science unlocked the obstacles to progress. Following the work of the French mathematician and philosopher René Descartes and of the English physicist and mathematician Isaac Newton in the seventeenth and early eighteenth centuries, nature was increasingly seen as a complex mechanism, as mere matter in motion, and so was stripped of any deeper significance. Land, plants and animals were increasingly seen as commodities. Later, the Scottish economist Adam Smith contributed self-interest and 'the invisible hand' as the route to wealth. Capitalism was unleashed into the world and began the process of cumulative economic and technological growth which has built up into the globalizing economy of today. On a world scale, this process was accompanied by unprecedented rates of population growth and environmental change.

modernity

The attempt to substitute rationalism for traditional authority, whether this had come from the monarchy or the church, was a crucial aspect of modernity. Science played a central role both in providing a new understanding of the laws which underlay the apparent chaos of nature and in developing new materials and technologies. Much of the inspiration for modern scientific thought came from a rediscovery of Greek science, especially via Islamic scholars and libraries. However, science could not answer ethical questions and many attempts were made to rethink these in the light of the new rationalism. Utilitarianism was formulated by the English philosopher and social reformer Jeremy Bentham in his *Principles of Morals and Legislation* (written in 1780, published in 1789) as a guide to policy which did not rely on divine or earthly authority but on maximizing pleasure and minimizing pain. However, although these latter were clear principles, they proved difficult to calculate or predict. In practice, economic power and demand came to be accepted as sufficient to guide policy-making.

In this process of change, some of the restraints of community and religion weakened: no longer was it widely believed that humanity had a duty of stewardship over nature or held land from God or the monarch. Increasingly, people regarded themselves as freehold owners with the right to deal with land, animals, plants and resources in their own interest.

Industrialization soon began to produce pollution and squalid cities, however, triggering new ideas which began to challenge the ideals of modernity even before it had displaced the theological world view. These ideas sprang initially from the heart of the emerging industrial societies, from German philosophy and English poetry, and were strengthened by crucial impacts from the emerging earth and life sciences. German philosopher Immanuel Kant, having celebrated the mechanistic scientific world view in his *Critique of Pure Reason* (1781) and, having attempted to reconcile it with ethics in the *Critique of Practical Reason* (1788), found himself unsatisfied. His *Critique of Judgement* (1790) went on to confront the issue of human experience and recognized a realm of the aesthetic.

Philosophers still debate the significance of the third *Critique*, but poets were quick to celebrate the aesthetic. English poets like Samuel Taylor Coleridge, William Wordsworth and Percy Bysshe Shelley went on to develop the Romantic school of poetry which celebrated direct experience of nature and the significance of intuitive rapport with the wild. Their sense of nature approached pantheism, and they went far beyond admiration of pleasant environments to confront *the sublime*: the awe-inspiring qualities of high mountains, waterfalls, snowfields and raging seas, that remind people of the insignificance of humankind in relation to the natural world.

Furst (1977), one of the leading scholars of Romanticism, argues that it was not a uniform movement, but one with a distinctive geography. For example, precursors to Romanticism started in England with works like Thomas Gray's famous 'Elegy Written in a Country Churchyard' of 1750, at a time when France was dominated by Neo-Classicism. From about 1770, Germany entered the Romantic period with works by Johann von Goethe, poet, novelist and playwright who led the German Romantic *Sturm und Drang* movement, and by Johann von Schiller, poet and historian and a principal dramatist of his time. True Romanticism was formulated in Germany from 1790 onwards, affecting history and philosophy as well as literature. From 1820, the English Romantic poets dominated the movement, followed from 1830 by French authors, notably Victor Hugo. Each country was influenced by its own traditions – the English more individualist, the Germans more idealist and the French dominated by the Revolution. The distinguishing features of the whole movement were increasing individualism, a stress on the creative imagination and an emphasis on feeling. While some early Romantics emphasized contact with nature, the movement tended towards introversion and fantasy, especially in Germany.

Simultaneously, the newer scientists were building views of the world which challenged the mechanistic view. The Scottish geologist James Hutton began to recognize a world in which geological and biological processes interacted in ways far more complex than a mere machine, and Charles Darwin's conception of evolution through natural selection not only suggested how these global processes could have evolved together, but directly contradicted the idea that the human species was separate from nature. Evolution and the new science of ecology increasingly insisted that human beings were a part of nature and that agricultural and industrial technology not only were dependent on natural materials and processes, but were also increasingly able to disrupt them. In the twentieth century, science has moved far beyond the clockwork universe recognized by Descartes and Newton – indeed, some aspects of contemporary science seem even more fantastic than the subterranean universes formulated by some of the Romantics. By the 1970s, experience of the problems brought about by population increase, developments in technology, and economic growth had generated a widespread reaction and a range of environmentalist responses. However, industrial society, and especially its capitalist version, continued to extend and consolidate its hold over society and nature worldwide.

Summary of section 1.4

mode of production

- This brief sketch of human society's impacts on and attitudes to nature has placed *mode of production* at centre stage. Hunter-gatherers, cultivators and industrialized societies have recognized different natural phenomena as resources, used different techniques to exploit them and made different kinds and degrees of impact on natural systems. Partly because of their reliance on geographically variable resources, each mode of production has its own forms of geographical unevenness.
- However, modes of production are not just technologies: they include the social values and structures that regulate behaviour. For much of human history, myth and religion have been a crucial influence; more recently science and political ideology have added their voices. These influences are never unequivocal but there do seem to have been two crucial changes: first, the move from animism for hunter-gatherers to gods more or less separate from nature in early agricultural societies; and, second, the displacement of gods as rulers by the rule of scientific law during the Enlightenment. Clearly, these are not monolithic changes, and values may persist which originated far back in history; indeed, global society is extremely heterogeneous in its values. Nevertheless, it is possible to identify dominant values and their environmental consequences.
- The notion of a change of dominant values at key points in social evolution is a particularly important one for us, as many social scientists argue that we are living now at such a time of change: the replacement of modernity by a new form of social (dis)organization.

Activity 5 Accounts like the one in this section can never be complete; indeed, they may omit or de-emphasize aspects which other people regard as crucial. Reflect for a moment on other implications of religious writings like the Genesis story and the expulsion from the Garden of Eden to which I referred in section 1.1.1. Then turn to Reading B by Carolyn Merchant, entitled 'Gender and environmental history', which you will find at the end of this chapter. As you read, think about the following question:

- How does Plumwood's focus on dualisms relate to Merchant's two ways of incorporating a gender perspective into environmental history?

1.5 Globalizing society and representations of nature

The second half of the twentieth century has seen a huge increase in human activity: population has doubled, industrial production has increased sevenfold and international trade even faster. The capitalist world economy is almost global in its penetration: resources are extracted from the ends of the earth and pollution is ubiquitous in atmosphere and ocean. Yet, as **Allen and Hamnett (1995)** make clear, the shrinking of the world is selective and uneven. Its effects are complex and contradictory and it both creates and reflects the differences between places. Rapid change has broken down many institutions: states seem decreasingly able to resist international pressures, political parties and trade unions have lost authority and even the family seems to be crumbling. Optimists point to economic growth and increasing numbers of people who have 'never had it so good': they point to increasing choice of individual lifestyles to suit every need. Pessimists point to increased inequality and growing crime and international conflict. No-one could possibly believe that all is for the best in the best of all possible worlds.

One of the paradoxes of the modern world is that the globalization of human society has left individuals with increasingly fragmented experience. This fragmentation is nowhere more apparent than in environmentalist thought: consciousness of environmental problems is increasingly widespread but the range of environmentalist diagnoses and prescriptions is baffling. One key distinction between approaches is the degree to which they are anthropocentric (focused on solving problems for humans) or ecocentric (accepting that environmental policy should consider the interests of nature as well as those of humans).

Activity 6 Turn now to Reading C, 'Exploring the environmental spectrum: from anthropocentrism to ecocentrism', by Robyn Eckersley. Consider the following questions as you read:

- Can you identify a parallel difference between Merchant's critique of Worster and the human welfare ecology view of resource conservation?
- If the authors Eckersley quotes are right in seeing multiple reasons for nature preservation but futility in demarcation of 'islands' of wilderness, how can the contradiction be resolved?
- Eckersley presents her spectrum of positions as if it ranges from a feasible *anthropocentrism* to a less feasible but more desirable *ecocentrism*. Do you accept that we must choose a particular position on the spectrum? If so, which would you choose? If not, how else can the problem be defined?

anthropocentrism
ecocentrism

My view is that each position on the spectrum makes a valid case: we do need efficient methods of production, a quality environment for reproduction and some kind of ecological preservation. Even extreme anthropocentrics want enough nature to support their life-support systems and even extreme ecocentrics want to live as humans and not as wild animals. Therefore, although contemporary environmentalist thought might be wide ranging, and although a demarcation of 'islands' of wilderness may be seen by some as a futile approach to the preservation of nature, there is a consensus that nature *should* be preserved and current thinkers are now formulating ways of expressing the need for wilderness in new terms, one of the most persuasive of which is the case for biodiversity (see Box 1.2).

Box 1.2 Biodiversity

The inclusion of a Biodiversity Convention within the 1992 Earth Summit, and the refusal of the USA to join the overwhelming majority of countries in signing it, alerted many people to the importance of this new concept and the possibility that it would be politically divisive.

The concept seems highly appropriate to a world of socially modified ecosystems. It does not concern itself with naturalness but with the diversity of species, of gene pools and of ecosystems, all of which have ecological value and potential economic value. Although only a minority of species are known to science, it is clear that species diversity is vastly greater in Equatorial regions than in temperate ones, with rainforest as the most diverse ecosystem of all. Consequently, rainforest destruction is the most serious threat to biodiversity. However, just as rainforest destruction has resisted environmentalist protest, the protection of biodiversity sets the possessors of diverse ecosystems, mostly less developed countries, at loggerheads with interests in the developed world.

A prime example is plant-based drugs. From quinine and cocaine to rosy periwinkle and reserpine, about half of modern drugs derive originally from tropical plants and sustain a global market worth over US $50 billion a year. Yet the companies active in this trade are all based in the industrialized world and no payment has ever been made for the original resource. The Biodiversity Convention suggests that future use of genetic resources should be subject to bilateral agreement and payment, but ignores the imbalance of economic and political power of the different sides.

Similar problems apply to crop genes. Almost all the crops in world agriculture originate from areas in the less developed world, and this is where most genetic variation exists in nature. This variability is essential to the huge plant-breeding industry, but most of the technological know-how and financial resources are in the developed world, as are most of the world's gene banks, where seeds are cold-stored for long periods. Existing collections are exempt from new regulations. A particular irony is that varieties produced by multinational plant-breeders or biotechnology companies are now being patented, while those produced by poor farmers are regarded as accidental.

In principle, biodiversity seems a valuable concept. In practice, natural biodiversity is accorded little value while humanly engineered varieties are of great commercial value.

(Further information can be found in CIIR, 1993.)

The analysis in this chapter goes a step beyond the anthropocentric–ecocentric debate: it recognizes that the dualism humankind–nature cannot be maintained, a conclusion also reached by Evernden (1992). There are two key problems. First, the dualism is factually inaccurate: humans are a part of nature, dependent on it for food, water and air and vulnerable to heat, cold, predators and infection. Second, nature provides key inputs to human society and technology, in the form of resources.

The key question being addressed by this book is: 'Are current lifestyles sustainable?' This chapter has shown that one aspect of past and current lifestyles – the existence of a surrounding wilderness – is only true now if we discount indigenous occupants and is unlikely to persist in any form for more than a further decade or two; this will pose serious problems for many environmentalists. However, wilderness seems not to be an essential for most people: what matters to the majority is the quality of life in humanized areas. Here, there are serious doubts for two reasons linked to the society–nature dualism. First, the growth of the human population is explosive and seemingly beyond social control: our natural propensity to reproduce is creating huge problems both for ourselves and for natural systems. Second, economic and technological growth requires increasing quantities of raw materials, many of which are available only in limited quantities. It is possible that shortages of supply, or pollution caused by use, could cut off the growth which seems vital for the future. Hence, a deeper analysis of human society's relation to nature requires a more detailed consideration of population growth and of natural resources – the two topics identified in the book introduction as central to the question of global overcrowding.

References

ALLEN, J. and HAMNETT, C. (1995) 'An uneven world', in Allen and Hamnett (1995).

ALLEN, J. and HAMNETT, C. (eds) (1995) *A Shrinking World? Global Unevenness and Inequality*, Oxford, Oxford University Press in association with The Open University.

BALASUBRAMIAN, R. (1991) 'The Hindu attitude to knowledge and nature', in Ravindra, R. (ed.) *Science and Spirit*, New York, Paragon House.

CATER, E. (1995) 'Consuming places: global tourism', in Allen and Hamnett (1995).

CIIR (1993) *Biodiversity: What's at Stake?*, London, Catholic Institute for International Relations.

DALBY, R. (1994) *The Alaska Guide*, Golden, CO, Fulcrum Publishing.

DE SILVA, P. (1991) 'Environmentalist ethics', in Fu, C.W. and Wawrytko, S.A. (eds) *Buddhist Ethics in Modern Society*, New York, Greenwood.

ECKERSLEY, R. (1992) *Environmentalism and Political Theory: Toward an Ecocentric Approach*, London, UCL Press.

EVERNDEN, N. (1992) *The Social Creation of Nature*, Baltimore, MD, Johns Hopkins Press.

FURST, L.R. (1977) *Romanticism in Perspective* (2nd edn), London, Macmillan.

GADGIL, M. and GUHA, R. (1992) *This Fissured Land: an Ecological History of India,* Delhi, Oxford University Press.

GLACKEN, C. (1967) *Traces on the Rhodian Shore,* Berkeley and Los Angeles, CA, University of California Press.

HASHIMOTO, A. (1993) 'Cultural values and the environment: the Japanese experience', paper presented to *Values and the Environment* conference, University of Surrey, 23–24 September.

KUMMER, D.M (1991) *Deforestation in the Postwar Philippines,* Chicago, IL, University of Chicago Press.

MASON, H. (1970) *Gilgamesh: a Verse Narrative,* New York, New American Library.

MAULL , H. (1992) 'Japan's global environmental policies', in Hurnell, A. and Kingsbury, B. (eds) *The International Politics of the Environment,* Oxford, Clarendon Press.

MERCHANT, C. (1990) 'Gender and environmental history', *The Journal of American History,* Vol. 76, No. 4, pp. 1117–21.

NUTTALL, M. (1993) 'Environmental policy and indigenous values in the Arctic: Inuit conservation strategies', paper presented to *Values and the Environment* conference, University of Surrey, 23–24 September.

OELSCHLAEGER, M. (1991) *The Idea of Wilderness,* New Haven, CT, Yale University Press.

PLUMWOOD, V. (1993) *Feminism and the Mastery of Nature,* London, Routledge.

SILVERTOWN, J. and SARRE, P. (1990) *Environment and Society,* London, Hodder and Stoughton in association with The Open University.

SIMMONS, I. (1989) *Changing the Face of the Earth,* Oxford, Basil Blackwell.

THE NEW ENCYCLOPÆDIA BRITANNICA (1985) (15th edn), Chicago, IL, Encyclopædia Britannica Inc.

TSUMI, S. and WEIDNER, M. (1989) *Environmental Policy in Japan,* Berlin, Edition Sigma.

TUAN, Y.F. (1974) *Topophilia: a Study of Environmental Perception, Attitudes and Values,* Englewood Cliffs, NJ, Prentice-Hall.

TURNER, F.J. (1947) *The Frontier in American History,* New York, Holt, Rinehart and Winston.

Reading A: *Mark Nuttall, 'Environmental policy and indigenous values in the Arctic: Inuit conservation strategies'*

Throughout the Arctic and sub-Arctic, non-native attitudes to the environment have long been underpinned by the authority of rational scientific knowledge, and by a frontier ideology that has shaped both the course of economic development and the post-contact history of its indigenous peoples. But industrial activity and the commercial exploitation of both renewable and non-renewable resources brings the real threat of severe environmental damage. This was most recently illustrated in the spring of 1989, when over 10 million gallons of crude oil spilled into Alaska's Prince William Sound from the tanker *Exxon Valdez*, which ran aground because of human error. Fears for similar environmental accidents in the future are not unfounded, especially given the extent of current oil and gas-based development strategies throughout the circumpolar north (e.g. see Nuttall, 1991; Vitebsky, 1990). Furthermore, attention is focused on atmospheric pollution and the effects of global warming and on damage caused to the Arctic ozone layer by increased emissions of man-made greenhouse gases, such as methane and carbon dioxide. As a result, the Arctic is regarded as a critical zone for global environmental change, and much energy is concentrated on designing appropriate resource management systems and environmental protection strategies.

However, science-based resource management systems often ignore the perspectives and values of the indigenous peoples, such as the Inuit, who live in the Arctic. The designation of wildlife refuges and national parks to safeguard animals and the environment, for example, often restricts the rights of Inuit to hunt in those areas, while international regulation has had an effect on subsistence whaling. Anthropologists have not only called for environmental agencies to recognize the value of indigenous knowledge (e.g. Freeman, 1979), but have argued for the increased participation of local people in the design of environmental projects and the implementation of environmental policy (e.g. Drijver, 1992). In this way, environmental management programmes can integrate conventional scientific approaches with traditional or indigenous knowledge. In Greenland, Canada and Alaska, Inuit groups have begun to outline and put into practice their own environmental strategies and policies to safeguard the future of Inuit resource use, and to ensure a workable participatory approach to the sustainable management of resources. This paper illustrates this, by discussing Inuit resource use, and by providing an overview of the conservation strategies of both the Inuit Circumpolar Conference and the Greenland Home Rule Authorities.

Inuit resource use: an indigenous environmental ethic

Across the Arctic, from East Greenland to the Bering Sea coast of Siberia, many Inuit communities continue to rely on the harvesting of terrestrial and marine resources for subsistence purposes. The species most commonly harvested are marine mammals such as seals, walrus, narwhals, beluga, fin and minke whales, and land mammals such as caribou and muskoxen. While there is individual appropriation of these species, this is grounded within and dependent upon notions of collective appropriation and communal rights. Access to resources is collective, and their exploitation is guided by community regulations, which are often unwritten.

Collective appropriation of a resource area by an Inuit community is reflected in sharing patterns and distribution networks, and by communal claims to catch-shares from large sea mammals, such as walrus, bearded seals, narwhals and whales. Just as the use of land and sea does not entail any individual ownership rights, so no one hunter can claim ownership over the animals he catches. The sharing and distribution of meat is guided by an obligation to give which is central to Inuit culture (Nuttall, 1992). While feelings of morality and altruism guide sharing practices, reasons for sharing are also underpinned by cosmology and religious belief. For Inuit, animals are believed to give themselves up to hunters, and it is incumbent on the hunter to acknowledge this through ritual, and to give them in turn to other people. Through the sharing and giving of meat social relationships are sustained and expressed, and fundamental values that guide attitudes towards animals and the environment are reaffirmed (Nuttall, 1992; Wenzel, 1991). Inuit identity is founded upon, and derives meaning from, a culturally embedded system of shared relations, not only between persons, but between persons and animals. As Wenzel (1991, pp. 60–1) has put it, 'people, seals, polar bear, birds and caribou are joined in a single community in which animals give men food and receive acknowledgement and revival'.

The centrality of this sense of social relatedness, and complexity of reciprocal rights and obligations to persons and the environment, has been noted by many anthropologists working among other hunting and gathering peoples. Bird-David (1990), for example, has argued that hunter-gatherers perceive their environment as 'giving' and that the metaphor of sharing provides us with a key to understanding how hunter-gatherers relate to their environments. Hunting is not simply an economic exploitation of resources, it involves purposive action and is guided by ideological and ritual responsibility to the environment. For Inuit and other hunting peoples, such ideas have been important for both community survival and the sustainability of resources.

From an Inuit perspective, threats to wildlife and the environment do not come from hunting, but from airborne and seaborne pollutants, such as cesium isotopes, lead and mercury, entering the Arctic biosphere from industrial areas far to the south. Threats to the resource areas which Inuit depend on collectively also come from the impact of non-renewable resource extraction, such as hydrocarbon development. In setting out to counteract such threats, Inuit have argued that common property regimes based on local knowledge and cultural values are the only adequate systems of environmental management. And by claiming the right for international recognition as resource conservationists, Inuit have begun to use indigenous knowledge as political action. The success of this approach has been made possible by and with the support of the Inuit Circumpolar Conference.

An Inuit regional conservation strategy

The Inuit Circumpolar Conference (ICC) is a pan-Arctic aboriginal people's organization that represents the interests of Inuit in Greenland, Canada, Alaska and Siberia. Formed in Alaska in 1977, in response to increased oil and gas exploration and development, the ICC has had non-governmental organization status at the United Nations since 1983. Challenging the policies of governments, multi-national corporations and environmental movements, the ICC has argued that the protection of the Arctic environment and its resources should recognize indigenous rights and be in accordance with Inuit tradition and cultural values. Since its formation, the ICC has sought to establish its own Arctic policy, based on indigenous knowledge, for an Inuit homeland that reflects Inuit concerns about future development, together with ethical and practical guidelines for human activity in the Arctic (Stenbaek, 1985).

In 1985 the ICC set up its own Environmental Commission (ICCEC), to formulate an Inuit conservation strategy, worked out along the lines of the IUCN's [International Union for the Conservation of Nature's] world conservation strategy. The resulting Inuit Regional Conservation Strategy sketched out how best to design and implement sustainable resource management programmes that take into account the subsistence and cultural needs of local communities. Basically, the Inuit Regional Conservation Strategy aims to secure for the Inuit the rights to hunt and fish in traditional areas, and to preserve and protect living resources and so ensure the subsistence needs of future generations. As Inuit resource use is grounded in community norms and rules, the ICC advocates the management of the environment as a common-property regime.

Such a conservation strategy can only be effective, however, if there is adequate knowledge of resources and their use, and of Inuit environmental values.

Recognizing this, the ICC gives priority to the collection of information from scientific sources and from hunters and fishermen, and has established a data base on renewable resources. Work on the data base has hardly begun, and it requires the participation of Inuit communities and organizations throughout the Arctic. Of particular concern is that research should focus on human–environmental relations, and on both scientific and local knowledge of wildlife and ecological conditions. This kind of research is well underway in Greenland, where the Inuit population has a greater degree of decision-making power concerning environmental management than elsewhere in the Arctic.

Environmental policy and conservation strategies in Greenland

When Greenland achieved Home Rule from Denmark in 1979, Greenlanders became the first population of Inuit origin to achieve a degree of self-government. Since then, the Greenland Home Rule Authorities (HRA) have not only set a precedent for aboriginal self-determination in the Arctic, they have also set a precedent for indigenous people worldwide by implementing environmental protection strategies. Full jurisdiction over environmental issues had been transferred from Denmark to the HRA by January 1989, and in 1991 the environment, which had previously been the responsibility of the Department of Fisheries and Industry, became the concern of the new Department of Health and the Environment. The Greenlandic Parliament passed the country's first Environmental Protection Act in 1988 and this became effective in January 1989. As a result there has been much recent environmental legislation, together with research on the human use of renewable and non-renewable resources. The HRA regard environmental issues as the concern of the indigenous Inuit and consider it essential that indigenous knowledge guides and informs the directing of research on natural resources. As a result, much of this work is participatory, involving local people and foreign researchers.

In 1988 the Greenland Parliament also passed legislation concerned with the protection of wildlife most commonly harvested by local communities. In particular, protected seasons for birds such as common eiders, Brunnich's guillemot and white-fronted geese have been imposed. Caribou reserves have been designated along parts of the west coast, while outside of these reserves there is an open caribou hunting season from the beginning of August until the end of September. Similarly, the hunting of muskoxen is confined to the same period (and for four weeks during early winter). The hunting of polar bears is strictly regulated and confined to subsistence purposes only, but there is an open season from the beginning of September until the end of June. The hunting of seals and small cetaceans is not regulated, although the International Whaling Commission (IWC) sets quotas for the hunting of minke and fin whales. This is monitored closely by the HRA, which distribute the IWC quotas allocated to Greenland to the various municipalities in the country. These are then distributed to individual whaling vessels, and it is the responsibility of the municipal authorities to inform the HRA about the numbers of whales killed, as well as struck but not killed, and about any infringements of the whaling regulations. The HRA provide the IWC with an annual report on all whaling activities in Greenland, and this is used as a reference for working out the quotas for the following year.

Subsistence whaling and its regulation throws into relief the conflicting environmental orientations of indigenous Inuit, on the one hand, and foreign scientists and environmental agencies on the other. The Greenland authorities challenge the scientific data on whale populations used by the IWC to work out its regulatory policies as inadequate. In fact, some scientists would even agree that there is insufficient scientific knowledge about the biology and dynamics of whale stocks to implement satisfactory management plans (e.g. Heide-Jørgensen, 1990). Yet, scientists would also argue that, owing to demographic changes in Inuit communities and changes in hunting technology and methods, the self-regulation of whaling by Inuit communities themselves is not an option. A compromise needs to be reached and agreed upon. As a start, aerial counts of fin and minke whales have been carried out by the Greenlandic authorities in collaboration with international scientists, including representatives from the IWC.

But the Greenland authorities also recognize that effective resource management depends on a comprehensive understanding of how local people harvest and use wildlife, and of the cultural aspects of subsistence. With this in mind, in 1988 the Greenland authorities launched their most ambitious environmental research initiative to date. This aims to gather as much information as possible on all living resources in Greenland and on their appropriation by local communities. This is backed up by intensive anthropological research on traditional and modern subsistence, on social relations in hunting communities, on sharing patterns and on the symbolic and cultural importance of local landscapes (e.g. Nuttall, 1992).

Conclusions

Indigenous Inuit perspectives often find themselves in conflict with other competing visions of the future of the Arctic environment. Foreign scientists, the Inuit Circumpolar Conference, the Greenland authorities and local communities, however, all agree on the necessity of habitat protection. Yet, all may not agree on the reasons. For conservationists and environmentalists, the absolute protection of certain species of wildlife, such as whales and polar bears, may be desirable. For Inuit hunting communities, protection of breeding areas and closed seasons on hunting may be acceptable if an adequate and sustainable yield is maintained and made possible for future generations. Worldwide, resource agencies need to take into account the environmental values of local people and adopt a participatory approach to resource management. In the Arctic, the Inuit have found themselves a new role as resource conservationists and have initiated user-based research. As such they provide a model for the inclusion of indigenous values and environmental knowledge in the design and implementation of environmental policy.

References

BIRD-DAVID, A. (1990) 'The giving environment: another perspective on the economic system of hunter-hatherers', *Current Anthropology*, Vol. 31, No. 2, pp. 189–96.

DRIJVER, C. (1992) 'People's participation in environmental projects', in Croll, E. and Parkin, D. (eds) *Bush Base: Forest Farm,* London, Routledge.

FREEMAN, M.M.R. (1979) 'Traditional land users as a legitimate source of environmental expertise', in Nelson, J.G. *et al.* (eds) *Canadian National Parks: Today and Tomorrow, Conference II, Vol. 1, Studies in Land Use History and Landscape Change No. 7,* Waterloo.

HEIDE-JØRGENSEN, M. (1990) 'Small cetaceans in Greenland: hunting and biology', *North Atlantic Studies,* Vol. 2, Nos. 1 and 2, pp. 55–8.

NUTTALL, M. (1991) 'Mackenzie Delta gas: possible developments in the 1990s', *Polar Record,* Vol. 27, No. 160, pp. 60–1.

NUTTALL, M. (1992) *Arctic Homeland: Kinship, Community and Development in Northwest Greenland,* London, Belhaven Press; Toronto, University of Toronto Press.

STENBAEK, M. (1985) 'Arctic policy – blueprint for an Inuit homeland', *Etudes Inuit Studies,* Vol. 9, No. 3, pp. 5–14.

VITEBSKY, P. (1990) 'Gas, environmentalism and native anxieties in the Soviet Arctic: the case of Yamal Peninsula', *Polar Record,* Vol. 26, No. 156, pp. 19–26.

WENZEL, G. (1991) *Animal Rights, Human Rights,* London, Belhaven Press.

Source: Nuttall, 1993, pp. 194–9

Reading B: Carolyn Merchant, 'Gender and environmental history'

[...]

Donald Worster's 'Transformations of the Earth' [Worster, 1990], while a rich and provocative approach to the field of environmental history, lacks a gender analysis. His conceptual levels of ecology (natural history), production (technology and its socio-economic relations) and cognition (the mental realm of ideas, ethics, myths, and so on) are a significant framework for research and writing in this emerging field. His use of the mode-of-production concept in differing ecological and cultural contexts and his account of the changing history of ecological ideas in his major books have propelled environmental history to new levels of sophistication.

A gender perspective can add to his conceptual framework in two important ways. First, each of his three categories can be further illuminated through a gender analysis; second, in my view, environmental history needs a fourth analytical level, that of reproduction, which interacts with the other three levels. What could such a perspective contribute to the framework Worster has outlined?

Women and men have historically had different roles in production relative to the environment. In subsistence modes of production such as those of native peoples, women's impact on nature is immediate and direct. In gathering-hunting-fishing economies, women collect and process plants, small animals, bird eggs and shellfish, and fabricate tools, baskets, mats, slings and clothing, while men hunt larger animals, fish, construct weirs and hut frames, and burn forests and brush. Because water and fuelwood availability affect cooking and food preservation, decisions over environmental degradation that dictate when to move camp and village sites may lie in the hands of women. In horticultural communities, women are often the primary producers of crops and fabricators of hoes, planters and digging sticks, but when such economies are transformed by markets, the cash economies and environmental impacts that ensue are often controlled by men. Women's access to resources to fulfil basic needs may come into direct conflict with male roles in the market economy, as in Seneca women's loss of control over horticulture to male agriculture and male access to cash through greater mobility in nineteenth-century America or in India's chipco (tree-hugging) movement of the past decade, wherein women literally hugged trees to protest declining access to fuelwood for cooking as male-dominated lumbering expanded [Marburg, 1984; Rothenberg, 1976; Shiva, 1988; Etienne and Leacock, 1980].

In the agrarian economy of colonial and frontier America, women's outdoor production, like men's, had immediate impact on the environment. While men's work in cutting forests, planting and fertilizing fields, and hunting or fishing affected the larger homestead environment, women's dairying activities, free-ranging barnyard fowl, and vegetable, flower and herbal gardens all affected the quality of the nearby soils and waters and the level of insect pests, altering the effects of the microenvironment on human health. In the nineteenth century, however, as agriculture became more specialized and oriented toward market production, men took over dairying, poultry-raising and truck farming, resulting in a decline in women's outdoor production. Although the traditional contributions of women to the farm economy continued in many rural areas and some women assisted in farm as well as home management, the general trend toward capitalist agribusiness increasingly turned chickens, cows and vegetables into efficient components of factories within fields managed for profits by male farmers [Merchant, 1989; Bush, 1982; Sachs, 1983].

In the industrial era, as middle-class women turned more of their energies to deliberate child rearing and domesticity, they defined a new but still distinctively female relation to the natural world. In their socially constructed roles as moral mothers, they often taught children about nature and science at home and in the elementary schools. By the Progressive era, women's focus on maintaining a home for husbands and children led many women such as those quoted above to spearhead a nationwide conservation movement to save forest and waters and to create national and local parks. Although the gains of the movement have been attributed by historians to men such as President Theodore Roosevelt, forester Gifford Pinchot and preservationist John Muir, the efforts of thousands of women were directly responsible for many of the country's most significant conservation achievements. Women writers on nature such as Isabella Bird, Mary Austin and Rachel Carson have been among the most influential commentators on the American response to nature [Merchant, 1984; Norwood, 1984].

Worster's conceptual framework for environmental history can thus be made more complete by including a gender analysis of the differential effects of women and men on ecology and their differential roles in production. At the level of cognition as well, a sensitivity to gender enriches environmental history. Native Americans, for example, construed the natural world as animated and created by spirits and gods. Origin myths included tales of mother earth and father sky, grandmother woodchucks and coyote tricksters, corn mothers and tree spirits. Such deities mediated between nature and humans, inspiring rituals and behaviours that helped to regulate environmental use and exploitation. Similar myths focused planting, harvesting and first fruit rituals among native Americans and in such Old World cultures as those in ancient Mesopotamia, Egypt and Greece, which symbolized nature as a mother goddess. In Renaissance Europe the earth was conceptualized as a nurturing mother (God's vice-regent in the mundane world) and the cosmos as an organism having a body, soul and spirit. An animate earth and an I/thou relationship between humans and the world does not prevent the exploitation of resources for human use, but it entails an ethic of restraint and propitiation by setting up religious rituals to be followed before mining ores, damming brooks, or planting and harvesting crops. The human relationship to the land is intimately connected to daily survival [Allen, 1984; Eisler, 1988; Berger, 1985; Bord and Bord, 1982; Merchant, 1980].

When mercantile capitalism, industrialization and urbanization began to distance increasing numbers of male élites from the land in seventeenth-century England and in nineteenth-century America, the mechanistic framework created by the 'fathers' of modern science legitimated the use of nature for human profit making. The conception that nature was dead, made up of inert atoms moved by external forces, that God was an engineer and mathematician, and that human perception was the result of particles of light bouncing off objects and conveyed to the brain as discrete sensations meant that nature responded to human interventions, not as active participant, but as passive instrument. Thus the way in which world views, myths and perceptions are constructed by gender at the cognitive level can be made an integral part of environmental history [Merchant, 1980].

While Worster's analytical levels of ecology, production and cognition may be made more sophisticated by including a gender analysis, ideas drawn from feminist theory suggest the usefulness of a fourth level of analysis – reproduction – that is dialectically related to the other three. First, all species reproduce themselves generationally and their population levels have impacts on the local ecology. But for humans, the numbers that can be sustained are related to the mode of production: more people can occupy a given ecosystem under a horticultural than a gathering-hunting-fishing mode, and still more under an industrial mode. Humans reproduce themselves biologically in accordance with the social and ethical norms of the culture into which they are born. Native peoples adopted an array of benign and malign population control techniques such as long lactation, abstention, coitus interruptus, the use of native plants to induce abortion, infanticide and senilicide. Carrying capacity, nutritional factors and tribally accepted customs dictated the numbers of infants that survived to adulthood in order to reproduce the tribal whole. Colonial Americans, by contrast, encouraged high numbers of births owing to the scarcity of labour in the new lands. With the onset of industrialization in the nineteenth century, a demographic transition resulted in fewer births per female. Intergenerational reproduction, therefore, mediated through production, has impact on the local ecology [Boserup, 1965, 1970; Harris, 1979; Merchant, 1990; Wells, 1985].

Second, people (as well as other living things) must reproduce their own energy on a daily basis through food and must conserve that energy through clothing (skins, furs, or other methods of bodily temperature control) and shelter. Gathering or planting food crops, fabricating clothing and constructing houses are directed toward the reproduction of daily life.

In addition to these biological aspects of reproduction, human communities reproduce themselves socially in two additional ways. People pass on skills and behavioural norms to the next generation of producers, and that allows a culture to reproduce itself over time. They also structure systems of governance and laws that maintain the social order of the tribe, town, or nation. Many such laws and policies deal with the allocation and regulation of natural resources, land and property rights. They are passed by legislative bodies and administered through government agencies and a system of justice. Law in this interpretation is a means of maintaining and modifying a particular social order. These four aspects of reproduction (two biological and two social) interact with ecology as mediated by a particular mode of production.

Such an analysis of production and reproduction in relation to ecology helps to delineate changes in forms of patriarchy in different societies. Although in most societies governance may have been vested in the hands of men (hence patriarchy), the balance of power between the sexes differed. In gatherer-hunter and horticultural communities, extraction and production of food may have been either equally shared by or dominated by women, so that male (or female) power in tribal reproduction (chiefs and shamans) was balanced by female power in production. In subsistence-oriented communities in colonial and frontier America, men and women shared power in production, although men played dominant roles in legal-political reproduction of the social whole. Under industrial capitalism in the nineteenth century, women's loss of power in outdoor farm production was compensated by a gain of power in the reproduction of daily life (domesticity) and in the socialization of children and husbands (the moral mother) in the sphere of reproduction. Thus the shifts of power that Worster argues occur in different environments are not only those between indigenous and invading cultures but also those between men and women [Merchant, 1989; Cott, 1977; Epstein, 1981; Bloch, 1978; Welter, 1966].

A gender perspective on environmental history therefore both offers a more balanced and complete picture of past human interactions with nature and advances its theoretical frameworks. The ways in which female and male contributions to production, reproduction and cognition are actually played out in relation to ecology depends on the particular stage and the actors involved. Yet within the various acts of what Timothy Weiskel has called the global ecodrama [Weiskel, 1987] should be included scenes in which men's and women's roles come to centre stage and scenes in which nature 'herself' is an actress. In this way gender in environmental history can contribute to a more holistic history of various regions and eras.

References

ALLEN, P.G. (1984) *The Sacred Hoop: Recovering the Feminine in American Indian Traditions,* Boston, MA, Beacon Press.

BERGER, P. (1985) *The Goddess Obscured: Transformation of the Grain Protectress from Goddess to Saint,* Boston, MA, Beacon Press.

BLOCH, R. (1978) 'American feminine ideals in transition: the rise of the moral mother, 1785–1815', *Feminist Studies*, Vol. 4, pp. 101–26.

BORD, J. and BORD, C. (1982) *Earth Rites: Fertility Practices in Pre-Industrial Britain*, St Albans, Granada.

BOSERUP, E. (1965) *The Conditions of Agricultural Growth: the Economics of Agrarian Change Under Population Pressure*, Chicago, IL, Aldine.

BOSERUP, E. (1970) *Women's Role in Economic Development*, New York, St Martin.

BUSH, C.G. (1982) 'The barn is his, the house is mine', in Daniels, G. and Rose, M. (eds) *Energy and Transport: Historical Perspectives on Policy Issues*, Beverly Hills, CA, Sage.

COTT, N.F. (1977) *The Bonds of Womanhood: 'Woman's Sphere' in New England, 1780–1835*, New Haven, CT, Yale University Press.

EISLER, R. (1988) *The Chalice and the Blade: Our History, Our Future*, San Fransisco, CA, Harper.

EPSTEIN, B.L. (1981) *The Politics of Domesticity: Women, Evangelism, and Temperance in Nineteenth-Century America*, Middletown, CT, Wesleyan University Press.

ETIENNE, M. and LEACOCK, E. (1980) *Woman and Colonization: Anthropological Perspectives*, New York, Praeger.

HARRIS, M. (1979) *Cultural Materialism: the Struggle for a Science of Culture*, New York, Random.

MARBURG, S. (1984) 'Women and environment: subsistence paradigms 1850–1950', *Environmental Review*, Vol. 8, pp. 7–22.

MERCHANT, C. (1980) *The Death of Nature: Women, Ecology, and the Scientific Revolution*, San Fransisco, CA, Harper and Row.

MERCHANT, C. (1984) 'Women of the progressive conservation movement', *Environmental Review*, Vol. 8, pp. 34–56.

MERCHANT, C. (1989) *Ecological Revolutions: Nature, Gender, and Science in New England*, Chapel Hill, NC, University of North Carolina Press.

MERCHANT, C. (1990) 'The realm of social relations: production, reproduction, and gender in environmental transformations', in Turner II, B.L. (ed.) *The Earth as Transformed by Human Action*, New York and Cambridge, Cambridge University Press.

NORWOOD, V. (1984) 'Heroines of nature: four women respond to the American landscape', *Environmental Review*, Vol. 8, pp. 34–56.

ROTHENBERG, D. (1976) 'Erosion of power: an economic basis for the selective conservativism of Seneca women in the nineteenth century', *Western Canadian Journal of Anthropology*, Vol. 6, pp. 106–22.

SACHS, C.E. (1983) *The Invisible Farmers: Women in Agricultural Production*, Totowa, NJ, Rowman.

SHIVA, V. (1988) *Staying Alive: Women, Ecology, and Development*, London, Zed.

WEISKEL, T. (1987) 'Agents of empire: steps toward an ecology of imperialism', *Environmental Review*, Vol. 11, pp. 275–88.

WELLS, R. (1985) *Uncle Sam's Family: Issues and Perspectives on American Demographic History*, Albany, NY, State University of New York Press.

WELTER, B. (1966) 'The cult of true womanhood, 1820–1860', *American Quarterly*, Vol. 18, pp. 151–74.

[WORSTER, D. (1990) 'Transformations of the earth: toward an agroecological perspective in history', *Journal of American History*, Vol. 76, No. 4, pp. 1087–106.]

Source: Merchant, 1990, pp. 1117–21

Reading C: *Robyn Eckersley, 'Exploring the environmental spectrum: from anthropocentrism to ecocentrism'*

[...]

Major streams of environmentalism

In presenting the following overview of the major streams of environmentalism, I have drawn on the pioneering typologies of environmentalism developed by John Rodman and, more recently, Warwick Fox, who elaborates the most exhaustive classification scheme in the ecophilosophical literature [Rodman, 1978, pp. 45–56; 1983, pp. 82–92; Fox, 1986, pp. 21–9; 1990, ch. 6]. [...] Whereas Rodman has sought to crystallize the major currents in the history of the environmental movement in order to uncover their complexities and ambiguities, Fox has developed a more general, analytical map that is intended to provide a close to exhaustive categorization of the range of ecophilosophical positions (i.e. whether or not they are represented by a particular historical movement). The approach here will be primarily historical, since my main concern is to relate clusters of particular environmental ideas to particular movements and to point out the contribution of, ambiguities in, and potential for alliance between, these various movements. [...]

Moving from the anthropocentric toward the ecocentric poles, the major positions that I will be discussing are resource conservation, human welfare ecology, preservationism, animal liberation, and ecocentrism. [I do not discuss the most blatant anthropocentric environmental position, which Fox characterizes as 'unrestrained exploitation and expansionism', since no Green activist or emancipatory ecopolitical theorist would support this position.] This spectrum represents a general movement from an economistic and instrumental environmental ethic toward a comprehensive and holistic environmental ethic that is able to accommodate human survival and welfare needs (for, say, a sustainable 'natural resource-base', a safe environment, or 'urban amenity') while at the same time respecting the integrity of other life-forms. However, since part of my concern is to draw out the ambiguities in, and the potential for forming alliances between, some of these historical currents of environmentalism, the general movement from anthropocentrism to ecocentrism will not appear as a strict linear progression. For example, some of the arguments for preservationism are *more* ecocentric than those for animal liberation, while other preservationist arguments represent a variation of some of the arguments used by the human welfare ecology stream.

This general overview of environmentalism will also help to explain how some currents of environmentalism have had more influence in some countries than others and how this has influenced both the nature and goals of the Green movement and the expression of Green theory in those countries. For example, the human welfare ecology stream has played a relatively more prominent role in Europe, whereas the preservationist stream has had more influence in 'New World' regions such as North America, Australia and New Zealand (where there are considerably more areas of wilderness to preserve) [see Hay and Haward, 1988; Eckersley, 1992]. This has given rise to different emphases in Green theory and practice in those regions. [...]

[...]

Resource conservation

Although the idea of conservation, in the sense of the 'prudent husbanding' of nature's bounty, can be traced back as far as Plato, Mencius, Cicero and the Old and New Testaments, its twentieth-century scientific and utilitarian manifestation is intricately bound up with the rise of modern science from the sixteenth century [see Drengson, 1990, p. 3; Glacken, 1965, p. 158]. Those who have inquired into the historical roots of the modern conservation doctrine have generally traced its popularization in North America to Gifford Pinchot, the first chief of the United States Forest Service, described by Devall as the 'prototype figure in the [conservation] movement'

[Devall, 1979, p. 140; see also McConnell, 1971; Hays, 1959; Rodman, 1983]. Central to Pinchot's notion of conservation was the elimination of waste, an idea that the environmental historian Samuel P. Hays has dubbed 'the gospel of efficiency', which he sees as lying at the heart of the doctrine of conservation. Yet Pinchot's ideas were also deeply imbued with the ethos of the Progressive era to which he belonged; indeed, in his book *The Fight for Conservation,* he identified 'development' as the first principle of conservation, with 'the prevention of waste' and development 'for the benefit of the many, and not merely the profit of the few' forming the second and third principles, respectively [Pinchot, 1910, p. 46]. Moreover, as McConnell observes, it was taken for granted that the principle of waste prevention meant 'maximizing output of economic goods per unit of human labor' [McConnell, 1971, p. 430]. According to Devall, the Pinchot-led conservation movement in the United States helped to 'professionalize "resource management"' and further the centralization of power in large public bodies (such as the US Forest Service) based on principles of 'scientific management' [Devall, 1979, p. 140].

[...]

The resource conservation perspective may be seen as the first major stop, as it were, as one moves away from an unrestrained development approach. Not surprisingly, it is the least controversial stream of modern environmentalism – indeed, it has become somewhat of a foe to more radical streams of environmentalism. The general acceptability of the resource conservation perspective arises from the fact that it proceeds from a human-centred, utilitarian framework that seeks the 'greatest good for the greatest number' (including future generations) by reducing waste and inefficiency in the exploitation and consumption of non-renewable 'natural resources' (e.g. oil) and ensuring a maximum sustainable yield in respect of renewable resources (e.g. fisheries, soil, crops and timber). As such, it is a perspective that is inextricably tied to the production process and, by virtue of that fact, necessarily regards the non-human world in use-value terms. This is reflected, among other things, in the *language* used by adherents of this stream of environmentalism; after all, 'resources' are, as Neil Evernden points out, 'indices of utility to industrial society. They say nothing at all of experiential value or intrinsic worth' [Evernden, 1984, p. 10; see also Livingston, 1981, pp. 43–6]. Similarly, Laurence Tribe has argued that to treat human material satisfaction as the *only* legitimate referent of environmental policy analysis and resource management leads to 'the dwarfing of soft variables' such as the aesthetic, recreational, psychological and spiritual needs of humans and the different needs of *other* life-forms [Tribe, 1974; see also Livingston, 1981, pp. 24–34]. While the recognition of the use value of the non-human world must form a necessary part of any comprehensive environmental ethic, resource conservation is too limited a perspective to form the *exclusive* criterion of even a purely anthropocentric environmental ethic.

Human welfare ecology

[...]

[...] Whereas the resource conservation movement has been primarily concerned with improving economic productivity by achieving the maximum sustainable yield of natural resources, the major preoccupation of the human welfare ecology movement has been the health, safety and general amenity of the urban and agricultural environments – a concern that is encapsulated in the term 'environmental quality' [Livingston, 1981, pp. 34–41]. In other words, the resource conservation stream may be seen as primarily concerned with the *waste* and *depletion* of natural resources (factors of production) whereas the human welfare ecology stream may be seen as primarily concerned with the *general degradation,* or overall state of health and resilience, of the physical and social environment. For the human welfare ecology stream, then, 'sustainable development' means not merely sustaining the natural resource-base for human *production* but also sustaining biological support systems for human *reproduction.* Moreover, by focusing on both the physical *and* social limits to growth, the human welfare ecology stream has done much to draw attention to those 'soft variables' neglected by the resource conservation perspective, such as the health, amenity, recreational and psychological needs of human communities.

More significantly, the human welfare ecology stream, unlike the resource conservation stream, has been highly critical of economic growth and the idea that science and technology alone can deliver us from the ecological crisis (although the human welfare ecology stream has, of course, been dependent on the findings of ecological science to mount its case). The kind of ecological perspective that has informed this stream of environmentalism is encapsulated in Barry Commoner's 'four laws of ecology': everything is connected to everything else, everything must go somewhere, nature knows best (i.e. any major human intervention in a natural system is likely to be detrimental to that system), and there is no such thing as a free lunch [Commoner, 1971, pp. 29–44]. These popularly expressed ecological insights have challenged the technological optimism of modern society and the confident belief that, in time, we can successfully manage all our large-scale interventions in natural systems without any negative consequences for ourselves. The realization that there is no 'away' where we can dump our garbage, toxic and nuclear wastes, and other kinds of pollution has given rise to calls for a new stewardship ethic – that we must protect and nurture the biological support system upon which we are dependent. Practically, this has led to widespread calls for 'appropriate technology' and 'soft' energy paths, organic agriculture, alternative medicine, public transport, recycling and, more generally, a revaluation of human needs and a search for more ecologically benign lifestyles.

Since it is in urban areas that we find the greatest concentration of population, pollution, industrial and occupational hazards, traffic, dangerous technologies, planning and development conflicts, and hazardous wastes, it is hardly surprising that cities and their hinterlands have provided the major locale and focus of political agitation for human welfare ecology activists. Nor is it surprising that human welfare ecology has been the strongest current of environmentalism in Green politics in the most heavily industrialized and domesticated regions of the West, most notably Europe. In particular, the many different popular environmental protests or 'citizen's initiatives' in West Germany that provided the major impetus to the formation of *Die Grünen* have primarily been urban ecological protests falling within this general rubric. Not surprisingly, the ecological pillar in *Die Grünen's* platform is generally couched in the language of human welfare ecology [*Die Grünen*, 1983, p. 7].

By virtue of its primary concern for *human welfare* in the domestic environment, however, this stream has generally mounted its case on the basis of an anthropocentric perspective. That is, the public justification given for environmental reforms by human welfare ecology activists has tended to appeal to the enlightened self-interest of the human community (e.g. for *our* survival, for *our* children, for *our* future generations, for *our* health and amenity). Indeed, the human welfare ecology stream has no need to go any further than this in order to make its case: it is enough to point out that 'we must look after nature because it looks after *us*'. Moreover, defenders of this perspective can say to their ecocentric critics that human welfare ecology reforms would, in any event, directly improve the well-being of the non-human community as well. Why, they ask, should we challenge the public and lose the support of politicians with perplexing and offbeat ideas like 'nature for its own sake' when we can achieve substantially the same ends as those sought by ecocentric theorists on the basis of our own mainstream anthropocentric arguments? The ecocentric rejoinder, however, is that if we restrict our perspective to a human welfare ecology perspective we can provide no protection to those species that are of no present or potential use or interest to humankind. At best wildlife might emerge as an indirect beneficiary of human welfare ecology reforms

[Livingston, 1981, p. 42]. More generally, an anthropocentric framework is also likely to wind up reinforcing attitudes that are detrimental to the achievement of comprehensive environmental reform in the long run, since human interests will always systematically prevail over the interests of the non-human world. As Fox puts it, employing only anthropocentric arguments for the sake of expediency might win the occasional environmental battle in the short term. However, in the long term 'one is contributing to losing the ecological war by reinforcing the cultural perception that what is valuable in the non-human world is what is useful to humans' [Fox, 1990, p. 186; see also Evernden, 1984].

Preservationism

[...] In North American environmental history, the conflict between Gifford Pinchot of the US Forest Service, on the one hand, and John Muir of the Sierra Club, on the other hand, is generally taken as the archetypical example of the differences between resource conservation and preservation. In short, whereas Pinchot was concerned to *conserve* nature *for* development, Muir's concern was to *preserve* nature *from* development.

[...]

It is noteworthy that, whereas wilderness was once feared by the early European colonists in New World regions such as Australia and North America as a hostile force to be tamed, to an increasing number of Westerners wilderness has become, for a complex range of reasons, a subject of reverence, enlightenment, and a locus of threatened values. [...] Indeed, it is arguably the campaigns for wilderness preservation, more than any other environmental campaigns, that have generated the most radical philosophical challenges to stock assumptions concerning our place in the scheme of things, thereby forcing theorists to confront the question of the moral standing of the non-human world. Despite John Muir's pious and outmoded vocabulary, his public defence of 'wild nature' has made a lasting impression on the modern environmental imagination. As Stephen Fox shows, Muir found:

a divergence between Christian cosmology and the evidence of nature. 'The world we are told was made for man,' he noted. 'A presumption that is totally unsupported by facts. ... Nature's object in making animals and plants might possibly be first of all the happiness of each one of them, not the creation of all for the happiness of one. Why ought man to value himself as more than an infinitely small composing unit of the one great unit of creation? ... The universe would be incomplete without man; but it would also be incomplete without the smallest transmicros[c]opic creature that dwells beyond our conceitful eyes and knowledge.

[Fox, 1981, pp. 52–3]

The link between Muir's particular pantheistic world view and the ecocentric philosophy of more recent times is widely acknowledged, although there are important differences. Rodman, for example, has argued that Muir's egalitarian orientation toward other species was 'faint in comparison to the religious/esthetic theme' in his life and writings – and that an ethic that is primarily based on awe has significant limitations [Rodman, 1978, p. 51; 1983, pp. 84–6]. In so far as wilderness appreciation has developed into a cult in search of sublime settings for 'peak experiences' or simply places of rest, recreation and aesthetic delight – 'tonics' for jaded Western souls – it tends to converge with the resource conservation and human welfare ecology positions in offering yet another kind of human-centred justification for restraining development. [...]

Moreover, J. Baird Callicott – a thoroughgoing non-anthropocentric philosopher – has criticized the very concept of wilderness on the grounds that it enshrines a bifurcation between humanity and nature; is ethnocentric and sometimes racist (e.g. it overlooks the ecological management by the aboriginal inhabitants of New World 'wilderness' regions through such practices as fire lighting) and ignores the dimension of time in suggesting that the ecological status quo should be 'freeze-framed' [Callicott, 1991, especially pp. 14–15]. As Callicott rightly argues, we need to be wary of reinforcing this human/nature bifurcation and to develop, where possible, a more dynamic and symbiotic approach to land management that acknowledges that humans can live alongside wild nature. However, as Callicott acknowledges, there also remains a strong case for the reservation and protection of large areas of representative ecosystems as the best means of conserving species diversity and enabling ongoing speciation.

[...]

More recently, environmental philosophers have pointed to the wide range of anthropocentric arguments that have been advanced in favour of wilderness preservation (some of which have already been canvassed above). Fox provides the most exhaustive classification of these arguments to date [Fox, 1990, ch. 6]. Building on and adding to work by William Godfrey-Smith and George Sessions, Fox identifies nine kinds of argument for preserving the non-human world on the basis of its instrumental value to humans. He refers to these as the 'life-support', 'early warning system', 'laboratory' (i.e. scientific study), 'silo' (i.e. stockpile of genetic diversity), 'gymnasium' (i.e. recreational), 'art gallery' (i.e. aesthetic), 'cathedral' (i.e. spiritual), 'monument' (i.e. symbolic) and 'psychogenetic' (i.e. psychological health and maturity) arguments. He also divides these nine arguments into five general categories of argument that emphasize the 'physical nourishment value', the 'informational value', the 'experiential value', the 'symbolic instructional value', and the 'psychological nourishment value' of the non-human world to humans.

It is easy to see how many of the more tangible arguments for the preservation of wilderness can be quite persuasive politically, especially the more economically inclined arguments, such as those that refer to the recreational potential of wilderness or those that demonstrate the importance of maintaining genetic diversity to provide future applications in medicine and agriculture. However, it is important not to underestimate the political potency of some of the less tangible arguments for wilderness preservation. For example, the preservation of wild nature is seen by many as both a symbolic act of resistance against urban and cultural monoculture and the materialism and greed of consumer society *and* a defence (both real and symbolic) of a certain cluster of values. These include freedom, spontaneity, community, diversity and, in some cases, national identity.

Part of the political potency of arguments of this latter kind lies in the fact that the defence of wild nature is at the same time a defence of a certain cluster of values of *social* consequence. That is, they represent not only a defence of biological diversity and of 'letting things be' but also a renewed assault on the one-dimensionality of technological society. In this respect, Thoreau's oft-quoted dictum – 'in wilderness is the preservation of the world' – may be seen as taking on both an ecological *and* political meaning.

[...]

Animal liberation

Alongside the three major streams of environmentalism discussed above is a fourth stream that has developed relatively independently and has its origins in the various 'humane' societies for the prevention of cruelty to animals that emerged in the eighteenth and nineteenth centuries. The modern animal liberation movement, unlike the resource conservation, human welfare ecology, and preservation movements, has from its inception consistently championed the moral worthiness of certain members of the non-human world. However, while the animal liberation movement might have been one of the first streams of environmentalism to have stepped unambiguously over what might be called the 'great anthropocentric divide', such a step, as many ecophilosophical critics have recently pointed out, was not as momentous as it might first appear. In the view of these critics, the philosophical foundations of the animal liberation movement are unduly limited and fall well short of a rounded ecocentric world view [Rodman, 1977; 1983, pp. 86–8; Shepard, 1974; Tribe, 1974, pp. 1344–5; Callicott, 1980;

Livingston, 1984, pp. 61–72; Fox, 1985, pp. 26–8].

The popular case for the protection of the rights of animals is a relatively straightforward revival of the arguments of the modern utilitarian school of moral philosophy founded by Jeremy Bentham. In enlarging the conventional domain of ethical theory, Bentham had argued that human moral obligation ought to extend to all beings capable of experiencing pleasure and pain, regardless of what other characteristics they may possess or lack. The important question for Bentham in respect of whether beings were morally considerable was 'not, Can they *reason*? nor, Can they *talk*? but, *Can they Suffer*?' [Bentham, 1789, quoted in Singer, 1975, p. 8].

In drawing on Bentham's moral philosophy [i.e. utilitarianism], the contemporary animal rights theorist Peter Singer has argued in favour of the moral principle of equal consideration (as distinct from treatment) of the interests of all sentient beings regardless of what kind of species they are. The criterion of sentience is pivotal. For example, Singer has insisted that the 'capacity for suffering and enjoyment is *a prerequisite for having interests at all,* a condition that must be satisfied before we can speak of interests in a [morally] meaningful way' – indeed, he has argued that the criterion of sentience [...] is the 'only defensible boundary of concern for the interests of others' [Singer, 1975, pp. 8–9]. [...]

[...]

The implication of Singer's argument is that, where practicable, we must avoid inflicting any suffering on sentient beings. Accordingly, supporters of animal liberation advocate the prohibition of the hunting and slaughtering of all sentient beings (the corollary of which is vegetarianism), the prohibition of vivisection, and the prohibition of 'factory farming'. Although Singer's major focus has been the abuse of domestic animals, his argument also provides a justification for the protection of the habitat of wild animals, fish, birds and other sentient fauna. That is, forests and wetlands ought to be protected where it can be shown that they are instrumentally valuable to sentient beings for their 'comfort and well-being' in providing nesting sites, breeding habitat, and sustenance.

[...]

Ecocentric philosophers, however, have been critical of Singer's moral philosophy for regarding *non-sentient* beings as morally inconsequential. As Rodman has observed, Singer's philosophy leaves the rest of nature in:

> *a state of thinghood, having no intrinsic worth, acquiring instrumental value only as resources for the well-being of an elite of sentient beings. Homocentrist rationalism has widened out into a kind of zoocentrist sententism.*
>
> *[Rodman, 1977, p. 91; see also Callicott, 1980, p. 318]*

Trees, for example, are considered to be valuable only in so far as they provide habitat, can be turned into furniture, or otherwise rendered serviceable to the needs of sentient life-forms. To the extent that synthetic substitutes can be made to perform the services of non-sentient life-forms, then the latter will be rendered dispensible.

Some environmental philosophers have also mounted a more subtle critique of the animal liberation perspective. According to John Rodman, not only does this approach render *non*-sentient beings morally inconsequential but it also subtly degrades *sentient non*-human beings by regarding them as analogous to 'defective' humans who likewise cannot fulfil any moral duties [Rodman, 1983, p. 87; see also Fox, 1985, p. 27]. Rodman sees this tendency to regard sentient non-humans as having the same standing as defective or inferior human beings as analogous to (and as ridiculous as) dolphins regarding humans 'as defective sea mammals who lack sonar capability' [Rodman, 1977, p. 94]. The result is that the unique modes of existence and special capabilities of these non-human beings are overlooked.

A further criticism levelled against Singer's moral philosophy is that it is atomistic and therefore unsuitable for dealing with the complexities of environmental problems, which demand an understanding and recognition of not only whole species but also the interrelationships between different natural cycles, systems and populations. According to Rodman, the progressive extension model of ethics (which includes Christopher Stone's argument for the legal protection of the rights of non-sentient entities, discussed in the following section) tends:

to perpetuate the atomistic metaphysics that is so deeply imbedded in modern culture, locating intrinsic value only or primarily in individual persons, animals, plants, etc., rather than in communities or ecosystems, since individuals are our paradigmatic entities for thinking, being conscious, and feeling pain.

[Rodman, 1983, p. 87; see also Callicott, 1980]

Finally, critics have pointed to the tension between Singerian justice and an ecological perspective by noting that animal liberation, when pressed to its logical conclusion, would be obliged to convert all non-human animal carnivores to vegetarians, or, at the very least, replace predation in the food chain with some kind of 'humane' alternative that protects, or at least minimizes the suffering of, sentient prey. As Fox argues, besides representing 'ecological lunacy', animal liberation:

would serve, in effect, to endorse the modern project of totally domesticating the nonhuman world. Moreover, it would also condemn as immoral those 'primitive' cultures in which hunting is an important aspect of existence.

[Fox, 1990, p. 195]

Singer has in fact admitted that the existence of non-human carnivores poses a problem for the ethics of animal liberation. Despite this concession, he nonetheless counsels a modification to the dietary habits of at least some domestic animals in referring his readers to recipes for a vegetarian menu for their pets! [Singer, 1975, pp. 238–9].

To conclude, then, animal liberation has mounted a compelling challenge to anthropocentrism in pointing to its many logical inconsistencies. However, Singer's alternative criterion of moral considerability (i.e. sentience), while a *relevant* and significant factor, is too limited and not sufficiently ecologically informed to provide the exclusive criterion of a comprehensive environmental ethics. As we shall see, ecocentric theorists have identified broader, less 'human analogous' and more ecologically relevant criteria to determine whether a being or entity has 'interests' deserving of moral consideration.

Ecocentrism

Ecocentric environmentalism may be seen as a more wide-ranging and more ecologically informed variant of preservationism that builds on the insights of the other streams of environmentalism thus far considered. Whereas the early preservationists were primarily concerned to protect wilderness as sublime scenery and were motivated mainly by aesthetic and spiritual considerations, ecocentric environmentalists are also concerned to protect threatened populations, species, habitats and ecosystems *wherever situated* and irrespective of their use value or importance to humans. (This kind of concern is well illustrated by the activities of the international environmental organization Greenpeace.) In particular, ecocentric environmentalists strongly support the preservation of large tracts of wilderness as the best means of enabling the flourishing of a diverse non-human world. Accordingly, in what I refer to as New World regions such as North America, Australia and New Zealand (where significant areas of wilderness still remain), it is not surprising that the greatest concentration of ecocentric activists can usually be found in organizations, campaigns, or movements that promote the protection of wilderness. Two noteworthy examples here are the Earth First! movement in the United States and The Wilderness Society in Australia.

Much of the basic outline of an ecocentric perspective has already been foreshadowed [...] in the criticisms made of resource conservation, human welfare ecology, preservationism and animal liberation. An ecocentric perspective may be defended as offering a more encompassing approach than any of those so far examined in that it (i) recognizes the full range of human interests in the non-human world (i.e. it incorporates yet goes beyond the resource conservation and human welfare ecology perspectives); (ii) recognizes the interests of the non-human community (yet goes beyond the early preservationist perspective); (iii) recognizes the interests of future generations of humans and non-humans; and (iv) adopts a holistic rather than an atomistic perspective (contra the animal liberation perspective) in so far as it values populations, species, ecosystems and the ecosphere *as well as* individual organisms.

Now defenders of an animal liberation perspective might argue that their perspective is quite adequate to secure the protection of many non-sentient entities, such as ecosystems, and that for all *practical* purposes it is as good as an ecocentric perspective. This is because, as we saw in the previous section, if we attribute intrinsic value to all sentient beings, then we must also protect whatever is instrumentally valuable to *them* (e.g. *their* habitats and food sources). This would provide a case for the protection of forests, wetlands and any other habitat upon which sentient non-human beings depend for their survival and wellbeing. However, as we also saw in the previous section, ecocentric theorists have argued that this kind of approach not only leaves the rest of nature in a state of 'thinghood' – the only purpose of which is to service an élite of sentient beings – but that it is also too atomistic and, therefore, 'unecological' in the way in which it distributes intrinsic value in the world. For example, this kind of approach would attribute equal intrinsic value and, hence, equal moral consideration to the individual members of a native species or an endangered species as it would to the individual members of an introduced species or an abundant species (assuming the degree of sentience of each species to be roughly equivalent). This approach to the distribution of intrinsic value means that it would be considered no worse to kill, say, twenty members of a native species than it would be to kill ('weed out') twenty members of an introduced, feral species; for the same reason, it would be considered no worse to kill the last twenty members of a sentient endangered species than it would be to kill twenty members of an equally sentient species that exists in plague proportions. Similarly, an animal liberation perspective would attribute the same value to the individual animals that inhabit a flourishing, wild ecosystem as the equivalent number of domesticated or captive wild animals that might be managed by humans on a farm or in a zoo.

Even if one extends intrinsic value to all living organisms (i.e. animals, plants and micro-organisms) the same general kinds of problems apply. This is because such an approach still remains atomistic (i.e. it attributes intrinsic value only to *individual* living organisms) and therefore does not extend any moral recognition to populations, species, ecosystems and the ecosphere considered as entities in their own right. (I do not discuss a specifically 'life-based' stream of environmentalism in this survey of the major streams of environmentalism for the simple reason that, sociologically and politically speaking, this approach does not represent a major stream of environmentalism. Environmentalists who have moved beyond anthropocentrism tend, on the whole, to gravitate toward either the animal liberation approach or a straight-out ecocentric approach.)

Ecocentric theorists are concerned to develop an ecologically informed approach that is able to value (for their own sake) not just individual living organisms but also ecological entities at different levels of aggregation, such as populations, species, ecosystems and the ecosphere (or Gaia). [...]

References

BENTHAM, J. (1789) *An Introduction to the Principles of Morals and Legislation.*

CALLICOTT, J.B. (1980) 'Animal liberation: a triangular affair', *Environmental Ethics*, Vol. 2, pp. 311–38.

CALLICOTT, J.B. (1991) 'The wilderness idea revisited', MS.

COMMONER, B. (1971) *The Closing Circle: Nature, Man and Technology*, New York, Knopf.

DEVALL, B. (1979) 'Reformist environmentalism', *Humboldt Journal of Social Relations*, Vol. 6, pp. 129–57.

Die Grünen, Programme of the German Green Party (1983) London, Heretic.

DRENGSON, A.R. (1990) 'Forests and forestry practices: a philosophical overview', *Forest Farm Journal*, Vol. 2, p. 3.

ECKERSLEY, R. (1992) 'Environmental theory and practice in the Old and New Worlds: a comparative perspective', *Alternatives*.

EVERNDEN, N. (1984) 'The environmentalist's dilemma', in Evernden, N. (ed.) *The Paradox of Environmentalism*, Downsview, Ont., Faculty of Environmental Studies, York University.

FOX, S. (1981) *John Muir and his Legacy*, Boston, MA, Little Brown.

FOX, W. (1985) 'Towards a deeper ecology?', *Habitat Australia*, August, pp. 26–8.

FOX, W. (1986) 'Ways of thinking environmentally (and some brief comments on their implications for acting educationally)', in Wilson, J., Di Chiro, G. and Robottom, I. (eds) *Thinking Environmentally...Acting Educationally: Proceedings of the Fourth National Conference of the Australian Association of Environmental Education*, Melbourne, Victorian Association for Environmental Education.

FOX, W. (1990) *Toward a Transpersonal Ecology: Developing New Foundations for Environmentalism*, Boston, MA, Shambhala.

GLACKEN, C.J. (1965) 'The origins of the conservation philosophy', in Burton, I. and Kates, R.W. (eds) *Readings in Resource Management and Conservation*, Chicago, IL, Chicago University Press.

HAY, P.R. and HAWARD, M.G. (1988) 'Comparative green politics: beyond the European context?', *Political Studies*, Vol. 36, pp. 433–48.

HAYS, S.P. (1959) *Conservation and the Gospel of Efficiency*, Cambridge, MA, Harvard University Press.

LIVINGSTON, J. (1981) *The Fallacy of Wildlife Conservation*, Toronto, McClelland and Stewart.

LIVINGSTON, J. (1984) 'The dilemma of the deep ecologist', in Evernden (1984).

McCONNELL, G. (1971) 'The environmental movement: ambiguities and meanings', *Natural Resources Journal*, Vol. 11, pp. 427–35.

PINCHOT, G. (1910) *The Fight for Conservation*, New York, Doubleday Page and Co.

RODMAN, J. (1977) 'The liberation of nature?', *Inquiry*, Vol. 20, pp. 83–145.

RODMAN, J. (1978) 'Theory and practice in the environmental movement: notes towards an ecology of experience', in *The Search for Absolute Values in a Changing World: Proceedings of the Sixth International Conference on the Unity of the Sciences*, Vol. 1, San Fransisco, CA, The International Cultural Foundation.

RODMAN, J. (1983) 'Four forms of ecological consciousness reconsidered', in Scherer, D. and Attig, T. (eds) *Ethics and the Environment*, Englewood Cliffs, NJ, Prentice-Hall.

SHEPARD, P. (1974) 'Animal rights and human rites', *North American Review*, Winter, pp. 35–41.

SINGER, P. (1975) *Animal Liberation: a New Ethics for Our Treatment of Animals*, New York, The New Review.

TRIBE, L. (1974) 'Ways not to think about plastic trees: new foundations for environmental law', *The Yale Law Journal*, Vol. 83, pp. 1315–48.

Source: Eckersley, 1992, pp. 33–47

Population and environmental change: the case of Africa

Chapter 2

by Matthew Lockwood

2.1 Introduction

population growth

'I believe that population growth is *the* development and environmental problem facing the world in the 1990s' (Population Concern, 1991, p. 5). When Lynda Chalker, the UK's Minister for Overseas Development, said these words in an interview in 1990, she was putting forward a widely held view. In the same year, The Duke of Edinburgh, Patron of the International Union for the Conservation of Nature, spelled out what he saw as the direct dangers to wildlife and wilderness from population growth: 'There can be no doubt at all that the bulk of past and future extinctions [of species] is the direct consequence of the massive increase in the human population and the growth in the per capita demand for natural resources' (Population Concern, 1991, p. 5). World population growth thus looms large in the minds of many, with fears that the weight of humanity on resources is leading to a 'global timebomb', which will explode into famine, environmental disaster and war.

Others have criticized these views. The real problem, they argue, is not numbers of people, but rather the amount of resources they consume per head. Looking at it this way, it is the West which is 'overpopulated', rather than developing countries. Furthermore, population growth is high in developing countries precisely because the poor, especially women, need many children. A large family is the only way in which they can be guaranteed security in old age and help in time-consuming daily tasks.

This chapter examines some of these questions, looking particularly at Africa, south of the Sahara desert. One reason for this focus is that population in Africa is now growing more quickly than anywhere else in the world. It is often perceived as trapped in an environmental crisis caused by this runaway growth, and is thus a 'test case' for looking at ideas of overcrowding. However, there is another reason for choosing Africa for this chapter. Two of the central themes of this book are interdependence and uneven development. Africa, more than any other region, has been brought into the world economy through colonialization, with the establishment of exploitative relationships between Africa and Europe. The colonial period has left Africa an inheritance of extremely unevenly developed economies and natural resources (typically organized to provide crops, minerals and metals for export), and poverty. As the trends towards greater interdependence in the rest of the world continued after its independence from colonial rule, Africa has become one of the least integrated regions. This has intensified poverty in many countries.

This history has consequences for debates about population and the environment. The poverty of people in Africa means that they have low levels of consumption and more sustainable lifestyles than do people in Europe and North America. However, poverty also means that people may lack the means to protect and invest in their natural resource base. In solving the problems which greater populations pose, Africans cannot draw on the resources of a large part of the world, as Europe did during the colonial period.

The chapter is also designed to encourage you to reflect on how we think about population and resources in different cultural settings. Of whom do we think when we worry about the 'global timebomb' of population? As a preliminary exercise, have a look at Activity 1.

Activity 1 Look at the two photographs reproduced below and, as you do so, think about the following questions. Without counting, in which photograph do you think there are more people? Now count them carefully. Were you right? If not, why was your impression different? Which group seems more needy to you? Which group do you think uses more resources in meeting its needs? Why? Which group do you think would be more able to protect and transform its environment? Why?

This exercise reminds us of several basic points. The impact of people's needs on resources depends *both* on how many people there are *and* on how much they consume. We tend to think about people who are 'different', or who live in distant places, as an abstract mass – a 'population' – often in terms of their needs. But we think about people who are familiar and live close to us as groups of individuals, with skills and abilities. There can thus be a bias in the way in which we think about population–environment links in different settings.

In section 2.2.1, we briefly examine patterns of population growth in the world today. In section 2.2.2, we look at some of the ways in which population, resources and development have been linked by various thinkers. Section 2.3.1 tackles the question of *why* population growth in Africa is so high, looking at factors ranging from poverty, to the position of women, to cultural organization. The central question of the environmental impact of population growth in Africa is addressed in sections 2.3.2 to 2.4.2. Finally, section 2.5 concludes by considering the implications for policies in Africa – not only on the environment, but also policies on family planning and broader development policy.

2.2 Global population growth and environmental change

2.2.1 Global population patterns

When people think of the world being overpopulated, or of the 'population bomb', they usually think of a huge and rapidly growing mass of population in the countries of Latin America, Africa and Asia. But how true is this picture? The first step in understanding and assessing arguments about population growth and the environment is to get a grasp of contemporary trends in population patterns.

We start with *population growth.* There are several different aspects to world population growth, and we can only get a full picture by considering all of these together (Figures 2.1 and 2.2).

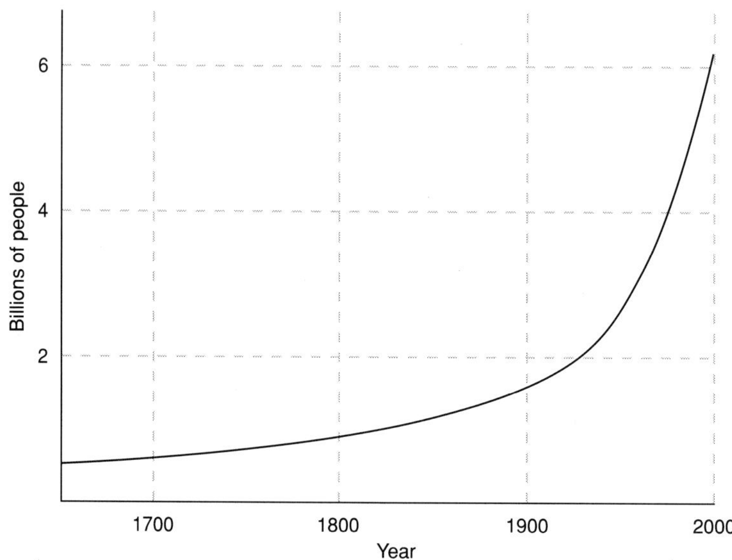

Figure 2.1 World population – the long run (Source: Meadows et al.*, 1992, p. 4, Figure 1.1)*

Note: World population has been growing exponentially since the beginning of the industrial revolution. In 1991 the world population growth rate was estimated to be 1.7 per cent, corresponding to a doubling time of 40 years.

The first point is the uniqueness of the twentieth century: total world population is growing at a rate, and to a size, unprecedented in human history (Figure 2.1). The second point, no less crucial than the first, is that the rate of growth is now slowing down, so that world population is expected to reach a maximum of about 12 billion people at some point soon after the year 2100 (Figure 2.2). Although the rate of growth is already slowing in most countries, the numbers added to world population each year are still growing and will continue to do so until shortly after the year 2000 (Demeny, 1989).

Most of the growth since 1950 has been in the developing world, and in particular in Africa and South Asia. Africa has already overtaken Europe in size of population, and by 2100 will be roughly five times as large. Some countries in Africa have had growth rates as high as 4 per cent per year, which implies a doubling of population about every 17 years.

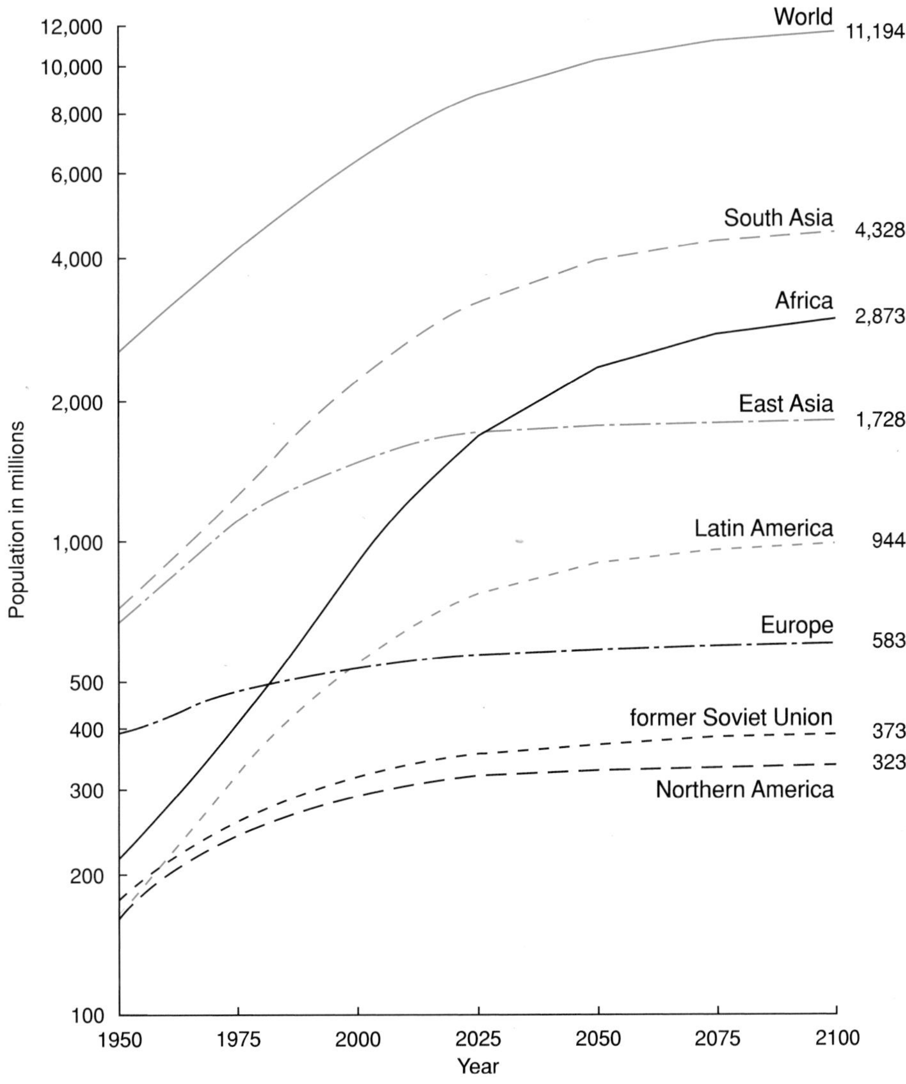

Figure 2.2 *World population – the future (Source: Demeny, 1984, p. 113, Figure 5)*

What determines population growth, and how do we know the path of future growth? Growth is defined simply as births minus deaths plus net migration. For large regions *fertility* and *mortality* are the key determinants of population growth. Of these, fertility, or the rate at which births occur, is the dominant influence.

fertility
mortality

We have reasonable estimates of birth and death rates for the main regions only from about 1950 (Figures 2.3 and 2.4). In 1950, death rates in developing countries were higher than in the developed world, but they fell steadily over the next 40 years. The difference in fertility was even more marked, and the timing of its decline has been more varied. In Latin America and the Middle East, fertility was already falling by the 1960s. Asia saw sustained decline in the 1970s, but at a much quicker rate in East Asia than in South Asia. Fertility in Africa has not shown signs of falling at all until very recently. This explains why population growth rates in Africa are currently the highest in the world.

There are many factors which bear on fertility rates, including the age at which people first have sex, and 'traditional contraceptives' such as abstinence and breast-feeding. However, the main immediate explanation for falling world fertility is the spread in use of modern contraceptives and sterilization (Figure 2.5). This process has been far from straightforward, and has involved controversies over the issues of motivation and coercion, especially in the promotion of sterilization.

Demographers make population projections from current population figures, plus assumptions about mortality and fertility. Sometimes our estimates of current population are badly inaccurate. For instance, the United Nations' (UN) estimate of Nigeria's population in 1990 was around 110 million, but the 1991 census gave a figure of 88 million. Generally, however, the major source of variation in population projections is in assumptions about the future path of fertility and mortality. The projections shown in Figure 2.2 are based on the assumption that the falls in fertility already under way will continue into the next century.

Activity 2 The character of writing about population trends is often strongly influenced by two features – the manner in which the figures are presented, and the assumptions made about the future. One of the most emotive writers on population is Paul Ehrlich. Turn now to the extract from his 1968 book *The Population Bomb*, which you will find as Reading A, 'Too many people', at the end of this chapter. As you read, ask yourself these questions:

- What feelings is Ehrlich trying to evoke in his readers?
- What assumptions is he making?
- At what points is he simply presenting information, and at what points is he making value judgements?
- Is it possible to write about population in a neutral way?

We have seen that population growth in Africa is currently higher than in other regions, and in section 2.3.1 below we explore why that is so. However, it is also true that population *density* in most African countries is much lower than in other parts of the world (Figure 2.6).

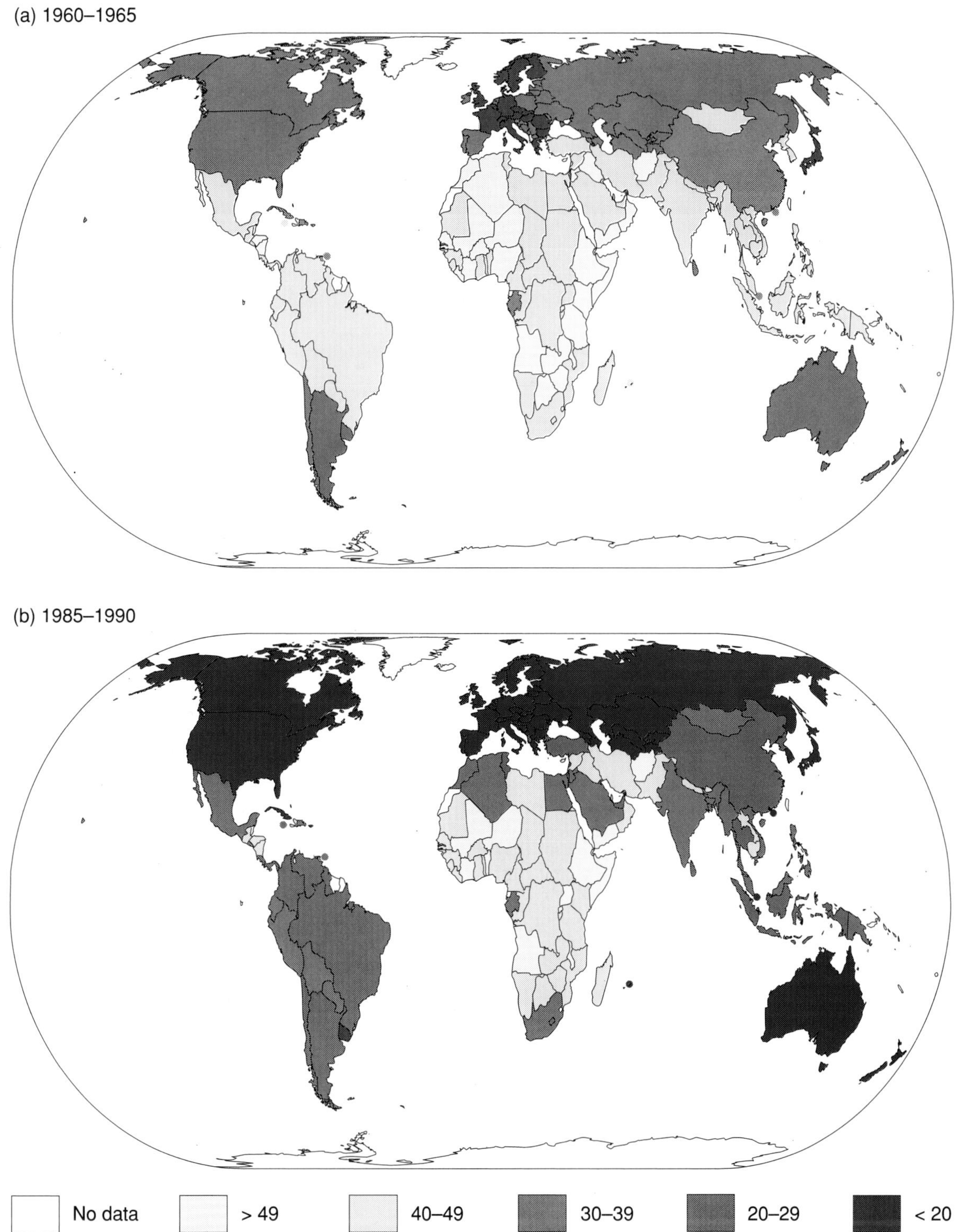

Figure 2.3 *Declines in crude birth rates (Source: compiled from data contained in Table 2 of Ross* et al., *1993, pp. 14–18)*

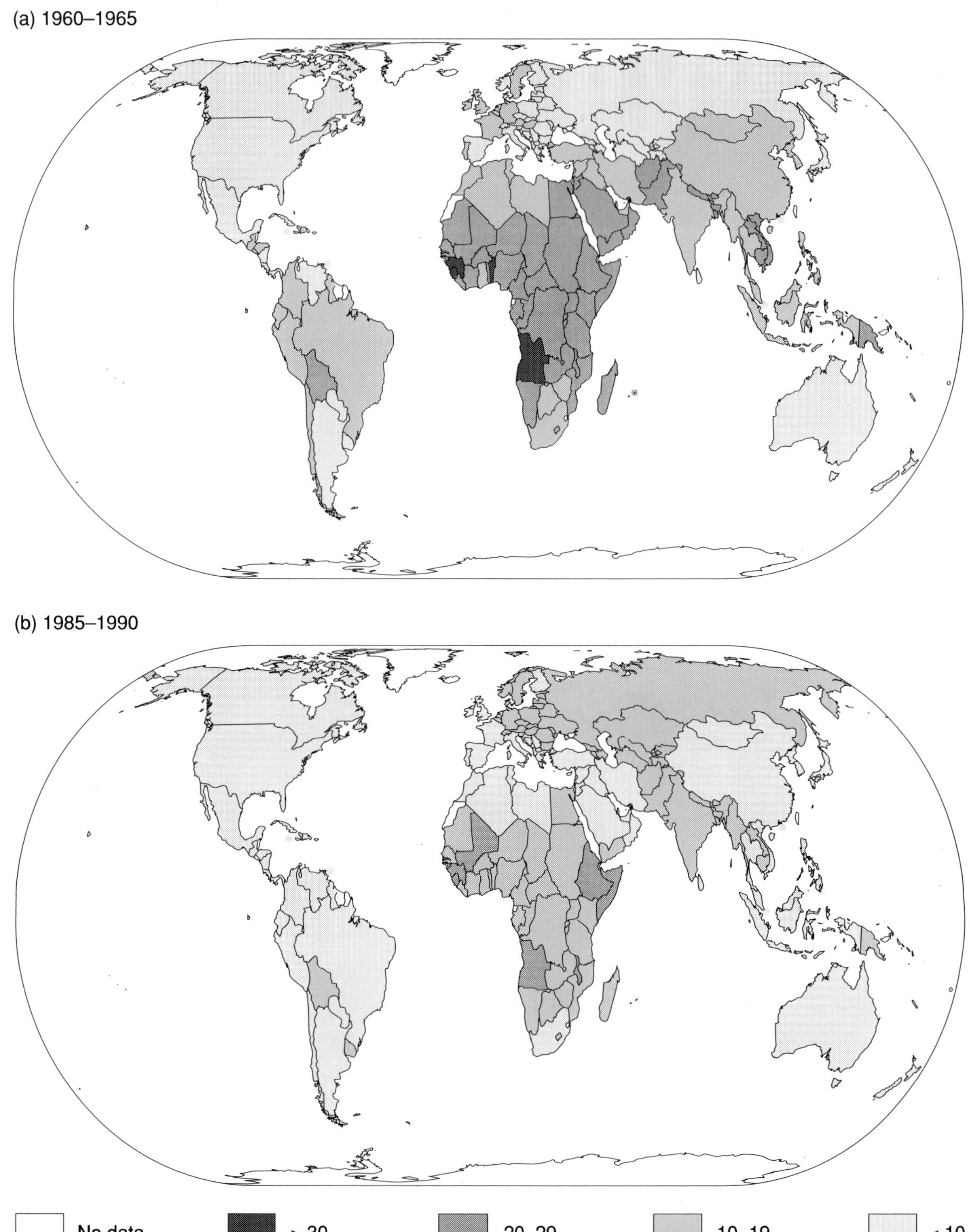

Figure 2.4 *Declines in crude death rates (Source: compiled from data contained in Table 2 of Ross* et al., *1993, pp. 14–18)*

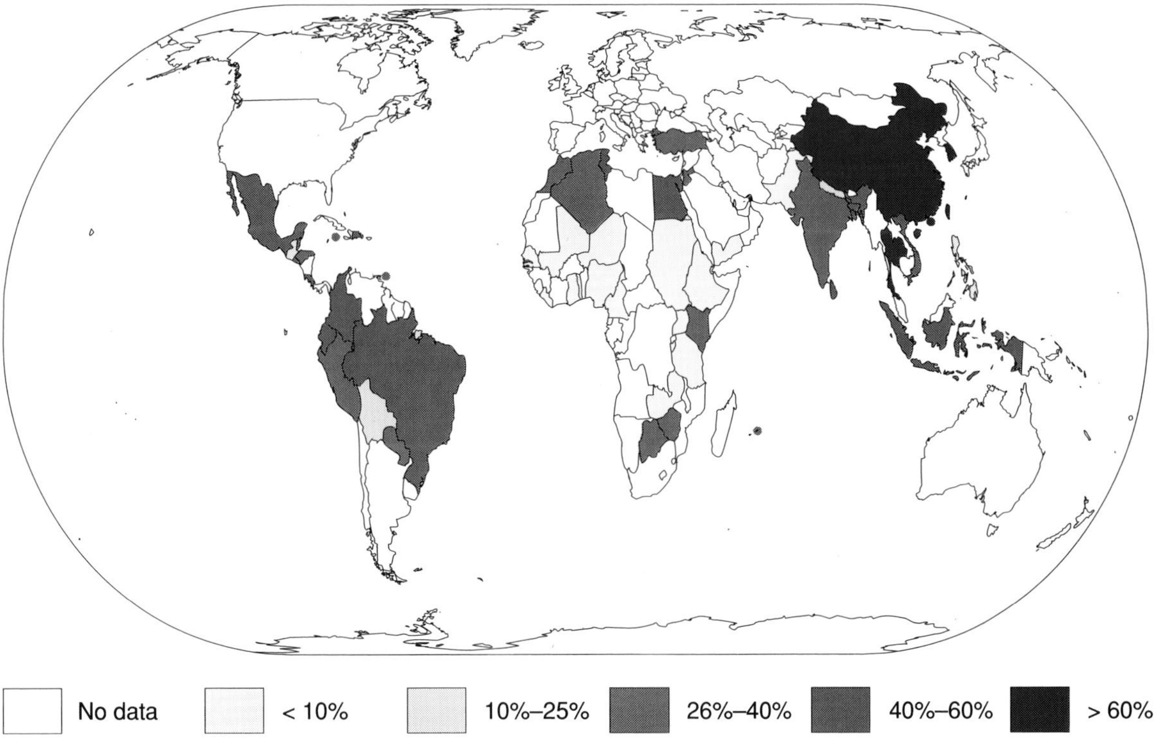

No data | < 10% | 10%–25% | 26%–40% | 40%–60% | > 60%

Figure 2.5 *The use of modern methods of contraception (Source: compiled from data contained in Table 8 of Ross* et al., *1993, pp. 38–46)*

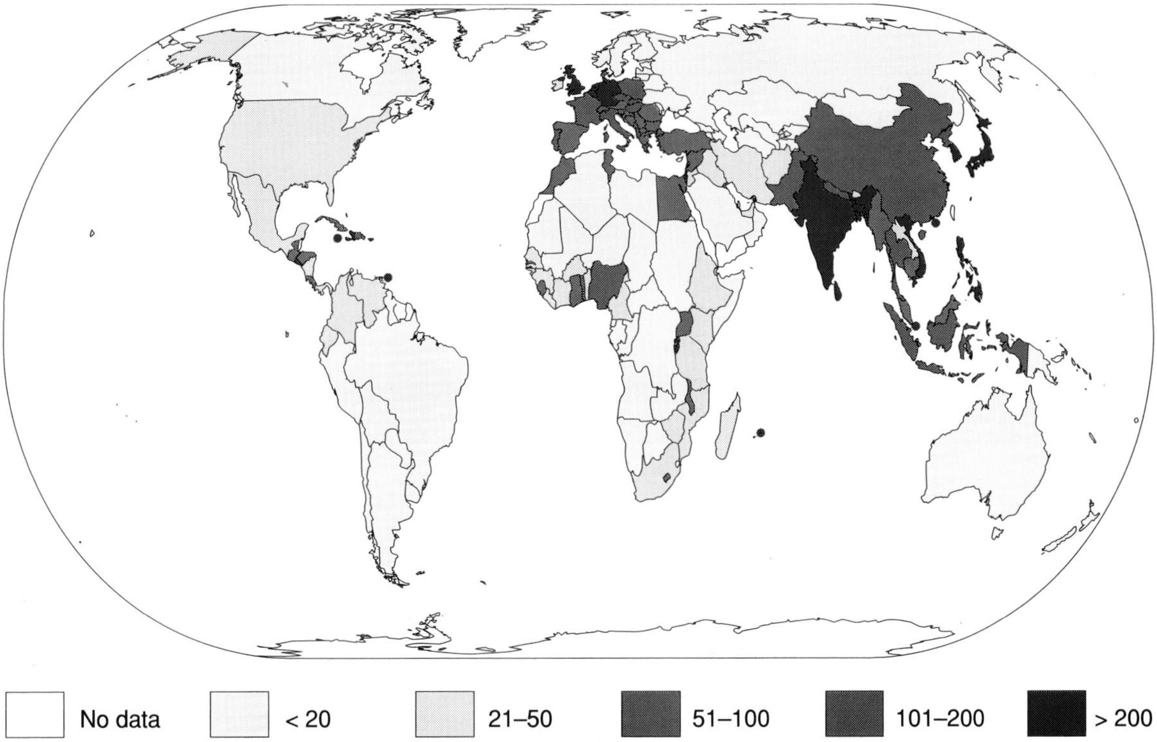

No data | < 20 | 21–50 | 51–100 | 101–200 | > 200

Figure 2.6 *Population density km², 1995 (Source: compiled from data contained in Table 1 of Ross* et al., *1993, pp. 9–13)*

Europe, Asia and parts of Latin America have the areas of highest population density in the world. Compare Bangladesh (2,266 people per square mile) and Holland (1,164 people per square mile) with Tanzania (81 people per square mile), Ethiopia (133 people per square mile) or Botswana (6 people per square mile). Even the most densely settled countries of Africa, such as Burundi and Rwanda, have only 500–600 people per square mile. This has led some to argue that Africa is in fact *under*populated, and needs more people. For example, this is one reason put forward to explain why much of Africa's farming systems rely on fallow and rotation of land, rather than on intensive methods involving manuring, fertilizing and irrigation. Others have criticized this argument by pointing out that much of the land area in African countries is not suitable for farming, so that 'real' densities, measuring numbers per area of arable land, are much higher. These points will be considered in more detail in sections 2.3.2 and 2.4.2 below.

So far the focus has been on how population is changing. At this point we need to turn to the question of what the environmental consequences of population change are. The next section explores some of the different answers given to this question over the last 200 years.

2.2.2 Population growth and the environment: three approaches

What are the environmental consequences of the unprecedented burst of population growth in the twentieth century? Although world population growth may eventually stop, can global resources accommodate 12 billion people? Will we run out of minerals, or sources of energy? Can we grow enough food to feed everyone? Will such large numbers produce too great a volume of greenhouse gases for the planetary ecosystem to absorb? These are large, controversial and complex questions, and we may never have decisive answers to them. However, what we can do as a first step is to look at the often radically opposed ways of thinking about population and the environment which have been put forward.

neo-Malthusianism

Cornucopians

Broadly speaking, there are three approaches. A *pessimistic* view is that population growth is a major threat to world economy, environment and political stability. Pessimists (sometimes called *neo-Malthusians*) argue strongly for population control. A second group are *optimists* (or *Cornucopians*), who argue that population growth, far from being a cause of economic and environmental problems, is in fact a positive stimulus to innovation and problem-solving. At their strongest, this group argues that the more people the better. A third, more *neutral* view is that population growth is neither strongly positive nor negative; rather, environmental problems have other social, political and economic causes.

The pessimistic view has its intellectual roots in the work of Thomas Malthus, a political economist writing just after the French Revolution. Many observers at the time were optimistic about the prospects for an improved standard of living for the rural peasantry, as power of the rural aristocracy was broken. Malthus wrote in opposition to this, arguing that any increases in incomes would lead to earlier marriages. This would fuel a population growth which in turn would undercut rural wages and exhaust the land. His most famous statement is that the supply of food increases more slowly than population, so that unrestricted population growth must eventually lead to famine.

In the current era, the Malthusian mantle has been donned most prominently by the socio-biologist Paul Ehrlich. In 1968 he published *The Population Bomb,* an extract of which you have just read in Activity 2, and in 1990 he updated his argument in *The Population Explosion,* published jointly with his wife. Another important neo-Malthusian work was *The Limits to Growth: a Report for the Club of Rome's Project on the Predicament of Mankind* by Meadows *et al.*, published in 1972. Pointing to the rapid growth of population in the twentieth century, these writers have modernized and extended Malthus' arguments. They argue that population growth will lead to the exhaustion of mineral resources, leading to scarcities of key inputs and the collapse of economic growth. They associate malnutrition with the inability of food production in underdeveloped countries to keep pace with population growth. They draw attention to the role of population growth in increasing pollution and environmental degradation in the form of deforestation and desertification. Perhaps most of all, they warn that the activities of rapidly growing populations are destroying the diversity of species of plants and animals on the earth.

For all these issues, the impact on resources, food stocks and the environment depends somewhat on levels of consumption and on technologies used in production. This is sometimes represented in the equation:

$$I = P \times A \times T$$

(where I = impact; P = population; A = affluence; T = technology). However, in Ehrlich's words, 'overpopulation is now the dominant problem' (Ehrlich and Ehrlich, 1990), so that growth in P is much more important than any changes in A or T.

The neo-Malthusian position has been widely criticized. Some critics of the pessimistic view take the argument and turn it on its head. One of the most vigorous of these has been the economist Julian Simon. Simon argues that, far from being a source of economic and environmental disaster, population growth is a good thing. He noticed that, despite the fast growth of world population since the nineteenth century, the prices of scarce resources such as metals and minerals had actually gone down. He came to argue that, partly precisely because it creates problems, and partly because it provides more labour and brain power, population growth stimulates solutions. In terms of the equation $I = P \times A \times T$, growth in P will bring forth advances in T which lead to improvements in I. For example, shortages of minerals, metals and fuels stimulate substitution and exploration for new sources. Shortages of food and land stimulate the development of new, more productive types of agriculture. Increases in pollution stimulate the development of better, cleaner technology, and new methods of cleaning up the mess. Simon's arguments have been applied mainly to issues such as macro-economic growth and natural resources. Another economist, Ester Boserup, has made very similar arguments about agriculture, which are discussed in section 2.3.2 below.

Simon, Ehrlich and others have tried to back up their arguments with empirical data on population growth and various indicators of resource use and environmental change.

Activity 3 The newspaper article 'Betting the planet', reproduced below, describes a wager made between Paul Ehrlich and Julian Simon. The bet was based on what would happen to the scarcity of some key resources – metals and minerals – over the 1980s. If a resource becomes more scarce, its price usually rises, so Ehrlich bet that the prices of these metals and minerals would rise. Simon believed the opposite. Read the article to find out who won. After you have read it, think about the following questions:

- Does the fact that one of the two won the bet mean that he is right about other issues? If not, why not?
- Much of the argument between the two is about what will happen in the future, as well as what has happened up until now. What do you think will happen, as population growth continues into the next century?

Betting the planet

John Tierney reports

In 1980 an ecologist and an economist chose a refreshingly unacademic way to resolve their differences. They bet $1,000, specifically over the future price of five metals, but at stake was much more – a view of the planet's limits, a vision of humanity's destiny. It was a bet between the Cassandra and the Dr Pangloss of our era.

They led two intellectual schools – sometimes called the Malthusians and the Cornucopians, sometimes the doomsters and the boomsters – that use the latest in computer-generated graphs and foundation-generated funds to debate whether the world is getting better or going to the dogs. The argument has generally been as fruitless as it is old, since the two sides never seem to be looking at the same part of the world at the same time. Dr Pangloss sees farm silos brimming with record harvests; Cassandra sees topsoil eroding and pesticide seeping into ground water. Dr Pangloss sees people living longer; Cassandra sees rainforests being decimated. But in 1980 these opponents managed to agree on one way to chart and test the global future. They promised to abide by the results exactly 10 years later and to pay up out of their own pockets.

The pair – who have never met in all the years they have been excoriating each other – are both 58-year-old professors who grew up in the Newark, New Jersey suburbs. The ecologist, Paul R. Ehrlich, has been one of the world's better-known scientists since publishing The Population Bomb in 1968. More than three million copies were sold. When he is not teaching at Stanford University or studying butterflies in the Rockies, Ehrlich can generally be found on a plane on his way to give a lecture or to collect an award. He is the pessimist.

The economist, Julian L. Simon of the University of Maryland, often speaks of himself as an outcast, which isn't quite true. His books carry jacket blurbs from Nobel laureate economists, and his views have helped shape policy in Washington for the past decade. But Simon has certainly never enjoyed Ehrlich's academic success or popular appeal. He is the scourge of the environmental movement.

[...]

Ehrlich decided to put his money where his mouth was by responding to an open challenge issued by Simon to all Malthusians. Simon offered to let anyone pick any natural resource – grain, oil, coal, timber, metals – and any future date. If the resource really were to become scarcer as the world's population grew, then its price should rise. Simon wanted to bet that the price would instead decline by the appointed date. Ehrlich derisively announced that he would "accept Simon's astonishing offer before other greedy people jump in." He then formed a consortium with two colleagues at the University of California at Berkeley specialising in energy and resources.

In October 1980 the Ehrlich group bet $1,000 on five metals – chrome, copper, nickel, tin and tungsten – in quantities that each cost $200 in the current market. A futures contract was drawn up obligating Simon to sell Ehrlich and his colleagues these same quantities of the metals 10 years later, but at 1980 prices. If the 1990 combined prices turned out to be higher than $1,000, Simon would pay them the difference in cash. If prices fell, they would pay him. The contract was signed and Ehrlich and Simon went on attacking each other throughout the 1980s. During that decade the world's population grew by more than 800 million, the greatest increase in history, and the store of metals buried in the earth's crust did not get any larger.

[...]

The bet was settled last month without ceremony. Ehrlich did not even bother to write a letter. He simply mailed Simon a sheet of calculations about metal prices – along with a cheque for $576.07. Simon wrote a thank you note, adding that he would be willing to raise the wager to as much as $20,000, pinned to any other resources and to any other year in the future.

Each of the five metals chosen by Ehrlich's group, when adjusted for inflation since 1980, had decline[d] in price. The drop was so sharp, in fact, that Simon would have come out slightly ahead overall even without the inflation adjustment called for in the bet. Prices fell for the same Cornucopian reasons they had fallen in previous decades – entrepreneurship and continuing technological improvements. Prospectors found new lodes, such as the nickel mines around the world that ended a Canadian company's near monopoly of the market. Thanks to computers, new machines and new chemical processes, there were more efficient ways to extract and refine the ores for chrome and the other metals.

For many uses the metals were replaced by cheaper materials, notably plastics, which became less expensive as the price of oil declined (even during this year's crisis in the Persian Gulf, the real cost of oil remained lower than in 1980). Telephone calls went through satellites and fibre-optic lines instead of copper wires. Ceramics replaces [*sic*] tungsten in cutting tools. Cans were made of aluminium instead of tin, and Vogt's fears about America going to war over tin remained unrealised. The most newsworthy event in the 1980s concerning that metal was the collapse of the international tin cartel, which gave up trying to set prices in 1985 when the market became inundated with excess supplies.

Is there a lesson here for the future? "Absolutely not," says Ehrlich. "The bet doesn't mean anything. Julian Simon is like the guy who jumps off the Empire State Building and says how great things are going so far as he passes the 10th floor. I still think the price of those metals will go up eventually, but that's a minor point. The resource that worries me the most is the declining capacity of our planet to buffer itself against human impacts. Look at the new problems that have come up, the ozone hole, acid rain, global warming. It's true that we've kept up food production – I underestimated how badly we'd keep on depleting our top soil and ground water – but I have no doubt that sometime in the next century food will be scarce enough that prices are really going to be high even in the United States. If we get climate change and let the ecological systems keep running downhill, we could have a gigantic population crash."

Simon was not surprised to hear about Ehrlich's reaction. "Paul Ehrlich has never been able to learn from past experience," he said, then launched into the Cornucopian line on the greenhouse crisis – how, even in the unlikely event that doomsayers are right about global warming, humanity will find some way to avert climate change or adapt, and everyone will emerge the better for it. But Simon did not get far into his argument before another thought occurred: "So Ehrlich is talking about a population crash," he said. "That sounds an even better way to make money. I'll give him heavy odds on that one."

Source: *The Guardian*, 28 December 1990, p. 25

In the case of metals and minerals, then, the optimism of Julian Simon won out over the pessimism of Paul Ehrlich, at least for now (you can find out more about metal and mineral resources in Chapters 4 and 5). However, there are several points to make about this.

First, what appears to be a good solution for the users of resources – Western consumers – is also a disaster for the producers of resources. For example, since the late 1970s the world market for copper has collapsed, which has led countries which relied on copper exports, such as Zambia, into debt and economic collapse. So a solution in terms of *resources* may not always be a solution in terms of *people*.

Second, there can be cases where the solution to one resource problem leads to further problems with a different resource. Let us take the example of food. Up to the 1960s, the rapid population growth of the twentieth century looked like posing a world food problem, especially for poor populous countries like India. However, the development of new technologies in agriculture, including the Green Revolution in South Asia, meant that per capita world food production increased over the next 30 years, despite a doubling of the population. This may look like support for the Simon argument that population growth provides its own solutions, but this must be qualified. First, malnutrition has not disappeared in Asia, since it is obviously the distribution of food, rather than the total supply, which determines this. Distribution is determined by social, economic and political factors, not population growth. Second, intensive farming, both in underdeveloped countries and in the North, has brought environmental problems, such as salination through irrigation and the build-up of nitrates and pesticides in soils. Finally, there are signs that, partly because of these environmental problems, the limits to intensive food production may have been reached.

A third point about the optimist versus pessimist debate is that 'impact' issues tend to get grouped together. The pessimists assert that population growth is always a dominant and negative factor, while the optimists argue that it is always a positive influence. However, it is possible that the significance of population varies according to the issue.

Atmospheric change is a good example of this. Chlorofluorocarbon (CFC) gas, which destroys ozone, comes primarily from industrial products, such as refrigerators, aerosols and packaging. The rapid rise in CFC emissions, and their subsequent fall, was mainly to do with industrial production processes, and most releases come from rich industrial countries with low population growth. Greenhouse gases such as carbon dioxide and methane, by contrast, are produced in greater volumes by a wider range of countries. Richer countries can still produce a lot more greenhouse gas per head; for example, cumulative per capita emissions of carbon dioxide up to 1986 were about 30 times higher in Canada than they were in China. This is because people in richer industrial countries use more energy per head. However, the role of population growth is still significant. According to Paul Harrison, about half the growth in carbon dioxide emissions over the period 1960–1988 in Africa and Latin America was due to population growth, and about 40 per cent in Asia (Harrison, 1992, Figure 4).

So far, whether it is seen as having a negative or a positive effect, the focus is still on the consequences of population growth. But this debate between the optimists and the pessimists is seen by many observers as missing the point. They take the view that environmental and resource problems are more to do with economic and social structures than with population growth. This third approach also goes back a long way. Karl Marx, writing in the nineteenth century, was an early critic of Malthus. Marx was principally concerned with the development of capitalism, and he saw how the agricultural revolution, and later the industrial revolution, meant that the supply of resources and the use of resources could be transformed at a rapid pace. Food supply, for example, could be greatly increased in a short space of time by the use of new technologies, which made population growth largely irrelevant, except as an extra source of potential workers to keep wages down. Furthermore, technological change was not driven along by population growth, but by the competition inherent in the capitalist system.

Many present-day economists would agree with much of Marx's analysis of the role of population. In terms of the $I = P \times A \times T$ equation above, this view means that A and T can change relatively quickly, compared with P. Further, they do not change primarily in response to an increase in population, as Julian Simon argued, but rather in response to such factors as economic policy, profitability, growth in demand, and investment in research and development by individuals, governments and corporations. Equally, A and T do not always change in a favourable way, to reduce the impact of population growth (see Box 2.1).

Box 2.1 Population, energy use and carbon dioxide emissions in China

A major determinant of carbon dioxide emissions is energy use. For a long period, total energy use in China grew rapidly, since high population growth and the development of heavy industry coincided. According to the UN, carbon dioxide emissions from industrial sources in China went up from 215 million tonnes of carbon in 1960 to 406 million tonnes in 1980.

Since the late 1970s, however, population growth in China has slowed down, because the government adopted a population policy where each couple was allowed only one child. The rate of natural increase in 1990 was down to about one per cent per year. But over the same period, there have also been dramatic reforms in China, leading to high economic growth. Industrial output grew rapidly, and although there was greater energy efficiency over this period, carbon dioxide emissions increased even faster, up to 678 million tonnes of carbon in 1990.

The outlook for emissions of carbon dioxide, and indeed general pollutants, is not encouraging. Energy efficiency gains in the 1980s have been taken as far as they can (Smil, 1993). At the same time, industrial growth has speeded up as the economic reforms continue apace. While the highest population growth rates ever recorded in the world have been in the region of 4 per cent per year, industrial output can grow much faster. In both 1992 and 1993, the economy of China grew by 13 per cent, while industrial output has been growing at 25 per cent per year or more. Energy demand is outstripping supply, and China has been slow to move away from reliance on coal as a primary source.

In China, then, economic growth has had a rapid impact on the environment largely independently of population dynamics.

Summary of section 2.2

- World population is much higher than it has ever been, and has grown faster in the twentieth century than ever before.
- However, the rate of growth is slowing, and world population will probably stabilize at about 12 billion early in the next millennium.
- Most of the growth has been and will be in Africa and Asia, especially South Asia.
- Growth has been high because fertility has been higher than mortality for a considerable time, but is falling now principally because fertility is dropping rapidly as modern contraception is adopted.
- There is controversy over whether Africa is over- or underpopulated compared with the rest of the world, in terms of densities of people on land.
- Thinking about population and the environment has developed largely within traditions: neo-Malthusian, Cornucopian and the neutral approach.
- The neo-Malthusian position is the best known of these, perhaps because it is intuitively appealing – population growth is something simple and tangible to which to attribute global problems.
- However, closer inspection often reveals that it simplifies too much, and underestimates the adaptability of people, as the case of metals and minerals shows.
- We shall also see in section 2.4.1 below that, although it appears almost obvious, a negative effect of population on the environment is in fact quite difficult to demonstrate.

2.3 Africa: overcrowded or underpopulated?

For neo-Malthusians, Africa is the proof that overpopulation causes poverty and environmental catastrophe. With its high growth rates, it is already seen as trapped in decline, with the inevitability of much more growth summoning up the spectres of the ultimate Malthusian price for overpopulation – famine and war. At the same time, Africa should be a test case for the views of the optimists. Africa is the poorest region of the world, and yet the most rapidly growing. More people pose problems, but according to analysts like Julian Simon, they also provide solutions. In this sense perhaps Africa needs more people, since it has been seen in the past as underpopulated. But the key question here is whether the *economic* resources needed to solve problems will be available.

In this section we will explore these issues at some length. The first thing we need to do, however, is understand why Africa's population is growing so fast.

2.3.1 High fertility in Africa

We saw in section 2.2.1 that current rates of population growth in Africa are high compared with other parts of the world. We also saw that this was mainly due to high fertility. To understand fully the population–environment picture in Africa, we therefore need to look at why fertility in Africa has been and remains so high.

A word of caution, however, about looking at Africa as a unit. This can be useful in certain ways, but there are also important differences *within* Africa. Fertility rates are not the same everywhere on the continent; nor are economic structures or cultural factors.

Early attempts to explain high fertility cast it in terms of outdated 'customs'. In the pre-colonial period, infant mortality was high, and so Africans had many children to ensure family survival. However, with the availability of modern medicine, mortality had fallen and high fertility was unnecessary. It only continued because it was propped up by 'irrational' religious and cultural values.

This approach, associated with modernization theory, was challenged in the 1970s by an increasing body of evidence that children in Africa (and elsewhere in the developing world) made significant contributions to work on family farms in rural areas. This led to the view that a large family was in fact economically rational. The demographer John Caldwell captured this in the notion of *wealth flows*. In the rural areas of Ghana and Nigeria in West Africa, where Caldwell had worked, children contributed more through their labour than they cost their parents in terms of food and clothing, so there was a net wealth flow up from child to parent. By contrast, in low-fertility Britain, a child is expensive to raise, and will not usually directly support his or her parents when they are old. In this case, the net flow is downwards from parent to child (Caldwell, 1982).

wealth flows

Activity 4 Turn now to Reading B, 'Children: activities, costs and returns', by John C. Caldwell, which you will find at the end of the chapter. When you have read this extract on the costs and benefits of children in Yoruba society (southern Nigeria), think about raising children in the UK. What are the main differences from the Yoruba case in terms of costs and returns? What factors do you think might change the situation in Yoruba society?

The idea of wealth flows has been used widely to explore the value of children to parents, in terms of their labour, and also in their role of providing security and care for parents in their old age. It has been particularly important in understanding women's attitudes to large families. Having a large number of children places a great physiological strain on women, as well as a great burden of daily work in feeding, cleaning and raising them. However, children can also be a key source of help to women in these tasks, and of especial importance in providing security to women when they are old.

This argument is of great relevance to the population–environment question, because some kinds of environmental degradation may increase the need for children's labour. For example, deforestation may mean that women (who are usually responsible for domestic fuel) have to go further in search of fuelwood, and the help of children becomes even more important (see, for example, Dankelman and Davidson, 1988).

One implication of this theory is that education will have a large impact. First of all, schooling raises the costs of having children, both directly in terms of fees and uniforms, and so on, and also indirectly in that they will no longer be available to help their parents in farm work. A second factor (which Caldwell stresses) is that education, by introducing Western concepts of the nuclear family where children are dependent, will undermine the *cultural* basis of the upward wealth flow. A third point, emphasized by economists, is that when women become educated, they can begin to move out of traditional roles into paid work, and their time becomes more valuable. This means that raising children becomes more costly in terms of potential earnings lost.

The importance of education, especially the education of women, for falling fertility rates has been widely confirmed by surveys. However, the wealth flows theory has not been universally accepted. One problem is that, although education for both sexes has seen a great expansion in Africa since the 1960s, fertility rates have actually increased. This has led Caldwell to amend his theory somewhat, to provide an explanation of why African fertility is different (Caldwell and Caldwell, 1987). His answer shifts the emphasis from the economic role of children to their cultural role. High fertility, he argues, is an outcome of African religious traditions, which are based on the worship of ancestors. Ancestors are 'reborn' in children, so this tradition ensures a generally pro-natalist, or pro-birth, attitude. It is this attitude which also limits the spread in the use of contraception seen in Latin America and Asia.

Perhaps the most significant development for all these theories is that, very recently, there are signs that fertility is starting to decline in certain African countries, such as Botswana, Zimbabwe and Kenya. The reasons for these declines are still debated.

Controversy over the causes of Africa's high fertility thus continues. However, one of the few issues on which most of the theories agree, and which may have the most significant implications for the future, is the status of women. Here, maintaining the expansion of education for women is key.

2.3.2 Two views on population and rural environments

So far, we have seen that, for many countries in Africa, rates of population growth are currently higher than in other regions of the world, and this is due specifically to high fertility rates. We have also noted the beginnings of a fall in fertility in a few countries, heralding a likely slow down of growth in the future, though not for some time still.

What is the implication of these patterns for African environments? Does the rapid expansion of population, and hence agriculture and livestock rearing, lead to the exhaustion of soils and rangelands and hence to desertification? Does the expansion of agriculture in tropical Africa and a burgeoning demand for fuelwood lead to deforestation?

There are two opposing views on population and local environmental change, closely related to the optimists' and pessimists' views described in section 2.2.2 above. One approach, again very much associated with the name of Malthus, sees population growth leading to an increased demand for food, and therefore to an increased use of land for farming and livestock rearing. As land reserves are used up, farming becomes more intensive. 'Traditional' farming systems in developing countries were 'extensive', using a lot of land. They involved long periods of fallow, whereby an area would be farmed for a while, then abandoned as the nutrients in the soil were used up. New areas would be cleared and the old area left to fallow. As populations built up and land ran out, this system could not survive, and farmers began to shorten the length of the fallow period, until eventually they started to farm the land every year. According to this view, with shorter fallow periods, soils have less time to recover, and soil degradation begins. The physical structure of the soil begins to break down, and most of its nutrients are exhausted or leach out. Incomes and food production fall, and poverty increases. A similar process occurs with the overstocking of rangeland with cattle; too many cattle will clear land of vegetation, leaving it vulnerable to sheet erosion and, ultimately, to desertification (see Plate 5).

The opposing idea, that population growth might have a positive influence on the productivity of environments, is associated with the name of Ester Boserup. Drawing on a wide range of examples from different countries and periods of history, she developed an alternative argument. As populations on land grow, the demand for food and marketable crops does indeed increase, and fallow periods shorten. But rather than this leading to declining productivity and environmental degradation, it spurs the application of more labour and complementary inputs per unit of land. New technologies which enhance yields are sought out and adopted. Labour-intensive investments in soil are made, since these help to maintain yields. The quality of the land is improved (Boserup, 1965).

Activity 5 Turn to Reading C by Michael Mortimore, entitled 'Population growth and land degradation', at the end of the chapter. As you read, think about the following questions:

- At what stage do the two 'pathways' diverge?
- What factors do you think will determine whether a particular rural area follows one pathway or the other?

There is another useful, but simple, way of thinking about the differences between these two views. The notion of 'pressure' comes from the physics of gases. As the volume of gas builds up in a container, its pressure against the walls of the container increases. Analogously, as a total population in an area builds up, the population pressure on the environment increases. Here, however, the views diverge. According to a Malthusian position, the environment is like a gas bottle, with a fixed ability to absorb human activity. This fixed ability is usually called the *carrying capacity* of a resource. Pushed beyond that point, it degrades very rapidly, similar to a container buckling or exploding under pressure. The Boserupian conception, by contrast, is that the environment is more like a balloon – its shape and capacity is actually changed as pressure builds up. Carrying capacity is therefore not fixed, but can be influenced or shaped by the application of labour and technology (Zaba and Scoones, 1994).

carrying capacity

Activity 6 Below are two sets of questions to get you to think critically about the notion of carrying capacity – an idea which seems simple at first, but on second thoughts looks more problematic.

1 Think of one example in which you have altered the carrying capacity of an environment (a garden for example). What resources did you put into this environment? How 'natural' was the environment before you started?

2 Zaba and Scoones (1994, p. 197) write: '... most of us have no problem with the notion of the carrying capacity of Botswana (pop. 1.3 m, area 567,000 sq km), but would be incredulous at the idea of calculating the carrying capacity of Birmingham (pop. 1.1 m, area 300 sq km).' Why is it inappropriate to calculate the carrying capacity of Birmingham? Do these reasons also make it actually inappropriate to apply the notion to Botswana?

Summary of section 2.3

- Various explanations have been put forward to explain high fertility in Africa.
- An important development has been the recognition that children in African societies provide benefits to parents that are absent in Western societies, including providing security.
- The value of children to women is particularly important.
- One factor likely to change fertility is the status of women, and here education is key.
- Rising population densities can lead either to degradational or to conservation pathways.

2.4 Africa: a unique case?

Boserupian pathway

In theory, either of the two pathways – to degradation or to conservation – discussed in section 2.3.2 above, might develop from population growth. The *Boserupian conservation pathway* is obviously the desirable one, so the important question then becomes, *under what conditions does the Boserupian path develop, and under what conditions is it blocked?* This path has often been seen as blocked in Africa, for three reasons.

The first of these reasons is that population growth has been so rapid at national levels. This has been seen as a major threat to the possibility of Boserupian effects coming into play:

> *The relevant question for Africa is whether the catalyzing factor of population is ahead of or behind the pace of farmer-based innovation. The question reveals a major limitation of the Boserup model. High rates of population in an initially high density area jeopardize the perceived benefits of autonomous intensification ...*
>
> *(Lele and Stone, 1989, p. 9)*

Second, Boserupian investments in land and water resources are argued to be blocked in Africa because of types of *social organization* which differ from those in other regions (such as Europe or South East Asia), where societies have adapted successfully to high population densities.

A third reason often put forward for the lack of hope in Africa is what is argued to be the *unique fragility of African environments*:

Africa is different. Her climate, her soils, her geology, her patterns of disease, all pose problems of a severity that most of Asia and Latin America do not have to face.

(Harrison, 1987, pp. 27–8, original emphasis)

Because the great majority of Africa's soils are 'difficult to manage', the outcome of intensification – permanent cultivation – is impossible (Timberlake, 1985, p. 65). As a result, as populations build up, the resulting overcultivation, overgrazing and deforestation inevitably bring desertification:

Africa's environment is so sensitive that when a certain level of population density is reached traditional methods begin to degrade the soil on which all future production depends.

(Harrison, 1987, p. 27)

The combination of rapid population growth, a lack of institutions capable of coping with the crisis, and a fragile natural resource base lead to a poor prognosis for Africa:

Africa is overpopulated now because … its soils and forests are rapidly being depleted …

(Ehrlich and Ehrlich, 1990, p. 39)

Of course, population growth is not the only factor. Also to blame are the generally harsh environment of [sub-Saharan Africa], its unreliable climate, its adverse trade terms, its weak infrastructure, and its faulty development policies. But there is much agreement that the list of problems is headed by population growth.

(Myers, 1989, p. 214)

2.4.1 Assessing 'population pressure'

How can we assess these arguments? The first step is to think about what kind of evidence we need in order to show that population pressure *causes* 'environmental degradation'.

The answer to this depends partly on what *environmental degradation* actually means. This term is usually used very loosely to cover a variety of equally loosely defined processes such as deforestation, desertification and soil erosion. In the case of desertification, one review of the literature identifies over one hundred definitions (Glantz and Orlovsky, 1983). Although it need not be defined this way, many uses of the term 'degradation' imply *irreversible destruction of biological potential.* If this is the case, identifying degradation means identifying irreversible damage. This is actually quite difficult, because many ecosystems have considerable capacity to recover from natural cycles of climatic change, or human intervention.

environmental degradation

However degradation is defined, a second problem is proving that the important causal relationship is the one between degradation and growing numbers of people, as opposed to other factors. Such factors might be shifts in the activity of people, in technology, or in ownership of resources. There is also the pitfall of attributing environmental change to population pressure when it is in fact the outcome of cycles or trends in natural processes.

Again, the example of desertification is interesting. Despite the definitional problems mentioned above, Ehrlich and Ehrlich (1990) are of the view that desertification is a widespread process in the Sahel, the region of Africa just south of the Sahara. Let us look at an important part of their argument:

It is no accident that the most serious desertification is found in areas where burgeoning human populations are contributing to rapidly changing land-use patterns. For instance, the 1950s and 1960s were a period of unusually favourable rainfall in the Sahel. As a result, cash-crop agriculture expanded along with the human population. Specifically, the population of Niger increased from 2.5 to 3.8 million from 1954 to 1968, and peanut farming expanded from just over 500 square miles to some 1700 square miles. Nomadic herders of the Sahel who previously grazed animals on land that had disappeared under cash crops, were displaced to the north. They stocked new lands (which tribal traditions taught were undependable for the long term), and their herds increased during the moist phase. Then as tradition predicted, the climate turned dry again. The vegetation was completely removed by cattle, camels and goats, and millions of animals died. An unknown number of people, probably around 100,000, perished in the resulting famine.

(Ehrlich and Ehrlich, 1990, p. 130)

Let us assess this argument. The first important step is the association made between the increase in population in Niger with an increase in areas under peanuts (or groundnuts). The arithmetic of the figures given suggests that this is unlikely to be the whole story. Over the period, population increased by a factor of about 1.5, while the area under groundnuts increased more than three-fold. Other factors, including price and a greater demand for cash incomes, were clearly involved. These factors in turn must be situated in a broader colonial history of West Africa, in which the groundnut was promoted as an export crop for European markets, and adopted by farmers as a means of paying colonial taxes. Most of all, however, the groundnut boom was made possible by *suitable rainfall conditions*. Until the wetter years of the 1950s and 1960s, commercial groundnuts had largely been grown only in Senegal and Nigeria, not in Niger (Grainger, 1990, p. 74).

During this phase of groundnut expansion, herders moved north and were able to expand their herds. Groundnut expansion, caused by whatever factor, therefore did not damage the productivity of the pastoralists' herds, for otherwise they would not have increased in size, but rather decreased. This they did, of course, when drought struck in 1972 and 1973, although the figures for both animal and human deaths in the Sahel drought and famine are simply rough estimates. But the main reason why cattle died in large numbers was that the cattle population built up during a series of wet years which then ended abruptly in drought, killing off the vegetation on which the cattle fed. It had little to do with groundnuts or human population growth.

Ehrlich and Ehrlich would prove a decisive population–desertification link only if they could show that population growth caused cropping increases, which in turn forced pastoralists to graze herds where they caused an irreversible loss of biological potential. In fact, the evidence from Africa is that grazing ecosystems – both plants and animals – are relatively adaptable, and can recover quickly from drought (Behnke and Scoones, 1992). Assessments of pastoralism in Niger in the 1980s – which had recovered from the 1970s drought – found little evidence of damage from overgrazing.

All of this points to a central problem for proving population–desertification links – namely the fact that rainfall levels in the Sahel have been falling over the last 30 years. Lower rainfall means less moisture, and thus drier soil. Establishing downward trends in the productivity of crops or animals and linking these to population growth *above and beyond* the influence of rainfall patterns is very difficult. Analogous problems arise in other cases, such as soil erosion in the mountain areas of Ethiopia, where it is not easy to separate the impact of human activity from natural processes (Campbell, 1991).

Even if we can separate out the impact of rainfall or other factors, linking population growth in an area to degradation or soil erosion may be hard. As Paul Harrison says, 'Soil erosion involves so many factors, from the smallest detail of farm implements, through crop patterns to soil type, that it would be practically impossible to document the population link. The results would vary from one field to the next' (Harrison, 1992, p. 135). At the very least, then, making good causal links between population growth, cultivation or grazing and desertification is difficult. We have to pay careful attention to the other factors which may be involved, including sudden shifts in activity, or methods of farming, due to economic policy, and above all to rainfall patterns and other natural processes. We also have to show that farmers and pastoralists are actually following practices which are leading to damage, rather than just assuming that they are.

2.4.2 Social organization and agricultural intensification

A common theme in writings on African environments is that Africa is caught in a 'technology trap' (Harrison, 1992, p. 105), unable to move into Boserupian intensification. How true is this? In one sense, there is a great irony here, since much of the early study of agricultural intensification, around the time of the publication of Boserup's work, was done in West Africa. Many of these investigations showed an association between population growth and agricultural intensification, including the move from long fallow to permanent cultivation systems. However, the pessimistic view of African social organization would expect such intensifications to lead to environmental and productivity decline. Increased frequency of cultivation would lead to soil erosion, and demand for fuelwood would lead to deforestation.

What reasons are there for being so pessimistic? One issue is how control over resources is organized. In 1968 the socio-biologist Garrett Hardin put forward the view that where access to resources is free, people will tend to overuse them. He argued that only by making all natural resources privately owned would people have an incentive to moderate their use and start investing in those resources.

Activity 7 Hardin's paper 'The tragedy of the commons' has become a classic in the population–environment literature. He portrays a situation where, although each individual person behaves rationally, everyone is worse off, and no-one has any incentive to change. Turn now to Reading D, an extract of this paper, which you will find at the end of the chapter. As you read, think about the following questions:

- Do you agree with Hardin?
- Can you think of an example of a communal 'good', like a national park, which has collapsed under pressure of population?
- Are all communal resources managed in the way assumed by Hardin?
- What do you think of Hardin's view of human nature?

Hardin has been criticized on the grounds that he has misunderstood the true nature of common property resource systems, which do not involve uncontrolled free access, but rather are communally managed (for example, Shepherd, 1989). However, the view often still prevails that Africa has less private ownership of agricultural and range lands, and that this is a major cause of overcultivation and overgrazing.

In many African contexts, people are said to hold 'usufructuary' rights to land, meaning that, although they do not own the land, they have rights to use it. Such rights are usually seen as managed through kinship groups called lineages, since these are the primary social grouping in African society. This contrasts with Asia, where social and political allegiance is based on territory. This difference between lineage, on the one hand, and territory, on the other, is another reason put forward for the particularly bad prognosis for the population–environment relationship in Africa. The logic of the lineage is that more people are always good, since this increases the power and capacity of the lineage. In territorial-based systems, there is more concern for the balance between people and resources (Cain and McNicoll, 1986).

Do these arguments mean that agricultural intensification in Africa will necessarily lead to environmental degradation? One relatively well-documented example suggests the contrary, that existing social institutions *can* adapt in response to population growth in successful ways. This example is set in a semi-arid area, and lies near the large city of Nairobi in Kenya. It is not the kind of area which would be expected to respond to rapid population growth by adaptations bringing about stability or improvement in the environment, but it did.

Activity 8 The best introduction to this example, which comes from Machakos District in Kenya, is to look at the two photographs of a particular hillside in the district, reproduced opposite. The top photograph was taken in 1937, the bottom one in 1991. Over this period, population in Machakos had increased more than five-fold. What differences can you see? In which year do you think more soil erosion would have occurred? Why?

Soil and water resources in Machakos have improved greatly since the 1930s, and deforestation has been reversed. As Tiffen *et al.* say: '... the movement of farming households into the drier ecological zones of Kenya is sometimes claimed to have a deleterious impact on the environment. It need not' (Tiffen *et al.*, 1994, p. 7).

How did such a remarkable transformation take place? The full story is fascinating, though complex, and is told in *More People, Less Erosion* by Tiffen *et al.* (1994). However, the key underlying elements were the roles of markets and changing institutions, as Reading E, 'The economics of recovery' by Mary Tiffen and Michael Mortimore shows us. Turn to the end of the chapter and read this now, and then think about the arguments about institutional limits I have just described. Why did they not apply? How did Machakos escape the 'technology trap'?

Kiima Kimwe Hill (north-east slope) in 1937

Kiima Kimwe Hill (north-east slope) in January 1991

Machakos is an example of the basic Boserupian principle – that improvements in land and other natural resources actually *need* population pressure. There are several lessons from the Machakos experience. One obvious one is that investment in resources such as land is not possible without labour and capital, and these are only available if people receive a reasonable income for what they produce or do.

Another basic message is that existing social institutions were capable of adapting to higher population pressures, and acted to facilitate conservation. Customary rights to land for the Akamba, the main group in Machakos, guaranteed security of tenure, and thus an incentive to invest in new technologies and in terracing. Akamba family organization in the 1930s was of a 'traditional African' type: based on clans, with polygyny and a large family as a social ideal. However, the society was open to change, and people came to see more value in education and opportunities through migration. Women took on new roles in farm management, as men increasingly migrated to Nairobi.

However, one of the most significant factors involved in facilitating conservation in Machakos has been the persistence and adaptation of a 'traditional' social institution: mutual 'self-help' groups composed of relatives, but also of friends and neighbours. These groups, called *mwethya*, have been important for allowing women and younger people to have an input into community development, a role formerly monopolized by older men. The *mwethya*, and other institutions such as the church and traders associations, have provided channels for new information and ideas to enter Akamba society.

The gender aspects of the Machakos story are crucial. As in most farming communities, men and women interact with the environment in ways which are highly gender specific. This can be seen, for example, in their different interests in tree management (Box 2.2). However, women are usually responsible for tasks particularly affected by environmental change, such as water and fuelwood collection (see Plate 6). Thus whether or not women have the resources and opportunities for investing in the environment, will make the difference between sustainability and collapse.

Box 2.2 Women, men and trees in Machakos

People in Machakos have managed tree resources both as individuals and in self-help groups such as the *mwethya* and later women's groups. Management involves planting, tending and harvesting fruit, bark, wood and other products. Trees are thus investments of money and labour, and require both the resources to invest in them, and also reasonably secure rights to the products.

Men and women plant trees for different reasons, and often this means different kinds of trees. Men plant trees for timber for construction, for fuel, for charcoal-burning and brick-making, for shading crops and for wind-breaks, whereas women favour fruit trees. Women have secure rights to the products of the trees, and access to the market to sell fruit. They therefore have a good incentive to plant and maintain fruit trees. Early forestry projects seemed to miss this altogether – they ignored the fact that women plant and tend most trees, and focused on conservation rather than looking at trees as a source of income.

Source: adapted from Tiffen *et al.*, 1994, pp. 221–2

How representative is the Machakos case? It is certainly not unique in Africa. The densely settled zone around the northern Nigerian city of Kano is another example of a sustainable farming system under population pressure in a semi-arid area near a big city (Mortimore, 1970; Mortimore *et al.*, 1990). Studies of the Kano zone also provide important evidence that, contrary to the view that the areas near big cities are denuded of trees, tree density actually increases as one approaches the city (Cline-Cole *et al.*, 1990).

However, the 'Machakos effect' may not be applicable in all cases. In the semi-arid, densely settled northern areas of the Mossi plateau in Burkina Faso, deforestation and sheet erosion are serious problems, and it has taken direct interventions by government and non-governmental organizations (NGOs) to stabilize soils through the construction of stone barriers. Paradoxically, one factor in the failure of previous conservation efforts in Burkina Faso may have been that French colonial administrations took people's time away from their land through forced labour.

The two examples of Kano and Machakos provide us with a new perspective on the view that Boserupian-type intensification will be blocked in Africa. Both areas are close to cities, with access to urban labour markets, or markets for agricultural produce which give farmers a high enough price to realize a surplus. People in areas that are this well integrated into a national economy may thus be in a better position to invest in the environment than those in distant, poor regions. However, this obviously depends also on the terms on which that integration takes place; low prices for their goods, or much extraction through taxation, will deprive farmers of the resources they need to invest.

2.4.3 Are African environments really uniquely fragile?

So far in section 2.4, we have looked at two issues. First, we saw that, although it appears 'obvious', it is in fact often difficult to establish evidence that rapid population growth is a major causal factor in environmental change. Second, we saw in two important cases that rapid population growth did not lead to environmental disaster, but rather to improvement and stability.

Why then has there been such a pessimistic view of the capacity of African environments and people to adapt to population growth? One factor, mentioned above, is the pervasive opinion that tropical and semi-arid environments are excessively fragile. As a result, observers (typically European observers looking at African environments which they understood imperfectly) have tended to project catastrophic change onto what they see at any one time. Impressions become hardened into facts and recycled by generations of policy-makers.

There are many examples of this, including the case of the forestry expert E.P. Stebbing whose dire warnings of the 'encroaching Sahara' turned out to have little basis, but still have influence today (Mortimore, 1989, pp. 12–15). The main features of Stebbing's view were (i) that human activity (in this case shifting cultivation) in an inherently fragile, dry environment must cause damage; (ii) that this leads to a linear process of desertification; and (iii) that the observed spatial gradation from woodland to savanna to desert must be going on as a sequence through time. Thus, through human activity, the desert will march southwards (Stebbing, 1938).

Stebbing has been criticized for partial, seasonally biased data and leaps of logic, and his notion of a linear irreversible desert creep is not borne out by long-term evidence. Careful studies of the long-term record show that, rather than simple linear change from wet to dry, from forest to savanna, there has been a pattern of fluctuation back and forth over long periods of time (Nicholson, 1979). More recently, studies of the desert using remote sensing data have found that true desert boundaries are fragmentary, and move north as well as south (Forse, 1989; Binns, 1990).

The features of Stebbing's misperception listed above can also be seen in the case of the 'savanna–forest mosaic' found in Guinea and surrounding countries in West Africa (Fairhead and Leach, forthcoming). Such mosaic areas are made up of isolated islands of forest around villages in surrounding grasslands. The traditional view is that these islands are the relics of what was once continuous forest cover which has been degraded by bush burning and cultivation. Even contemporary large-scale afforestation projects in the area are based upon this view. Increased population pressure is destroying woodland and turning it into savanna. As with Stebbing's criticism of shifting cultivation in northern Nigeria, the implication is that local populations are ignorant of how they are destroying the environmental resource base.

Activity 9 Read in sequence the extracts from observers quoted in Reading F, 'Vegetation of the forest–savanna mosaic of Kissidougou Prefecture, Guinea: comparative quotations'. From 1893 onwards, the quotations assume that the forest was widespread in the only recent past, and that it will be totally destroyed in the imminent future. Why do you think it is that such views were repeatedly put forward, when in fact the region had not been covered in forest at any time in the last 100 years?

Again, a spatial pattern is read as history. Each generation of observers in the area has repeated the myth that the current mosaic pattern is the remains of a golden era, and represents an ecosystem on the verge of collapse.

It should be obvious that the mosaic pattern is unlikely to be the result of deforestation through human activity, since the islands of forest are found immediately around villages. Recent detailed research, using aerial photographs and anthropological fieldwork, has established that the forest islands are planted by local people, who manage them as part of a complex and sensitive cultivation system (Leach and Fairhead, 1993). The spread of settlement is leading to more, not fewer, forest islands.

There are other examples of how outsiders' (usually European) perceptions of African environments have projected onto these environments the notion of decline from a lost golden age into degradation and barrenness. They are not confined to savanna contexts, but extend even to tropical rainforest (Wood, 1993). This is often thought to be 'virgin' or 'primary' forest, an irreplaceable haven for biodiversity. However, much of Africa's tropical moist forest is secondary forest, and there is evidence from all over West and Central Africa that what is now forest was once cultivated (see Box 2.3 for the case of game reserves). The idea that cultivation in rainforest areas must lead inevitably to the destruction of soils and biodiversity has also been challenged (Wood, 1993).

Box 2.3 The reserving game

The imagination of the African ecological past as an unpopulated, virgin wilderness has had an important influence on contemporary perceptions of game reserves. Game reserves are relatively recent, being an innovation of the colonial period. The early colonial period probably saw an initial decline in human populations, as a result of war and disease (for example, Kjekshus, 1977). Areas which had been settled in the late nineteenth or early twentieth centuries were depopulated by the 1920s and 1930s. Wild animals multiplied. Game reserves were often set up in these areas, which even today bear traces of old village sites in the form of mangoes and other domesticated trees.

In some cases, the establishment of game reserves actually involved removing people from an area. For example, the Selous Reserve in south-eastern Tanzania, the largest game park in the world, involved the forced removal of large numbers of Ngindo-speaking people just after the Second World War (Cross-Upcott, 1956). Population growth has meant that people have started to repopulate the parks for farming and living. Traditions of village locations are still recent enough for inhabitants to see this as return, rather than invasion. Conflict with guardians of the reserves, who see the parks in terms of protection of game, is perhaps inevitable.

Still more examples could be explored, including that of rangelands, but the basic point is made: in the absence of evidence of change over time, degradation, like beauty, is in the eye of the beholder.

2.4.4 Future sustainability

Much of what we have explored so far should make us cautious about cataclysmic predictions of population growth causing the destruction of Africa's environments. However, high population growth cannot be sustained indefinitely, so what factors will determine whether existing patterns can be sustained in the future?

Population growth in Africa is still high, and will remain so for some decades. The limited and slow decline of fertility may be partly caused by environmental factors themselves. Even if population growth does not always hasten degradation, African environments are difficult and risky. Because it is not available through insurance or pensions, security can only be guaranteed through a large family.

Another important argument is that, as water resources dry up and forests are depleted, more work is created, especially for women who are usually responsible for gathering water and fuel. This creates a demand for more labour, and therefore more children. It is thus possible that environmental problems, even if they are not related to population growth, could support high fertility and, thus, population growth. A key policy to break this vicious cycle must be to increase the control that women have over resources.

The conditions for this population growth to lead to successful Boserupian adaptation are clear – that is, the resources, incentives and opportunities to invest in the natural resource base. These tend to be most disrupted by poverty, war and over-exploitation through markets or by governments. What happens to the capacity of people to adapt is therefore much dependent on these factors.

As populations build up on land, it is not inevitable that ecosystems are pushed to limits – rather, they may settle into managed stability. However, it is more than likely that the management of those ecosystems will not absorb everyone as populations grow. This means that more and more people will become involved in non-agricultural activities. As Mortimore points out: 'Ecological sustainability does not, therefore, and need not imply food sufficiency' (Mortimore, 1993, p. 19). More people can live in an area than can make a direct living off the land in that area.

It is worth noting that this leads to a very uneven picture in sub-Saharan Africa. Populations are concentrated in certain regions: on coasts, around cities such as Kano, around Lake Victoria and in highland areas. There is some migration from these areas, into less densely settled areas, but in many cases, as in Machakos, the non-agricultural economy is expanding, providing an alternative. However, in many countries the non-agricultural sectors have suffered contraction under economic programmes of structural adjustment.

Structural adjustment programmes have also involved cutbacks in health and education expenditure. These can only worsen the situation, as they will lead to more uneducated women and higher infant and child mortality, both probably contributing to high fertility, and therefore population growth. In poor and underdeveloped countries, economic policy becomes a key influence on future sustainability.

Summary of section 2.4

- Environmental degradation is hard to define.
- It is difficult to establish that it is population growth, rather than other factors, which cause it.
- The key lesson is to pay attention to a range of factors which may be at work, as well as population growth itself.
- There are different views of whether social organization in Africa acts to block the successful intensification of agriculture and management of the environment.
- Hardin's 'tragedy of the commons' argument would imply that it does.
- The case study of Machakos District in Kenya, however, serves as a counter-example, showing how traditional land tenure gave security.
- The Machakos example also points to some other factors in successful environmental management – economic resources available for investing in the land, openness to new ideas, and community groups which allowed women to gain control over natural resources.
- However, Machakos does not show that Boserupian-type intensification is inevitable in Africa, only that it is possible.
- Observers often project their own preconceptions about environmental decline on to situations which have quite different dynamics.

- Many assessments of African environments have been made on the basis of a 'snapshot': observers are all too ready to make the leap from what they see at any one time to the existence of a process of degradation.
- Population–environment relations in Africa are highly uneven.
- Successful adaptation requires resources, incentives and opportunities to invest in the natural resource base.

2.5 Conclusion

There is no doubt that world population is higher than ever before in history. This has enormous implications for our environment at both global and local levels. But one of the main conclusions of this chapter is that the relationship between population growth and environmental change is not a simple, inevitably negative one, even in an underdeveloped context such as Africa.

Many other factors – economic, political and social – also have a large impact on the environment. And the relationship between population and environment is itself affected by these forces, as the Machakos example from Kenya showed. This complexity means that we must appraise carefully any account which links the two. To some extent, this is an optimistic conclusion – we may yet escape the 'demographic trap'. But a growing population cannot be accommodated without the investment of *some* resources in the environment. The interaction of population and resources in a region therefore depends ultimately on the nature of interdependence of that region with the rest of the world. Poverty, unequal exchange and exploitation hamper the adaptability of societies to population growth. Equally, within societies, reducing population growth and enhancing environmental management means providing those most trapped in poverty with the resources and incentives to do so. Above all, this means empowering women.

If population growth does not inevitably lead to environmental degradation, if it can sometimes facilitate conservation and improvement, and is not a major cause of poverty on its own, does that mean that nothing should be done about it? Should family planning programmes be abandoned, or at least given a lower priority? Some Boserupian writers have indeed argued that there is a conflict between vigorously promoting family planning programmes at the same time as calling for environmental conservation, because the latter is very labour intensive (Mortimore, 1989, p. 209).

Even if population growth were never to have a negative impact on the environment, there are still good grounds, based on sound principles and evidence, for trying to provide family planning. The control of reproduction by individuals and couples is a basic human right, and the number of children a woman bears can have a profound impact on her health and that of the children (*Population Reports*, 1984). Further, placing the well-being and autonomy of parents at the centre of policy, rather than population growth, means that family planning programmes should emphasize informed choice, rather than simply lower fertility. This difference might be seen in terms of a distinction between a family planning policy and a population policy; one seeks to deliver a service, the other seeks to bring about a certain rate of population growth. The latter is far more susceptible to a slide into coercion.

A final point to think about is our perception of the issues. Many 'developed' countries have high levels of population density, but we do not think of population pressure as one of our own problems. Fear of high population growth can lead to a perception of fertility in developing regions as 'uncontrolled', and yet there have always been subtle social controls on fertility. The perception of African environments has been that they are much more fragile than our own. The perception of people's management of those environments has been that they are much less informed than our own. These fears are not always borne out: not all environmental disasters are only in the mind, but not all are real.

Summary of section 2.5

- The interaction of population and resources in a region depends ultimately on the nature of interdependence of that region with the rest of the world.

References

BEHNKE, R. and SCOONES, I. (1992) 'Rethinking range ecology: implications for rangeland management in Africa', Drylands Networks Programme Issues Paper 33, London, International Institute for Environment and Development.

BINNS, T. (1990) 'Is desertification a myth?', *Geography*, Vol. 75, Part 2, No. 327, pp. 106–13.

BOSERUP, E. (1965) *The Conditions of Agricultural Growth*, London, Allen and Unwin.

CAIN, M. and McNICOLL, G. (1986) 'Population growth and agrarian outcomes', Working Paper No. 128, New York, Population Council, Centre for Policy Studies.

CALDWELL, J. (1982) *Theory of Fertility Decline*, London, Academic Press.

CALDWELL, J. and CALDWELL, P. (1987) 'The cultural context of high fertility in sub-Saharan Africa', *Population and Development Review*, Vol. 13, No. 3, pp. 409–37.

CAMPBELL, J. (1991) 'Land or peasants? The dilemma confronting Ethiopian resource conservation', *African Affairs*, Vol. 90, No. 358, pp. 5–21.

CLINE-COLE, R., MAIN, H.A.C. and NICHOLS, J. (1990) 'On fuelwood consumption, population dynamics and deforestation in Africa', *World Development*, Vol. 18, No. 4, pp. 513–27.

CROSS-UPCOTT, A.R. (1956) 'A history of the Ki-Ngindo speaking peoples of Tanzania', unpublished PhD thesis, University of Cape Town.

DANKELMAN, I. and DAVIDSON, J. (1988) *Women and Environment in the Third World*, London, Earthscan.

DEMENY, P. (1984) 'A perspective on long-term population growth', *Population and Development Review*, Vol. 10, No. 1, pp. 103–26.

DEMENY, P. (1989) 'World population growth and prospects', Working Paper No. 4, New York, Population Council, Research Division.

EHRLICH, P. (1968) *The Population Bomb*, New York, Ballantine Books.

EHRLICH, P. and EHRLICH, A. (1990) *The Population Explosion*, London, Hutchinson.

FAIRHEAD, J. and LEACH, M. (forthcoming) 'Enriching landscapes: social history and the management of transition ecology in Guinea's forest–savanna mosaic', *Africa*.

FORSE, B. (1989) 'The myth of the marching desert', *New Scientist*, 4 February.

GLANTZ, M. and ORLOVSKY, N. (1983) 'Desertification: a review of the concept', *Desertification Control Bulletin* (UNEP), Vol. 9, pp. 15–22.

GRAINGER, A. (1990) *The Threatening Desert: Controlling Desertification*, London, Earthscan.

HARDIN, G. (1968) 'The tragedy of the commons', in Markandya, A. and Richardson, J. (eds) (1992) *The Earthscan Reader in Environmental Economics*, London, Earthscan.

HARRISON, P. (1987) *The Greening of Africa*, New York and London, Penguin and Paladin.

HARRISON, P. (1992) *The Third Revolution: Environment, Population and a Sustainable World*, London and New York, I.B. Tauris in association with The World Wide Fund for Nature.

KJEKSHUS, H. (1977) *Ecology Control and Economic Development in East African History: the Case of Tanganyika, 1850–1950*, London, Heinemann.

LEACH, M. and FAIRHEAD, J. (1993) 'Whose social forestry and why? People, trees and managed continuity in Guinea's forest–savanna mosaic', *Zeitschrift für Wirtschaftsgeographie*, Vol. 37, No. 2, pp. 86–101.

LELE, U. and STONE, S. (1989) 'Population pressure, the environment and agricultural intensification: variations on the Boserup hypothesis', MADIA Discussion Paper 4, Washington, DC, World Bank.

MEADOWS, D.H., MEADOWS, D.L. and RANDERS, J. (1992) *Beyond the Limits: Global Collapse or a Sustainable Future*, London, Earthscan.

MEADOWS, D.H., MEADOWS, D.L., RANDERS, J. and BEHRENS, W.W. III (1972) *The Limits to Growth: a Report for the Club of Rome's Project on the Predicament of Mankind*, London, Earth Island.

MORTIMORE, M. (1970) 'Population densities and rural economies in the Kano Close Settled Zone, Northern Nigeria', in Zelinsky, W., Kosinski, A.L. and Prothero, R.M. (eds) *Geography and a Crowding World,* New York, Oxford University Press.

MORTIMORE, M. (1989) *Adapting to Drought: Farmers, Famines and Desertification in West Africa,* Cambridge, Cambridge University Press.

MORTIMORE, M. (1993) 'Population growth and land degradation', *GeoJournal,* Vol. 31, No. 1, pp. 15–21.

MORTIMORE, M., ESSIET, E.U. and PATRICK, S. (1990) 'The nature, rate and effective limits of intensification in the small holder farming system of the Kano Close-Settled Zone', unpublished report, Federal Agricultural Coordinating Unit, Federal Government of Nigeria.

MYERS, N. (1989) 'Population growth, environmental decline and security issues in sub-Saharan Africa', in Ornas, A. and Salih, M.A. (eds) *Ecology and Politics: Environmental Stress and Security in Africa,* Uppsala, Scandinavian Institute of African Studies.

NICHOLSON, S. (1979) 'The methodology of historical climate reconstruction and its application to Africa', *Journal of African History*, Vol. 20, No. 1, pp. 31–49.

POPULATION CONCERN (1991) *Population Concern; Working for Change 1981–1991,* London, Population Concern.

POPULATION REPORTS (1984) 'Healthier mothers and children through family planning', Series J 27, Baltimore, MD, Johns Hopkins University.

ROSS, J.A., PARKER MAULDIN, W. and MILLER, V.C. (1993) *Family Planning and Population: a Compendium of International Statistics,* New York, The Population Council.

SHEPHERD, G. (1989) 'The reality of the commons', *Development Policy Review,* Vol. 7, pp. 51–63.

SMIL, V. (1993) *China's Environmental Crisis,* New York, M.E. Sharpe.

STEBBING, E.P. (1938) 'The man-made desert in Africa: erosion and drought', *Journal of the Royal African Society,* supplement for January.

TIFFEN, M. and MORTIMORE, M. (1992) 'Environment, population growth and productivity in Keyna: a case study of Machakos District', *Development Policy Review,* Vol. 10, No. 4, pp. 359–87.

TIFFEN, M., MORTIMORE, M. and GICHUKI, F. (1994) *More People, Less Erosion,* Chichester, John Wiley and Sons.

TIMBERLAKE, L. (1985) *Africa in Crisis: the Causes and Cure of Environmental Bankruptcy*, London, Earthscan.

WOOD, D. (1993) 'Forests to fields: resorting tropical lands to agriculture', *Land Use Policy*, Vol. 10, No. 2, pp. 91–109.

ZABA, B. and SCOONES, I. (1994) 'Is carrying capacity a useful concept to apply to human populations?', in Zaba, B. and Clarke, J. (eds) *Environment and Population Change,* Liege, Ordina Editions.

Reading A: Paul R. Ehrlich, 'Too many people'

Americans are beginning to realize that the undeveloped countries of the world face an inevitable population-food crisis. Each year food production in undeveloped countries falls a bit further behind burgeoning population growth, and people go to bed a little bit hungrier. While there are temporary or local reversals of this trend, it now seems inevitable that it will continue to its logical conclusion: mass starvation. The rich are going to get richer, but the more numerous poor are going to get poorer. Of these poor, a minimum of three and one-half million will starve to death this year, mostly children. But this is a mere handful compared to the numbers that will be starving in a decade or so. And it is now too late to take action to save many of those people.

In a book about population there is a temptation to stun the reader with an avalanche of statistics. I'll spare you most, but not all, of that. After all, no matter how you slice it, population is a numbers game. Perhaps the best way to impress you with numbers is to tell you about the 'doubling time' – the time necessary for the population to double in size.

It has been estimated that the human population of 6000 BC was about five million people, taking perhaps one million years to get there from two and a half million. The population did not reach 500 million until almost 8,000 years later – about 1650 AD. This means it doubled roughly once every thousand years or so. It reached a billion people around 1850, doubling in some 200 years. It took only 80 years or so for the next doubling, as the population reached two billion around 1930. We have not completed the next doubling to four billion yet, but we now have well over three billion people. The doubling time at present seems to be about 37 years. Quite a reduction in doubling times: 1,000,000 years, 1,000 years, 200 years, 80 years, 37 years. Perhaps the meaning of a doubling time of around 37 years is best brought home by a theoretical exercise. Let's examine what might happen on the absurd assumption that the population continued to double every 37 years into the indefinite future.

If growth continued at that rate for about 900 years, there would be some 60,000,000,000,000,000 people on the face of the earth. Sixty million billion people. This is about 100 persons for each square yard of the Earth's surface, land and sea. A British physicist, J. H. Fremlin [1964], guessed that such a multitude might be housed in a continuous 2,000-story *[sic]* building covering our entire planet. The upper 1,000 stories would contain only the apparatus for running this gigantic warren. Ducts, pipes, wires, elevator shafts, etc., would occupy about half of the space in the bottom 1,000 stories. This would leave three or four yards of floor space for each person. I will leave to your imagination the physical details of existence in this ant heap, except to point out that all would not be black. Probably each person would be limited in his travel. Perhaps he could take elevators through all 1,000 residential stories but could travel only within a circle of a few hundred yards' radius on any floor. This would permit, however, each person to choose his friends from among some ten million people! And, as Fremlin points out, entertainment on the worldwide TV should be excellent, for at any time 'one could expect some ten million Shakespeares and rather more Beatles to be alive'.

Could growth of the human population of the Earth continue beyond that point? Not according to Fremlin. We would have reached a 'heat limit'. People themselves, as well as their activities, convert other forms of energy into heat which must be dissipated. In order to permit this excess heat to radiate directly from the top of the 'world building' directly into space, the atmosphere would have been pumped into flasks under the sea well before the limiting population size was reached. The precise limit would depend on the technology of the day. At a population size of one billion billion people, the temperature of the 'world roof' would be kept around the melting point of iron to radiate away the human heat generated.

But, you say, surely Science (with a capital 'S') will find a way for us to occupy the other planets of our solar system and eventually of other stars before we get all that crowded. Skip for a moment the virtual certainty that those planets are uninhabitable. Forget also the insurmountable logistic problems of moving billions of people off the Earth. Fremlin has made some interesting calculations on how much time we could buy by occupying the planets of the solar system. For instance, at any given time it would take only about 50 years to populate Venus, Mercury, Mars, the moon, and the moons of Jupiter and Saturn to the same population density as Earth.

What if the fantastic problems of reaching and colonizing the other planets of the solar system, such as Jupiter and Uranus, can be solved? It would take only about 200 years to fill them 'Earth-full'. So we could perhaps gain 250 years of time for population growth in the solar system after we had reached an absolute limit on Earth. What then? We can't ship our surplus to the stars. Professor Garrett Hardin [1959] of the University of California at Santa Barbara has dealt effectively with this fantasy. Using extremely optimistic assumptions, he has calculated that Americans, by cutting their standard of living down to 18 per cent of its present level, could in *one year* set aside enough capital to finance the exportation to the stars of *one day's* increase in the population of the world.

Interstellar transport for surplus people presents an amusing prospect. Since the ships would take generations to reach most stars, the only people who could be transported would be those willing to exercise strict birth control. Population explosions on space ships would be disastrous. Thus we would have to export our responsible people, leaving the irresponsible at home on Earth to breed.

Enough of fantasy. Hopefully, you are convinced that the population will have to stop growing sooner or later and that the extremely remote possibility of expanding into outer space offers no escape from the laws of population growth. If you still want to hope for the stars, just remember that, at the current growth rate, in a few thousand years everything in the visible universe would be converted into people, and the ball of people would be expanding with the speed of light! [Cook, 1966]. Unfortunately, even 900 years is much too far in the future for those of us concerned with the population explosion. As you shall see, the next *nine* years will probably tell the story.

References

COOK, I.J. (1966) *New Scientist*, 8 September.

FREMLIN, J.H. (1964) 'How many people can the world support?', *New Scientist*, 29 October.

HARDIN, G. (1959) 'Interstellar migration and the population problem', *Heredity*, Vol. 50, pp. 68–70.

Source: Ehrlich, 1968, pp. 17–21

Reading B: *John C. Caldwell, 'Children: activities, costs and returns'*

[...] A majority [of Yoruba] believe that one needs a *very large* number of children (and many more merely believe that one needs a large number) to ensure first that some will grow up, secondly that some will be willing and able to help their parents once they are employed, and thirdly, so that they will have this ability, that some will be bright enough to win the educational qualifications that will secure them a job with a high income. There is almost unanimity in agreement about the need for this assistance in the parents' old age, a view held so strongly that it often obscures the profitability of such help at earlier times.

[...]

In the West an ever larger proportion of the possessions in the home do not supply aesthetic pleasure in themselves and cannot be consumed for direct satisfaction, but instead supply services. Telephones bring messages into the house and the family car brings purchases, the vacuum cleaner dusts the floor, the washing machine cleans the clothes and the washing-up machine the dishes, the refrigerator and freezer keep food so that fresh food has to be brought into the house less frequently, the mixer breaks up and mixes food while its grinding attachment grinds, the hot water system heats water and takes it to where it is needed, the electricity system brings fuel into the house, the incinerator removes waste and so on. In rural Yorubaland most of the consumer durables and other gadgets which provide these services cannot be afforded or even easily purchased or adequately serviced even where money is available; nor is electricity usually available, and, where it is, the cost is enormously high compared with that of human labour. But every one of these things can be done by human activity. Most involve carrying, picking up or pounding, all of which can be done by children. Indeed, in a traditional society they are done as a matter of course by children with little complaint. Nine out of every ten of our respondents agreed that 'small children save adults from menial tasks', and that 'children are important because of the help they give in the home'. Even outside the home such tasks abound. The Western observer often makes the mistake at first of assuming that most rural activity is clearing, digging, planting and harvesting, and that the great reductions in labour can be made by the introduction of tractors, saws, ploughs, mechanical cultivators and the like. In fact, more energy is probably consumed in the course of a year by carrying water, fuel, products for sale in the market and purchases from the market, and by walking to and from the nearest road or distant fields. Eventually, much of this activity will be reduced by the multiplication of feeder roads and vehicles, by the digging of wells closer to the village and even by the reticulation of water and electricity and the sedentarization of agriculture. In the meantime, there is much for children, even small children, to do. In a partly subsistence economy, children are not only producers for the market and for household consumption; they also provide subsistence services and make life for adults pleasanter and more gracious than it would otherwise be. In a different sense from that in which the term has previously been used, children are, indeed, 'consumer durables' [Blake, 1968].

[...] Children (and their parents as well) assume that children (at an early age) will do most of the things that adults of the same sex do. Around the age of five, imitative play seems to give way to activities that relieve others of some work. Amongst the very young there is little division by sex, and work is fairly evenly divided between doing housework and carrying things or messages outside the house. These are heterogeneous categories and an analysis based on them hides the changes in the balance of tasks that takes place with age: in the former a change from sweeping and light tasks to washing up or washing clothes to cooking and ironing among the older girls, and splitting wood among the boys, while in the latter the progression is from carrying messages and water to bearing heavier objects like firewood or fuel or to errands where responsibility is required such as when making purchases. Minding younger siblings depends more on birth order than age. In the traditional society (and still in rural areas amongst children who have not attended school), farming became an increasing proportion of all boys' work as they grew older, although there was a change in emphasis from tasks like weeding to such heavier labour as clearing and making yam mounds, and, amongst girls, marketing became more important although it by no means replaced housework. Schooling reduces the numbers undertaking such traditional work as farming and marketing, and urbanization has a similar impact in reducing the number of boys engaging in farming (although not the number of girls marketing). It is not the time in school that prevents educated boys from farming; it is a different attitude to life which is accepted, although often with misgivings, by their parents (Galletti *et al*, 1956, pp. 78–9; S.A. Aluko, personal communication).

Children from quite young ages help with market-oriented activities: the production of cocoa and foods eaten locally; the processing and preparation of foods derived from the oil palm, cassava and maize; spinning, weaving and dyeing; plaiting hair and so on. Children can also earn money on neighbouring farms during peak agricultural activity, and can collect firewood and wild plantain (the non-sweet banana) from the forest – in fact it appears that the possible earnings from the latter activity determine the minimum wage which can be paid for casual agricultural labour.[1]

The attempt to measure parents' feelings about the balance of cost and return on children has been made in both Ghana [...] and Nigeria. The results are unsatisfactory because many parents inevitably think of the question in terms of cash returns either from individual earnings or the expansion of the household's capacity for market production. There is little point in pressing the question whether the extra services in or outside the house make up for the extra costs and disadvantages of children, because most respondents state that additional children at no time result in additional costs or disadvantage. In terms of the current money balance of costs and returns, most respondents say that children who are not at school bring in more than the cost they impose by the time that they are about 15 years of age, while those still at school never do so. In terms of the cumulative return on children, most respondents over 50 years of age say that they have received back more than they spent. Indeed, 97 per cent of respondents felt that: 'The best investment is the education of one's children (or relatives)'.

An attempt was made to trace all money flows in each direction between parents and children in 300 families [...]. The investigation showed quite clearly that the present system of parent–child financial relationships – a system which does not penalize high fertility – depends for its working on the concept of extended family and communal help: those who can help usually do so. Most parents who did not give any assistance were unable to do so and were not blamed for their failure. At least they were not blamed sufficiently for it to result in any restriction of subsequent assistance from their children – if anything their greater need provoked a little more assistance. Similarly, those children who do not help are mostly those who can least afford to do so – the predicted revolt of the educated in favour of a Western type of system of obligations has not yet occurred on any significant scale. Investment in children is probably an investment in the real sense of the term. But this does not come about because there is any correlation between giving assistance to a specific child and receiving help back from that child. It comes about because greater financial help to capable children – and this qualification is clearly understood by the parents – may allow them to reach the urban white-collar occupations, and perhaps even the professions, with high and regular incomes. If they reach such heights they will in most cases return more money and remit it more regularly.

Children on average are reported to remit money to parents amounting to ten per cent, or a little more, of both fathers' incomes and household incomes; among those who remit at all the reported proportion is around 15 per cent. For perhaps half of all parents the number of remitters may eventually be three or more, thus increasing parental income by half. Such an analysis is illuminating but misleading. It hides much of the return on the investment. First, the monetary return is probably understated. Secondly, the non-monetary returns may well outweigh the monetary ones (speaking only in terms of goods and not of other gratifications). Thirdly, the successful investment buys security; the main aim is that the child should obtain employment with guaranteed tenure and regular salary that will permit help to be given to parents no matter what disaster should strike them – blighted cocoa crops, drought, flood, cholera, urban unemployment, a succession of funerals, or many other domestic or village crises.

There are, in fact, few other competitive sources of investment for rural populations. Peasant farming offers few outlets for extra money in contrast to extra labour; spraying equipment costs money but its acquisition is no longer thought of as optional. Usury is practised by some farmers but is more the speciality of the businessman.[2] Money is safely invested in a small business only if that business is run by a trusted and competent relative. More land can be acquired with money now that communal tenure has been modified, but it can only be profitably used if farmed by near relatives, usually children or a wife (Vanden Driesen, 1972, p. 53). These are not restrictions which favour low fertility.

One type of activity – once universal but now practised only by those grown-up children who remain in the village – has not been mentioned and does not appear on the kind of balance sheet just discussed. Sons tend to help their fathers on the farms throughout their fathers' lives. In fact, the most comfortably-off old man will be he who has some non-migrant children helping him locally and other migrant ones channelling the new forms of wealth from afar.

Notes

1 Personal communication from I.H. Vanden Driesen referring to the findings from his field research in the Ife Division of the Western State during 1968–69.

2 M.H. Peil (personal communication) on farmers; I.H. Vanden Driesen (personal communication) on businessmen.

References

BLAKE, J. (1968) *Population Studies*, Vol. 22, pp. 5–25.

GALLETTI, R., BALDWIN, K.D.S. and DINA, I.O. (1956) *Nigerian Cocoa Farmers: an Economic Survey of Yoruba Cocoa Farming Families*, Oxford, Oxford University Press.

VANDEN DRIESEN, I.H. (1972) *Africa*, Vol. 42, pp. 44–56.

Source: Caldwell, 1982, pp. 104–9

Reading C: Michael Mortimore, 'Population growth and land degradation'

[...]

The relevant parameter of population growth is rural density, or the population:land ratio. All observers agree that the effect of increasing population density is to bring about an increase in the frequency of cultivation. At low densities, long cultivation–fallow cycles are employed for fertility maintenance. Fallows and uncultivable land are used for grazing and for wood fuel collection. Increasing densities call for a larger percentage of cultivable land to be under cultivation at any one time (Allan, 1965). Long (forest or bush) fallow eventually gives place to short (or grass) fallow, annual cultivation, and (where technically feasible) multicropping (Boserup, 1965). The model is supported by comparative studies of substantial numbers of farming systems.

Thus far, observers of semi-arid farming systems have little cause for disagreement. However, at this point a divergence appears.

The degradational pathway

In a Malthusian outcome, shortening fallows leads to a loss of fertility of cultivated soils, declining crop yields, falling output per man-hour [*sic*] and, in the system as a whole, a fall in total output. As often stated, the sequence is modelled in Table C.1.

Table C1 Linkages between population growth and environmental degradation according to a Malthusian view

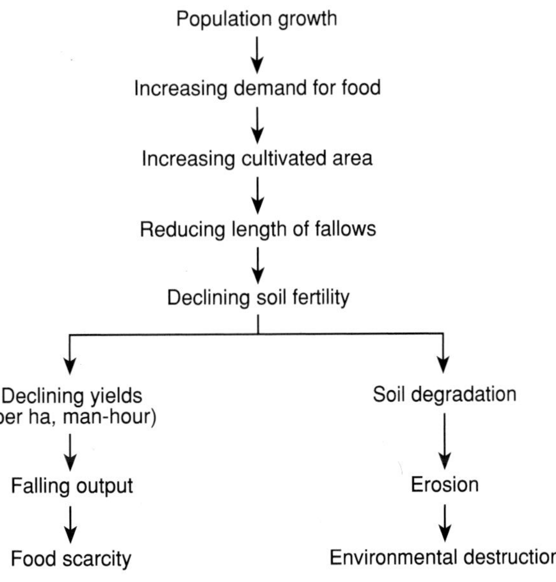

Food scarcity leads to further increases in the cultivated area and accelerated soil degradation. The outcome of the sequence is a food-starved community presiding over the irreversible destruction of its environmental resources.

In a semi-arid environment, owing to the inter-annual variability of the rainfall, there is an increasing frequency of harvest failures. The change in the ratio between cultivated and grazing land causes a decline in the amount of nutrient per hectare transferred by grazing animals from the natural ecosystem to the fields, accelerating the decline of their fertility (Gregoire and Raynaut, 1980). At the same time, as the area of grazing land declines, the peoples' efforts to maintain the same numbers of livestock lead to overgrazing, and pasture degradation. The growing population also requires more wood fuel, and so the woody component of the vegetation is subjected to a rising level of offtake that eventually exceeds the rate of natural regeneration. The effect of all these changes is to expose the soil to wind and water erosion, which removes the topsoil along with most of the nutrients.

Each piece of land has what we call its carrying capacity (for humans and animals) ... When that number is exceeded, the whole piece of land will quickly degenerate from overgrazing or overuse by human beings. Therefore, population pressure is definitely one of the major causes of desertification and the degradation of the land.

(Tolba, 1986)

Validation of the degradational pathway depends on establishing direct causal relationships between increasing population density and the degradation of land. But here there is a major difficulty. The data on land degradation are few, of poor quality, and disputed (Nelson, 1988; Warren and Agnew, 1988; Warren and Khogali, 1992). The hypothesis of a southward-advancing Sahara, for example, quoted in terms of so many kilometres a year, is no longer tenable. Rainfall change is admitted to have been as influential in some areas as management.

Average annual rainfall in most of the Sahel has fallen by 25–33 per cent since the mid-1960s (IUCN, 1989; Hulme, 1990), and there have been negative changes in the monthly distribution patterns. The frequency of droughts has increased. The effect of such changes is equivalent to moving most of the Sudano-Sahelian zone into a significantly drier ecological regime. The farming systems, people and livestock are confronted with the necessity for major adaptation (Boulier and Jouve, 1988; Mortimore, 1989). Rainfall deficiencies are a more important cause of the low crop yields that are widely reported, and of impoverished natural vegetation, than is management. This is shown by resurgent crop yields and pasture recovery during seasons of good rainfall.

In other parts of the semi-arid tropics, the rainfall record differs. But that for the Sahel shows that an unchanging climate should not be assumed, and the blame for degradation simply placed on the mismanagement of natural resources by farmers and pastoralists.

A disaster scenario is not a necessary outcome of Malthusian premises. Land productivity is a function not only of its intrinsic properties but also of the nature and level of labour and capital inputs, as recognized in the FAO/IIASA/UNFPA (1982) study of population-supporting capacities. The study discriminated between low inputs – a level assumed to represent present practice (in 1975) on smallholdings – intermediate, and high inputs. In 1975, of 25 sub-Saharan countries that we may consider have a 'desertification problem', 11 had densities below their potential supporting capacities, and in nine the densities exceeded their supporting capacities. But by raising the level of inputs from low to intermediate, even taking account of increases in population expected by the year 2000, the first number could be raised to 18 and the second reduced to four (Table C.2)

By contrast, of 21 countries in North Africa and South-west Asia, only one was estimated to have a supporting capacity in excess of its population density at low inputs in 1975, and even with improved inputs, only three by the year 2000. Yet much international concern about population-induced degradation focuses on sub-Saharan Africa.

Table C.2 Population-supporting capacities (PSC) and population densities, 1975 and 2000, according to the FAO

	Sub-Saharan Africa		N. Africa and S.W. Asia	
	1975	2000	1975	2000
At low inputs				
PSC > population density	11	7	1	3
PSC = population density	5	5	2	2
PSC < population density	9	13	18	16
Number of countries	25	25	21	21
At intermediate inputs				
PSC < population density	20	18	9	3
PSC = population density	2	3	1	3
PSC > population density	3	4	11	15
Number of countries	25	25	21	21

Source: FAO/IIASA/UNFPA, 1982

Note: Only those countries having substantial arid or semi-arid zones are included. [> = larger than; < = smaller than.]

The conservation pathway

In a 'Boserupian' outcome, declining yields lead to the substitution of fertilization for fallowing, beginning with increasing amounts of organic manure generated by the livestock. Labour inputs per hectare increase, not only because cultivation cycles are more frequent but because fertilization, intensive weeding and more careful husbandry become necessary.

Boserup (1965), in her intensification thesis, argued that land-conserving and yield-enhancing methods are introduced to maintain fertility, and increase yields per hectare and total output, as density rises. The effect of these and other technical changes is to increase output per hectare, but output per man-hour becomes subject to diminishing returns. Total output increases. The labour requirement rises so fast when short (grass) fallows and annual cultivation are adopted that the introduction of labour-saving technology (the ox-plough) becomes profitable. Land-conserving technologies (such as terracing on hillslopes) and field enclosures also become profitable.

Population density alone, however, cannot account for all of the changes that are being observed in land use systems. The penetration of the market into smallholder farming systems is everywhere pervasive and its effects cannot be distinguished from those of population density. It seems likely that some pre-colonial systems intensified under increasing population density, even without significant external market linkages. Such a development appears unlikely to have occurred in recent times, owing to the multiplication and intensification of such linkages.

Improved market access (higher prices) raises marginal returns to labour, so farmers cultivate larger areas (Pingali *et al.*, 1987). Also, the monetization of the household economy increases the need for income. Higher value market crops are partly substituted for lower value food crops. So the market amplifies the effect of population growth on intensification.

Market income may support capital investments in equipment, structures, or inputs and pay for hiring labour, which supplements family labour during bottlenecks. A less commonly noted form of investment, multipurpose trees, are planted and protected in increasing numbers, as they do not displace crops from farmland. They offer: (1) additional income, fuel and other useful output; (2) environmental benefits (shelter from wind and sun); and (3) – with some – enhanced soil fertility.

Livestock become increasingly dependent on labour-intensive feeding systems (crop residues, cut grass and browse) and less dependent on diminishing natural grazings. Their management is directed towards maximizing their current value for manure, draft and milk rather than their capital value. Crop and livestock sectors tend to become more integrated. The land tenure system reflects the need to secure access to resources and investment benefits, usually in the form of strengthened individual title to land. A 'Boserupian' model of agricultural intensification is consistent with an objective of sustainability – indeed, such an objective is intrinsic, as the importance placed on heritable title to land bears witness.

[...]

References

ALLAN, W. (1965) *The African Husbandman*, Edinburgh, Oliver and Boyd.

BOSERUP, E. (1965) *The Conditions of Agricultural Growth*, London, Allen and Unwin.

BOULIER, F. and JOUVE, P. (1988) *Etude Comparée de l'Evolution des Systémes de Production Saheliens et de leur Adaptation a la Sécheresse*, Montpellier, Département de Systémes Agraires.

FAO/IIASA/UNFPA (1982) *Potential Population Supporting Capacities of Lands in the Developing World*, Rome, FAO.

GREGOIRE, E. and RAYNAUT, C. (1980) *Présentation Génerale du Département du Maradi. Programme de Recherches sur la Région du Maradi*, Université de Bordeaux II.

HULME, M. (1990) 'The changing rainfall resources of Sudan', *Transactions of the Institute of British Geographers*, NS Vol. 15, pp. 21–34.

IUCN (1989) *Sahel Studies*, Gland, International Union for the Conservation of Nature.

MORTIMORE, M. (1989) *Adapting to Drought. Farmers, Famines and Desertification in West Africa*, Cambridge, Cambridge University Press.

NELSON, R. (1988) 'Dryland management: the desertification problem', Environment Department Working Paper 8, Washington, DC, The World Bank.

PINGALI, P.L., BIGOT, Y. and BINSWANGER, H.P. (1987) *Agricultural Mechanization and the Evolution of Farming Systems in Sub-Saharan Africa*, Washington, DC, The World Bank.

TOLBA, M.K. (1986) 'Desertification', *WMO Bulletin*, Vol. 35, pp. 17–22.

WARREN, A. and AGNEW, C.T. (1988) 'An assessment of desertification and land degradation in arid and semi-arid areas', Issue Paper 2, London, International Institute for Environment and Development.

WARREN, A. and KHOGALI, M. (1992) *Assessment of Desertification and Drought in the Sudano-Sahelian Region*, 1985–91, New York, United Nations Sudano-Sahelian Office.

Source: Mortimore, 1993, pp. 16–17, 20–1

Reading D: *Garrett Hardin, 'The tragedy of the commons'*

[...]

The tragedy of the commons develops in this way. Picture a pasture open to all. It is to be expected that each herdsman will try to keep as many cattle as possible on the commons. Such an arrangement may work reasonably satisfactorily for centuries because tribal wars, poaching and disease keep the numbers of both man and beast well below the carrying capacity of the land. Finally, however, comes the day of reckoning, that is, the day when the long-desired goal of social stability becomes a reality. At this point, the inherent logic of the commons remorselessly generates tragedy.

As a rational being, each herdsman seeks to maximize his gain. Explicitly or implicitly, more or less consciously, he asks, 'What is the utility *to me* of adding one more animal to my herd?' This utility has one negative and one positive component.

1 The positive component is a function of the increment of one animal. Since the herdsman receives all the proceeds from the sale of the additional animal, the positive utility is nearly +1.
2 The negative component is a function of the additional overgrazing created by one more animal. Since, however, the effects of overgrazing are shared by all the herdsmen, the negative utility for any particular decision-making herdsman is only a fraction of –1.

Adding together the component partial utilities, the rational herdsman concludes that the only sensible course for him to pursue is to add another animal to his herd. And another; and another ... But this is the conclusion reached by each and every rational herdsman sharing a commons. Therein is the tragedy. Each man is locked into a system that compels him to increase his herd without limit – in a world that is limited. Ruin is the destination toward which all men rush, each pursuing his own best interest in a society that believes in the freedom of the commons. Freedom in a commons brings ruin to all.

[...]

In an approximate way, the logic of the commons has been understood for a long time, perhaps since the discovery of agriculture or the invention of private property in real estate. But it is understood mostly only in special cases which are not sufficiently generalized. Even at this late date, cattlemen leasing national land on the western ranges demonstrate no more than an ambivalent understanding, in constantly pressuring federal authorities to increase the head count to the point where overgrazing produces erosion and weed-dominance. Likewise, the oceans of the world continue to suffer from the survival of the philosophy of the commons. Maritime nations still respond automatically to the shibboleth of the 'freedom of the seas'. Professing to believe in the 'inexhaustible resources of the oceans', they bring species after species of fish and whales closer to extinction [McVay, 1966].

The National Parks present another instance of the working out of the tragedy of the commons. At present, they are open to all, without limit. The parks themselves are limited in extent – there is only one Yosemite Valley – whereas population seems to grow without limit. The values that visitors seek in the parks are steadily eroded. Plainly, we must soon cease to treat the parks as commons or they will be of no value to anyone.

What shall we do? We have several options. We might sell them off as private property. We might keep them as public property, but allocate the right to enter them. The allocation might be on the basis of wealth, by the use of an auction system. It might be on the basis of merit, as defined by some agreed-upon standards. It might be by lottery. Or it might be on a first-come, first-served basis, administered to long queues. These, I think, are all the reasonable possibilities. They are all objectionable. But we must choose – or acquiesce in the destruction of the commons that we call our National Parks.

[...]

Freedom to breed is intolerable

The tragedy of the commons is involved in population problems in another way. In a world governed solely by the principle of 'dog eat dog' – if indeed there ever was such a world – how many children a family had would not be a matter of public concern. Parents who bred too exuberantly would leave fewer descendants, not more, because they would be unable to care adequately for their children. David Lack and others have found that such a negative feedback demonstrably controls the fecundity of birds [Lack, 1954]. But men are not birds, and have not acted like them for millenniums, at least.

If each human family were dependent only on its own resources; *if* the children of improvident parents starved to death; *if*, thus, overbreeding brought its own 'punishment' to the germ line – *then* there would be no public interest in controlling the breeding of families. But our society is deeply committed to the welfare state [Girvetz, 1950], and hence is confronted with another aspect of the tragedy of the commons.

In a welfare state, how shall we deal with the family, the religion, the race, or the class (or indeed any distinguishable and cohesive group) that adopts overbreeding as policy to secure its own aggrandizement [Hardin, 1963]? To couple the concept of freedom to breed with the belief that everyone born has an equal right to the commons is to lock the world into a tragic course of action.

[...]

References

GIRVETZ, H. (1950) *From Wealth to Welfare*, Stanford, CA, Stanford University Press.

HARDIN, G. (1963) *Perspectives in Biology and Medicine*, Vol. 6, p. 366.

LACK, D. (1954) *The Natural Regulation of Animal Numbers*, Oxford, Clarendon Press.

McVAY, S. (1966) *Scientific American*, Vol. 216, No. 9, p. 13.

Source: Hardin, 1968, pp. 60–6

Reading E: *Mary Tiffen and Michael Mortimore, 'The economics of recovery'*

[...] [B]ehind the physical recovery of the land lies a story of increasing value of production and investment per hectare. It is also demonstrated by the increase in the value of land, although this is also influenced by its increasing scarcity. Land prices have risen from a goat (current value about Ksh 350) per acre in the 1920s, to Ksh 40,000 without trees and Ksh 80,000 with trees in one AEZ 3[1] village in 1990.

The increase in the value of output per hectare took place by several routes:

- increased and more efficient use of the second, long rains. Almost all cultivated land is now double cropped, rains permitting;
- more careful husbandry and water retention in both seasons;
- increase in the ratio of cropped to pasture land (which has a lower output per ha), combined with greater integration of livestock, utilizing crop residues for feed and manure for crops;
- investment in fewer but higher quality and healthier livestock per household;
- a switch of some land into higher value crops, such as coffee, fruit trees and vegetables;
- planting and/or protection of trees in grazing lands, crop lands and hedgerows, to provide fuel, timber and fodder needs for own use or for sale.

This required investment, in capital and labour, at the farm level to terrace, hedge and fence, and to buy equipment, tree seedlings, improved livestock, etc., as well as increased working capital to finance the two seasons and to secure timely operations. At the community level it required investments in gully-stopping, dip construction, coffee-processing plants, roads, dams, etc. It also required a variety of services and complementary investments (in shops, stores, lorries, etc.) from traders who deliver inputs and incentive consumption goods and collect farm products.

When degradation was at its worst, the population density was around 50/km^2. Roads were few and transport expensive and time-consuming, so that in most of the District the only marketable commodity was livestock. In such a situation there is neither the incentive nor the means to invest. Improvements to degraded pasture land are high in cost relative to the benefit to be expected, especially from unimproved stock still very much subject to devastation by disease as well as drought. It is noticeable that the leading area of Machakos was the northern hills, the only area which already by the early 1950s had relatively good access to Nairobi. This was where fruit and vegetable and coffee production all started, and where the bench type of terracing was completed at an early stage.

Given the difficulty in a semi-arid, unpredictable climate to raise money from agriculture, the investment money had to come from outside the sector.

Investments made privately seem to have been more productive than those of ALDEV.[2] The most common source of monetary capital was off-farm work, although livestock sales also helped. The first priority was often a non-farm business (transport, a shop, etc); it was followed or accompanied by investment in schooling for the children and, if any investment were made in the farm at this stage, the money went into a plough or coffee trees. The investment in schooling was expected to secure, and did in many cases in the 1960s and 1970s, a well-paid job for the child, who in due course provided capital to develop the farm. Those who for one reason or another did not have a farm worth developing in the older heartlands invested by moving to new land, where they could develop a larger farm to compensate for the lower potential of the land. In both the new and the old farms there were considerable investments of family labour as well as paid and group labour. Accompanying all this was considerable community investment, through self-help, for dips, schools, feeder roads, etc., or through the co-operatives for coffee factories. Where once cattle had been the only possible investment and store of wealth against bad seasons there were now several. As leaders in the northern hills expressed it in 1990, 'Now, coffee is the cow'. Down in the lower-potential areas livestock were still valued: 'A family with cattle is the same as a family with a graduate'.

Notwithstanding the intense government activity in the ALDEV period, perhaps the most productive government investment directly into soil conservation has been on the advisory side, and through the provision of tools to groups in times of special hardship, a practice which has been found more effective than direct food aid (ODI, 1982). Tools were provided in the 1940s and 1950s, and during the late 1970s and the 1980s by both government and NGO programmes to groups working on a self-help basis. Many farmers reported to us that they had built their terraces without outside help, using labour which they had hired, but it is impossible to quantify the relative contributions of family labour, groups without subsidies, groups with subsidies, and paid labour to the process. In any case, it is certain that the labour value of the farmers' investments far outweighed the value of components provided by government or NGOs.

Enabling technologies

Soil and water conservation

The authorities always put the emphasis on soil conservation rather than water conservation, although some agricultural officers realized, and taught, that the bench terrace and cut-off also conserved water. Farmers preferred these labour-intensive works to the narrow-base terrace; they saved water and land, and were easier to maintain. The bench terrace was tried out in the 1930s but was not adopted (Gichuki, 1991). In 1948 it was being advocated by one party within the Agricultural Department (Throup, 1987), but was independently constructed by a farmer who had learnt the technique as a soldier in India and who was selling onions (KNA[3]: MKS/DC *Annual Report,* 1948 [Lambert, 1945]). By 1952 the authorities were calling it 'popular' (ALUS, 1953). By 1957 bench terrace construction surpassed the narrow-base terrace (Peberdy, 1961). A full review of soil and water conservation is given in Gichuki (1991). The history of soil and water conservation shows the importance of offering farmers economically and technically viable options if an introduced technology is to gain wide acceptance.

Examples of production technologies

Many new technologies were introduced in this period (see Mortimore and Wellard, 1992). Innovations came from multiple sources, including government research, farmers' observations and experiments, and through traders (cf. Biggs, 1989).

The plough at first enabled the extension of cultivated areas for market production. It also enables water-saving husbandry – ploughing on contour, and dry, early and row planting. First weeding by plough reinforces ridges as well as saving labour. The plough was adopted without extension effort, through purchases from traders. The early types were heavy, needing 4–8 oxen. Farmers have selected down to the current popular type, a light mouldboard drawn by two bulls or oxen. They use it so skilfully that they were uninterested in the much more expensive tool bar which the Machakos Integrated Development Programme tried to popularize in the early 1980s.

Katumani Composite B Maize was bred by government researchers in the early 1960s as a short-season, drought-escaping variety. It has been widely adopted, but farmers also cross breed it with local varieties, or plant both Katumani and a local variety that will outyield Katumani if the rains are good. It has added to the repertoire of choices. Together with the better husbandry practices which the plough enabled, it has led to better use of the long rains, as farmers now usually have enough time to prepare the land properly before replanting.

Coffee was important in making terracing profitable in hill areas. It brought in much more money and provided means to invest in other farm inputs and non-farm businesses. It also created employment. It was introduced in a tightly controlled manner; it spread more rapidly in the 1970s when controls relaxed and the price incentive was very high.

Fruit and vegetables were introduced by government, missions and traders. They have spread farmer to farmer, or trader to farmer, rather than via extension. Oranges and pawpaws are now major earners in AEZ 4 – replacing cotton, which has had all the government promotion. Crops such as French beans for the European markets are supplementing coffee in AEZ 2 and 3. Other farmers specialize in vegetables supplied to Indian communities in the United Kingdom as well as in Kenyan towns.

While profit motives were clearly important determinants of the adoption of new technologies it is significant that several of them (including the plough and Katumani maize) improved the efficiency of water use.

Changes in institutions

During the 60 years under review Akamba society has changed in ways which have facilitated the adoption of technological change and the accumulation and use of capital (Mbula Bahemuka and Tiffen, 1992). Four elements should be highlighted. Firstly, there are far more channels for information flows, both through formal services such as education and extension, and through informal means such as travel, meetings and meeting places. In 1930 the main information flow was between generations.

Secondly, there is a far wider base of leadership at the village level. This now includes women as well as men, and the enterprising or educated young as well as the old. In 1930 leadership was mainly in the hands of the older patriarchs. The emergence of women was partly due to male out-migration. Especially since Independence, the new leaders have been able to influence the government (mainly through party political channels) and to act together to secure desired amenities through self-help.

Thirdly, self-help groups at the local level, which derive from an older tradition of mutual help, have developed new features, including detailed project plans, elected leaders, and the ability to raise money as well as pool labour. This has been encouraged by a community development programme which originated in the 1950s and continues today. The groups not only pool local resources, but have learnt how to pull in capital and expertise from national and international sources, through their upward links to the political system and the churches and other NGOs. Most rural families belong to a self-help group; most are also church members. Most groups aim to develop a community amenity (a dam, dip, school, feeder road, etc.), or to develop a business (poultry, basket-making, buildings for rent). Others do turn-by-turn work to improve each other's farms or houses.

Fourthly, there is the family. Capital for farming improvements comes mainly from relatives. As we have seen, families aim to have at least one member of the family in a non-farm job and expected to contribute farm capital.

Factors that have assisted organizational ability, farm improvement and the uptake and development of new types of enterprise are:

(a) Education. While great sacrifices are made for secondary education, notable features in Machakos as compared with the rest of Kenya are the almost universality of primary education, the self-help village polytechnics turning out craftsmen, and adult education.

(b) The individualization of land tenure. Customary law that was already developing in this direction has been sanctioned by statute law.

(c) Community development support, initially provided through the government and the County Council, and now supported by many NGOs.

(d) The extension services have suffered many weaknesses, but have also been quite effective in spreading improved varieties of maize and certain husbandry techniques. Demonstrations, by the extension services or by neighbours, have been particularly effective. The Akamba have a saying 'Use your eye; the ear is deceptive'.

Community development and extension are both forms of adult education; they have contributed with schooling to increased information flows and awareness of alternatives.

[...]

Notes

[1 AEZ 3 is Agro-ecological zone 3. AEZs are a standard classification for Kenya, and AEZs 2 to 6 are represented in Machakos District. For further details see Tiffen *et al.*, 1994, p. 18.]

[2 ALDEV is the African Land Development Board, a Kenyan government programme which funded land improvement projects. For further details see Tiffen *et al.*, 1994, pp. 254–9.]

3 Kenya National Archives, Nairobi.

References

ALUS (1953) 'Annual report 1952', mimeo with photographs, Nairobi, African Land Utilisation and Settlement Board.

BIGGS, S.D. (1989) 'A multiple source of innovation model of agricultural research and technology promotion', Agricultural Administration (Research and Extension) Network Paper 6, London, Overseas Development Institute.

GICHUKI, F.N. (1991) 'Environmental change and dryland management in Machakos District, Kenya 1930–90: conservation profile', ODI Working Paper No. 56, London, Overseas Development Institute.

LAMBERT, H.E. (1945) DC/MKS/1/7/1 'Native land problems in the Machakos District with particular reference to reconditioning', MKS/DC *Annual Report 1948.*

MBULA BAHEMUKA, J. and TIFFEN, M. (1992) 'Akamba institutions and development, 1930–90', in Tiffen, M. (ed.) 'Environmental change and dryland management in Machakos District, Kenya, 1930–90. Institutional profile', ODI Working Paper No. 62, London, Overseas Development Institute.

MORTIMORE, M. and WELLARD, K. (1992) 'Profile of technological change: environmental change and dryland management in Machakos District, Kenya 1930–90', ODI Working Paper No. 57, London, Overseas Development Institute.

ODI (1982) 'Machakos integrated development programme: phase 1: evaluation', unpublished report (by Adams, M. *et al.*) prepared for the Ministry of Economic Planning and Development by Overseas Development Institute under assignment to the Commission of the European Communities.

PEBERDY, J.R. (1961) 'Notes on some economic aspects of Machakos District', mimeo, report for Ministry of Agriculture.

THROUP, D.W. (1987) *Economic and Social Origins of Mau Mau, 1945–53,* London, James Currey.

[TIFFEN, M., MORTIMORE, M. and GICHUKI, F. (1994) *More People, Less Erosion,* Chichester, John Wiley and Sons.]

Source: Tiffen and Mortimore, 1992, pp. 376–82

Reading F: *'Vegetation of the forest–savanna mosaic of Kissidougou Prefecture, Guinea: comparative quotations'*

Present-day

Around 1945, the forest, according to the elders, reached a limit 30 km north of Kissidougou town. Today, its northern limit is found at the level of Gueckedou-Macenta, thus having retreated about 100 km ... This deforestation is essentially the result of human action.

(Ponsart-Dureau, 1986, pp. 9–10; a study supporting a modern agricultural development project)

The opinion, quasi-general, is that the part north of Macenta, Gueckedou, Kissidougou will soon be no more than a vast poor savanna, the islands and gallery forests still present at risk of being rapidly destroyed.

(République de Guinée, 1988; national report for Tropical Forestry Action Plan)

In the green belts which surround the villages, one finds the relics of original primary forests ... (concerning certain 'indicator' savanna plants) one finds no individuals older than 35 years, supporting the thesis that this area has burned systematically only since this epoque.

(Green, 1986, p. 11; consultant on bush fire for Kissidougou's foreign-funded integrated rural development project)

The whole region was covered with forests about 99 years from 1992, which goes back to 1893 and corresponds with the Samorian period.

(Zerouki, 1993; large report on bush fire prepared by French and Guinean consultants for foreign-funded environmental programme)

Mid-twentieth century

The information obtained from the oldest inhabitants confirms what we supposed. The whole region was covered with forests around 75 years ago.

(Adam, French colonial forest botanist, in 1956 – published 1968, p. 926)

Early twentieth century

One has given the pompous name of forest in this country to simple thickets of wood, 2–3 km apart from each other and of an average area of 2–3 dozen hectares. To each of these thickets is attached a village which draws from it all the materials it needs ... almost treeless savannas joint these various thickets.

(Chauffaud, 1920; letter from Kissidougou Cercle administrator to Governor on the subject of coffee plantation)

Never, I believe, has a year so dry occurred in Kissidougou. I am left to say that from year to year, rain becomes more and more rare. And this I do not find extraordinary – even the contrary would astonish me – given the considerable and even total deforestation in certain parts of this region. From Kissidougou to Gueckedou, all has been cut ... the effects of this de-wooding are disastrous; one will soon see nothing more than entirely naked blocks of granite. A region so fertile become a complete desert. Now there rests no more than a little belt of trees around each village and that is all.

(Nicolas, 1914; report on local agriculture)

1893

The soil of the valleys has a more or less thick humus bed. This humus derives, as one understands it, from the immense forests which covered a large part of the soil, and which covered it entirely at an epoque relatively little distant from our own.

(Valentin, 1893; report on Kissidougou by its first French military administrator)

Sources of quotations

Archives

CHAUFFAUD (1920) Lettre du Commandant du Cercle à la gouverneur au sujet de 'Plantations de Caféiers', Guinean National Archives, Conakry, 2D 175.

NICOLAS, Etat de cultures indigènes (août 1914) Guinean National Archives, Conakry, 1R12.

VALENTIN, Rapport sur la Résidence du Kissi (1893) Senegalese National Archives, Dakar, 1G188.

Project reports/published articles

ADAM, J.G. (1968) 'Flore et végétation de la lisière de la forêt dense en Guinée', *BIFAN, T.XXX,* Series A, No. 3, pp. 920–52.

GREEN, W. (1991) 'Lutte contre les feux de brousse', rapport pour projet DERIK, Kissidougou, Dévèloppement Rural Intégré de Kissidougou.

PONSART-DUREAU, M-C. (1986) 'Le pays Kissi de Guinée forestière: contribution a la connaissance du milieu; problematique de developpement', memoire, Montpellier, Ecole Supérieure d'Agronomie Tropicale.

RÉPUBLIQUE DE GUINÉE (1988) Politique Forestière et Plan d'Action, Conakry, TFAP.

ZEROUKI, B. (1993) 'Etude Relative au Feu Aupres des Populations des Bassins Versants Ypes du Haut Niger', Conakry, Programme d'Amenagement des Bassins Versants Types du Haut Niger.

References to Fairhead and Leach study

FAIRHEAD, J. and LEACH, M. (1993) 'Degrading people? The misuse of history in Guinea's environmental policy', paper presented at the African Studies Association meeting, Boston, 4–7 December 1993. Forthcoming as Boston University African Studies Center Working Paper.

FAIRHEAD, J. and LEACH, M. (forthcoming) 'Enriching landscapes: social history and the management of transition ecology in Guinea's forest–savanna mosaic', submitted to *Africa.*

FAIRHEAD, J. and LEACH, M. with MILLIMOUNO, D. and KAMANO, M. (1992) 'Managed productivity: the technical knowledge used in local natural resources management in Kissidougou Prefecture', COLA Working Paper 2, Kissidougou and IDS, Brighton.

LEACH, M. and FAIRHEAD, J. (1993) 'Whose social forestry and why? People, trees and managed continuity in Guinea's forest–savanna mosaic', *Zeitschrift für Wirtschaftsgeographie,* Vol. 37, No. 2, pp. 86–101.

Source: quotations compiled by James Fairhead, School of African and Oriental Studies, University of London, and Melissa Leach, Institute of Development Studies, University of Sussex

Stabilizing population growth: the European experience

Chapter 3

by Ray Hall

3.1 Introduction

3.1.1 Why Europe?

Africa faces Europe across the Mediterranean sea, the continent with the world's fastest-growing population facing that of the world's slowest-growing population; the world's youngest population facing the world's oldest. The Mediterranean is a demographic 'fault line' of major importance, one which helps us to focus on the nature of population problems which face the world, since even low growth is not without problems. So from Africa, we can now turn to the other extreme demographically: Europe – densely populated, slow growing and rich.

Contrasts between Europe and Africa are even sharper if we compare the immediately adjacent regions: North Africa with Southern Europe. At present they are regions with similar numbers of people: 144 million in Southern Europe compared with 157 million in North Africa, and yet by 2025, Southern Europe's population will be only 148 million while North Africa's is projected to be almost double that – 280 million. The reason can be found in the very different birth rates of the two regions: Southern Europe has the lowest fertility, only 1.4 children per woman, with Italy (1.3) and Spain (1.2 children) having exceptionally low birth rates. By contrast, in North Africa the average number of children per woman is 4.5. Such low birth rates confound all demographic expectations, both about to what low levels birth rates can fall, as well as posing great problems of explanation. The proximity of these contrasting populations poses such questions as how permeable the frontiers between the regions should be, and what the impact will be of such proximity on both regions.

Europe, in sharp contrast to Africa, is usually represented as having reached an enviable stage in demographic terms. If this is true it also raises questions: how has Europe reached this point in its demography, and are there implications for the rest of the world? More critically, was the way in which Europe reached the current state problematic, and is the present so free of problems?

Within this book's overall examination of whether current lifestyles are sustainable, this chapter considers Europe's population and lifestyle within a global context. It asks three key questions:

1 How has Europe stabilized population growth?

2 What impacts has Europe's population had on the rest of the world?

3 Does Europe's experience of population change have implications for what will, or should, happen elsewhere?

The argument develops through four stages:

- Section 3.2 examines how Europe became the first continent to experience rapid population growth in the eighteenth and nineteenth centuries, and how both the reasons for growth and rates of growth compare with the rates of growth experienced in the less developed world in the twentieth century.

- Section 3.3 considers how Europe's population growth slowed and stabilized in the twentieth century, and again, whether this experience has any lessons for the rest of the world.
- Section 3.4 looks at the consequences for the rest of the world of Europe's population growth, resulting as it has in a large absolute number of people enjoying very high standards of living. To what extent did the rest of the world sustain Europe's population growth? How far does it sustain current European prosperity?
- Finally, section 3.5 examines Europe's current population growth rates and demographic trends. In particular, the changing structure of the population and other demographic changes are looked at, together with their implications for the future. Population growth may have stabilized in Europe, but there is still a range of population-related problems to be tackled, not least of which is how this no-growth population fits into a world of rapid population growth. Is a rich, no-population-growth society sustainable?

demographic transition

In Europe we can see the cycle of change that population has undergone (the *demographic transition*). The cycle starts with low growth rates which were the norm for many centuries, when the life of the average European was unpleasant by modern standards, and would certainly have been much shorter than that of today. It moves through a period of rapid rates of growth and then back to low rates of growth, albeit with much larger numbers at the end of the cycle than at the beginning, and vastly improved standards of living. From this cycle of growth and changing demography the model of demographic transition has been constructed which describes European population growth and underlies assumptions about population growth in the world (Figure 3.1) – although it is highly problematic that population growth elsewhere in the world can be expected to follow such a simplistic pattern.

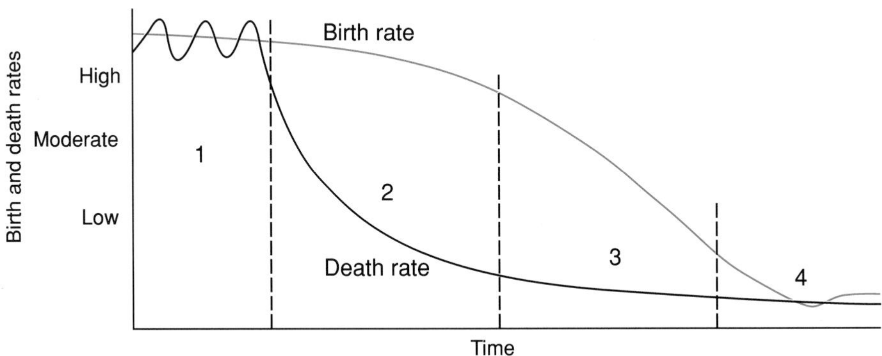

Figure 3.1 *The demographic transition model. The demographic transition model is based on a summary of European population change and underpins many assumptions about population change elsewhere in the world. It describes trends in European fertility and mortality as they were understood when the model was first developed in the first half of the twentieth century. It also includes an element of prediction – since it assumes other regions of the world, where population growth rates are still high, will follow a similar pattern of change, although this is highly problematic (Source: Hall, 1989, p. 4, Figure 1.2)*

3.1.2 How do populations grow?

Population grows or declines by a particular combination of births and deaths, growing when births exceed deaths and declining when deaths exceed births. This can be described by the key population equation:

$$P = (B-D) \pm M$$

(where P = population change; B = births; D = deaths, M = net migration).

When we are looking at population growth in any region we need to know the number of births (fertility) and deaths (mortality). There is a variety of indices used to measure fertility and mortality, including the crude birth rate and crude death rate which relate the number of births or deaths to the total population. More refined measures include the Total Fertility Rate (TFR) which is a measure of the average number of children per woman, and life expectancy at birth (e_0) which is a particularly useful summary measure of mortality giving the average age the population can expect to live to. The rate of growth of a population is a function of these two variables, births and deaths (see Chapter 2, section 2.2).

Activity 1 Turn now to Reading A by Massimo Livi-Bacci, entitled 'The space of growth', which you will find at the end of this chapter. As you read, examine the three diagrams carefully. Make sure you understand what is meant by *isogrowth curves*. In Figure A.1 note the contrast in distribution of growth rates between historical and present-day populations. Then look at Figures A.2 and A.3 which show in more detail the distribution of historical and modern-day populations along the isogrowth curves.

What should be evident is that countries may have the same or similar growth rates but achieve these by very different combinations of fertility and mortality. Select examples from Figures A.2 and A.3 to illustrate this.

Two types of demographic regime have been identified. A *low-pressure regime*, with moderate levels of fertility and mortality, and a *high-pressure regime* with high levels of fertility and mortality. It has been argued that populations which maintain equilibrium with resources by controlling fertility – that is, a low-pressure demographic regime – are in a more favourable position for economic growth than a high-pressure demographic regime.

low-pressure regime
high-pressure regime

3.2 How and why did Europe's population grow?

This section examines the course of European population growth and compares this growth with that found in the rest of the world in the twentieth century.

Populations appear always to have adjusted to resources available and for most of human history population growth rates were extremely low. As the productive capacity of land expanded, however, as a result of social and economic development, so populations began to grow (see Chapter 1, section 1.3.1). Even so, Europe's population grew only slowly and densities were low, with population numbers limited by available, locally produced food supplies and low agricultural productivity.

From the Middle Ages onwards, Europe underwent a range of dramatic changes that were to have repercussions for the whole world. There was a new approach to knowledge and learning and a desire to understand the world, as well as a desire for wealth and power. One particularly significant date among many was 1492 when Christopher Columbus set out to find a new route to the Spice Islands but landed in the previously 'unknown' continent of America instead. The implications of this discovery for both the indigenous peoples of America and for Europeans were profound. It meant that land and other resources available for European exploitation were immediately increased many times and European population growth need no longer be limited by the land and resources available within Europe itself.

At the same time, a range of technological and other developments were taking place in Europe that increased agricultural productivity. The so-called agricultural revolution was a long drawn-out process both temporally and spatially. The end result was a more efficient and productive agriculture and more reliable and increased food supplies which could support a larger population. The innovations in agriculture provided the essential foundation for those changes often described as the industrial revolution. Such a term, though, is far too limiting for the wide range of changes encompassed by it. Towns expanded, industries and commerce developed and the demand for labour grew apace. Previous environmental constraints evaporated and Europe's population began to grow rapidly. The more dramatic changes were focused on North-western Europe, with Britain leading the way in the agricultural and industrial changes and population growth.

3.2.1 Fertility and marriage

Pre-industrial Europe was characterized by moderate levels of fertility together with moderate levels of mortality. Look again at Figure A.2 in Reading A and compare England in 1750–1800 (E3) or Denmark, for example, with India in 1900. Most of the European populations exhibit the characteristics of a low-pressure demographic regime in contrast to other pre-industrial populations with a high-pressure demographic regime.

Pre-industrial Europe appears to have kept its population in line with environmental resources by strict controls on who could marry and when. 'No land, no marriage' seems to have been a medieval injunction (McLaren, 1990, p. 141). In Western Europe, especially, marriage entailed the founding of a new, separate household. The couple, therefore, had to be economically self-sufficient so marriage was strongly associated with control of land or completion of apprenticeships. Marriage took place at relatively late ages and long after puberty. The mean age of marriage for women would be between 23 and 28 years, with the mean age for men slightly older but in general with little age difference between spouses. At the same time, anything between 10 and 20 per cent of the adult population would remain unmarried. This pattern of marriage has been described as the *European marriage pattern* and was found west of a line from Trieste to St Petersburg (Hajnal, 1965). For such strict control over marriage to work, sexual self-control must have been an ingrained form of behaviour, especially as most data suggest that illegitimacy levels were low, particularly in rural areas – often only about 2 per cent of births (McLaren, 1990).

European marriage patterns

Was there birth control within marriage? There has been considerable debate about this and it was probably of only limited importance. It is almost certain, though, that some couples at least tried to limit fertility within marriage, mainly through coitus interruptus or sexual abstinence.

The European marriage pattern is in contrast to most non-European pre-industrial populations where marriage, for women at least, takes place soon after puberty, with a mean age of marriage between 16 and 18 years. Men also marry early, although at somewhat older ages than women. Here the couple usually begin married life within the household of the husband's parents and so do not need to be economically independent. Early marriage ensures higher fertility than in the European system. Both examples demonstrate that fertility is a social construct with geographical variations arising out of economic and social structures.

As agricultural and industrial developments accelerated in eighteenth-century Europe, population growth rates began to increase. The conventional reason for population growth in Britain at least, and which was incorporated into the model of demographic transition, was that growth was a result of declining death rates due to better food supplies and improved standards of living. More recently, however, very detailed work on parish registers in England and in other European countries has shown that population growth in the early stages of industrialization was the result of rising fertility due to earlier marriage rather than of falling mortality.

The new employment opportunities and growing towns enabled couples to marry earlier, thus giving rise to more children. At the same time, reductions in mortality were relatively slow, and in the growing urban areas mortality actually increased, so population growth rates were modest by today's standards in developing countries. The patterns of growth across Europe were varied and France in particular stands out as an exception to the model of demographic transition. Industrialization and urbanization were much slower in France than elsewhere in Western Europe, fertility was declining from the end of the eighteenth century and there was no period of rapid population growth.

3.2.2 Mortality decline

Mortality in pre-industrial Europe was characterized by low life expectancy and high infant and child mortality, with infectious endemic and epidemic diseases the principal killers. Mortality fluctuated under the influence of a variety of external (exogenous) factors such as the vagaries of climate, poor harvests and war. In general, though, death rates in pre-industrial Europe were lower than in non-European pre-industrial populations.

Activity 2 Analysis of English parish registers has shown how mortality levels in England fluctuated from the sixteenth to the nineteenth centuries. Periods of low mortality alternate with periods of high mortality.

Examine Figure 3.2. In (b), which periods had very low and very high life expectancy? Does there appear to be any relationship with the graph of changing fertility in (a)?

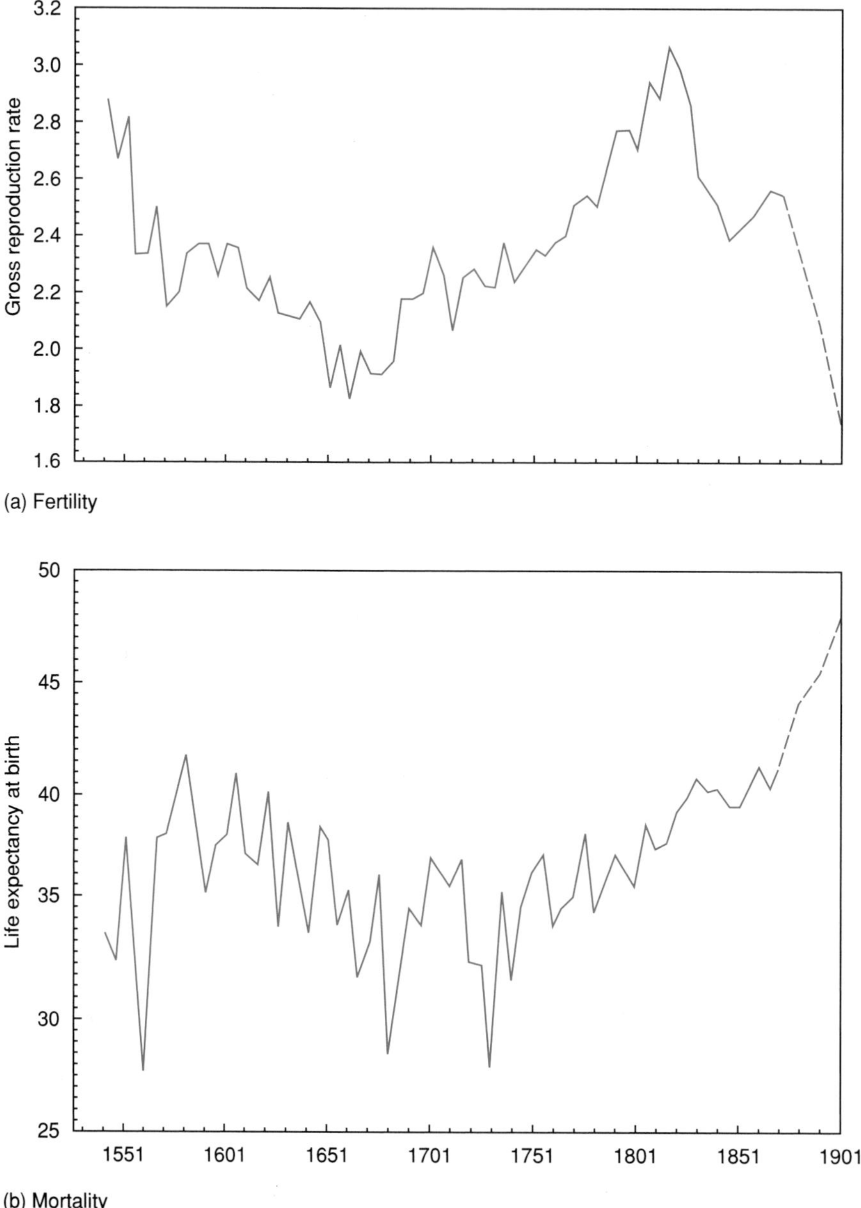

Figure 3.2 *Changes in English fertility and mortality, 1541–1901: quinquennial gross reproduction rates (GRR) and life expectancy at birth (e_o) (Source: Wrigley and Schofield, 1981, p. 231, Figure 7.6)*

Note: The data are centred on the years indicated. GRR is a measure of fertility which shows the average number of daughters that would be born to a woman during her lifetime.

With agricultural improvements from the eighteenth century onwards the food supply became more reliable. In turn, industrialization produced higher wages and rises in the standard of living which enabled individuals gradually to gain more control over their environment. This was reflected in a slow improvement in mortality from the 1780s to the 1820s: from a life expectancy of about 35 years to one of 40 years. Even so, by the period 1820 to 1870 life expectancy was still only about two years higher than at the end of the sixteenth century. Mortality fluctuated considerably in the early part of the industrialization process, and, although life expectancy increased, the levels achieved were little different from the late sixteenth century (Wrigley and Schofield, 1981).

Some diseases were particularly notorious in raising mortality levels. The control of killing disease was a slow process and only speeded up as local communities or national governments took concerted action. The elimination of plague as a killer in Europe illustrates the importance of such action in the control of disease.

Activity 3 Now turn to Reading B by Michael W. Flinn, 'The defeat of bubonic plague'. You will find this at the end of the chapter. Read the description of the action taken to control plague. What were the key factors in controlling the disease? Are there general lessons to be learnt from this example? One impact of increased migration and mobility has been the spread of infectious diseases around the world, or what has been described as the microbial unification of the world.

The elimination of plague as a major killer saw the rise of other diseases such as typhus, smallpox and tuberculosis to replace it and they found excellent breeding grounds in the rapidly growing towns. Smallpox was particularly significant in the eighteenth century, but was again brought under control from the late eighteenth to mid-nineteenth centuries through inoculation and vaccination. In the nineteenth century cholera spread along the world's seaways to Europe from its home in India and reached epidemic proportions in many cities for much of the century. Again, control was the result of institutional action, particularly the improvement of sanitation and water supplies towards the end of the nineteenth century. Figure 3.3 illustrates the relationship between improvements in life expectancy and public health in selected French cities from 1820 to 1900. Life expectancy in French towns rose from about 32 years in 1880 to about 45 years in 1900, and the changes corresponded closely to improvements in water supply and sanitation.

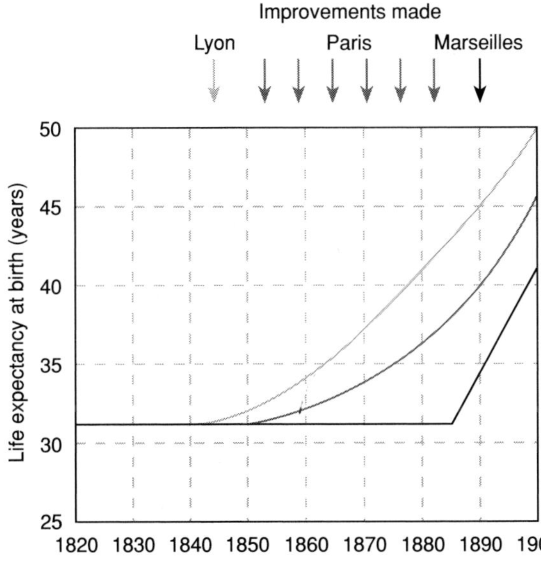

Figure 3.3 Life expectancy and improvements in water supply and sanitation in selected French cities, 1820–1900 (Source: World Bank, 1992, p. 99, Figure 5.1)

Mortality rates decreased slowly in industrializing Europe, and, although there were benefits from improved food supplies and higher wages, the unhealthy conditions of towns with poor housing, inadequate water supplies and poor sanitation almost certainly counterweighed these benefits. The rapid increase in numbers living in disease-ridden towns in the nineteenth century therefore cancelled out any improvements in mortality that might have followed from increased wealth. This partly explains at least why there was no significant change in life expectancy in England between 1820 and 1870.

Standards of living were improving but the early benefits were not experienced equally by all members of the population. In particular, there were no improvements in infant mortality until the twentieth century. There were also considerable regional variations in life expectancy for different ages. Infant mortality in Britain, for example, was much higher in northern industrial towns and inner London compared with rural areas and southern towns.

Activity 4 Turn to Reading C, 'Farr and the public health', by Ian Sutherland, at the end of the chapter. William Farr worked in the General Register Office for 40 years in the nineteenth century and wrote extensively about causes of death. The extract shows his concern with improving the environment in order to lower mortality. Even today, we still have much to learn about the relationships between environment and health, although the relationship between fog – and smog – and respiratory diseases is now well established.

Mortality levels started to fall towards the end of the nineteenth century, but rapid declines did not occur until the twentieth century. In 1900, life expectancy in Western Europe was still only around 50 years in spite of the economic changes that had taken place. Infant mortality was still over 100 per 1,000 live births. What is evident is that links between mortality and standards of living were much stronger in the nineteenth century than they are today. Globalization has resulted in a rapid spread of knowledge about mortality control, so that, for example, India in 1982 had a life expectancy of 55 years, while per capita income was below US $300 and the literacy rate below 40 per cent. By contrast, England, Sweden and the USA in 1900 had life expectancies below 50 even though per capita incomes were around US $1,000 (at 1982 values) and literacy rates over 80 per cent.

3.2.3 Historical rates of growth and comparisons with today

Overall, Europe's rates of population growth rose from around 0.5 per cent per annum in the eighteenth century to a maximum of around 1.5 per cent a year in the nineteenth century. In England and Wales, usually associated with very rapid population growth, the peak rate of growth was 1.6 per cent in the 1820s.

These rates of growth are much lower than those experienced during the twentieth century in many regions of the world. Even today, world population growth at just over 1.6 per cent per annum is higher than that of Europe at the peak of its growth, and individual countries in many parts of the world have experienced (and, indeed, are experiencing) rates of growth much higher than this.

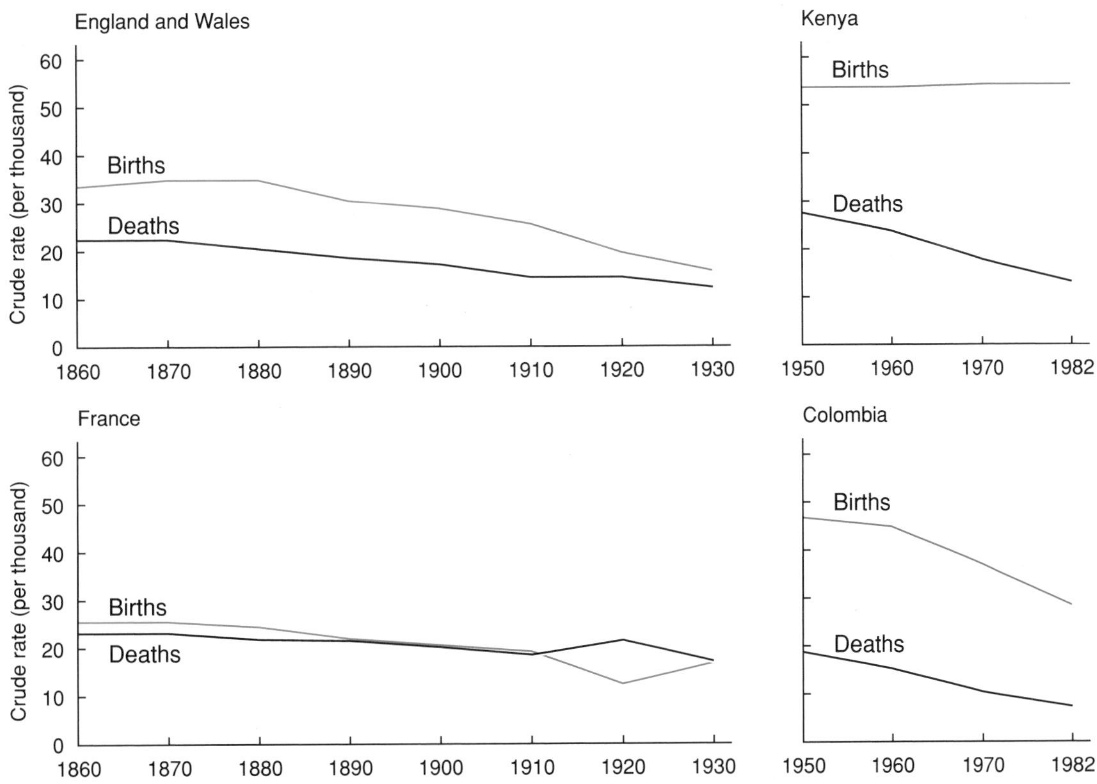

Figure 3.4 Fertility and mortality transition in developed and developing countries, 1860–1982 (Source: World Bank, 1984, p. 59, Figure 4.1)

Activity 5 Compare the peak birth and death rates of Kenya and Colombia with those of England and Wales and France in Figure 3.4. Note the contrasts between England and Wales and France and between Kenya and Colombia both in the birth and death rates experienced and the time-scale of changes in these rates.

Kenya's average annual rate of growth between 1970 and 1982 was 4 per cent per annum and today Africa averages 2.9 per cent per annum with many countries having growth rates around 3.5 per cent. These are more than twice the figure for Europe at its maximum rate of growth. Fertility, even though it increased in some parts of Europe in the early nineteenth century, reaching a maximum of around 36 births per thousand population in the 1870s in England and Wales, for example, was never as high as in less developed countries in the period after 1950. Crude birth rates of well over 40 per thousand population were common everywhere and even today some countries in Africa have rates over 50. At the same time, mortality rates declined relatively slowly for much of the nineteenth century in Europe, again helping to explain the slower rates of growth compared with today.

The final and important reason for Europe's slower rate of growth was its access to the new lands colonized by Europeans from the sixteenth century onwards. In the nineteenth century some of the surplus natural increase of European population growth could be siphoned off to these 'empty' lands waiting to be colonized. Emigration was of great importance in reducing the pressures of a growing population on the developing European economies.

3.2.4 Emigration

emigration *Emigration* provided Europe with a means of relieving demographic pressure during the period of most rapid population increase in the nineteenth century. It was particularly important as economies developed and diversified away from agriculture; emigration provided the unemployed rural populations with an alternative destination to the urban areas of Europe, especially in periods of economic depression. The only constraints on emigration were the cost to the individual, so the nineteenth century situation was very different to that operating in the world today where migration controls are enforced all over the world.

Something like 50 million people left Europe for the Americas and Australasia between 1846 and 1932. At the peak between 1881 and 1910 the numbers emigrating were equivalent to 20 per cent of Europe's population increase (see Figure 3.5 and note the relationship between rising rates of population growth and rising rates of emigration).

The areas of European settlement in the New World experienced very high rates of natural increase as a result of the young age structure of the migrants and their high fertility rates (Plate 7). The importance of European settlement beyond Europe is illustrated by the fact that, while Europe's share of world population remained reasonably constant between 1750 and 1950 at around 16 per cent, in those areas of European settlement outside Europe their share of world population increased from 24 per cent in 1750 to 36 per cent around 1900, remaining at that figure to 1950 (United Nations, 1973).

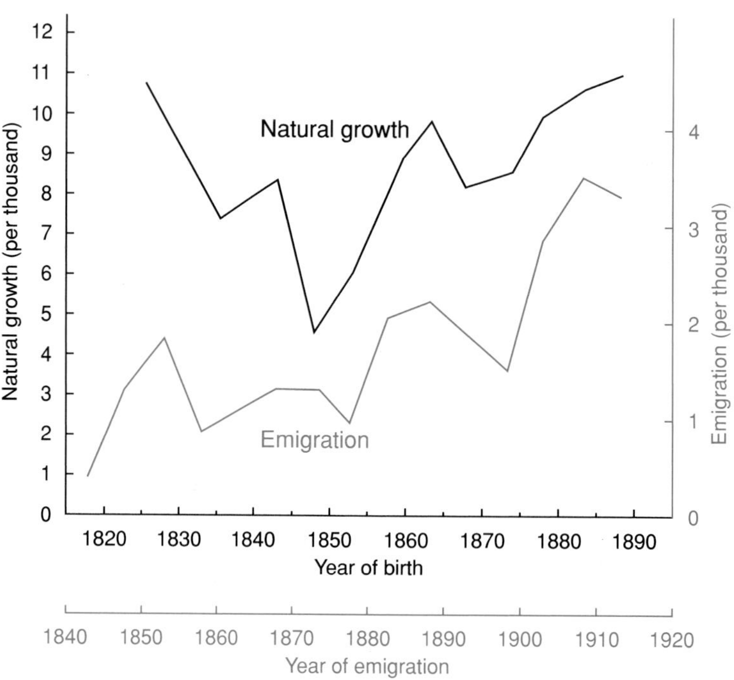

Figure 3.5 Emigration and natural growth for continental Europe (Source: Livi-Bacci, 1992, p. 125, Figure 4.9)

Summary of section 3.2

- Europe's population grew both as a result of a rise in the birth rate and slow declines in the death rate.
- Economic changes made it easier for couples to marry earlier and so have more children.
- Improvements in mortality took place slowly, especially as conditions in towns resulted in high death rates from infectious diseases.
- Overall growth rates in Europe were considerably lower than they have been in many parts of the world since 1950.
- Europe had the additional safety valve of emigration to the New World, which siphoned off excess population, especially in the later nineteenth century.

3.3 The slow-down of European population growth

Look again at the diagram of the model of demographic transition (Figure 3.1 in section 3.1.1). It shows mortality decline beginning in stage 2, but growth rates only really begin to slow down in stage 3 with the decline of fertility. The process of declining birth and death rates was completed in stage 4 in the twentieth century.

In this section we look at how European population growth slowed to the very low levels of today. We need to look in some detail at the reasons for the declines in both mortality and fertility which may help in understanding population change in the less developed world. We will also consider what other related changes have taken place which make contemporary European society very different from any time in the past.

Population growth in the eighteenth and nineteenth centuries resulted in an increase of over 200 million people in Europe, from 95 million in 1700 to 295 million in 1900, in spite of the very large numbers migrating overseas. Only Ireland was exceptional, with a decrease in population as a result of massive emigration in the second half of the nineteenth century. There were regional variations and Northern Europe experienced more growth than Southern Europe and the Balkans. The twentieth century, however, saw the end of this period of rapid growth, and up to the Second World War was characterized by falling rates of population growth.

3.3.1 Decline of mortality in the twentieth century

The most dramatic improvements in mortality have taken place in the twentieth century. Declines in infant and child deaths, especially, have been important in contributing to the slowing-down of growth rates since parents no longer feel impelled to have extra children as security against child deaths. The continued improvement in standards of living, together with medical discoveries – particularly the widespread use of antibiotics after the Second World War – resulted in a rapid decline in deaths, especially from infectious diseases. The extent of mortality improvements is illustrated by the figures for England and Wales. Here the life expectancy at birth in 1891–1900 was 44.1 years for a man and 47.2 for a woman, and 10.3 and 11.3 years at the age of 65; by 1990 these figures had risen to 73.2 for men and 78.5 for women at birth, and 17.6 years and 27.7 years at the age of 60. Apart from the overall increase in life expectancy, the widening difference in life expectancy between men and women is notable.

Throughout Europe mortality has continued to improve through to the present, and while there were considerable contrasts even as late as 1950 between Northern and Western Europe and Eastern and Southern Europe, today the differences are negligible in the areas outside Eastern Europe. Within Eastern Europe, though, life expectancies have stagnated or even deteriorated since 1960 and the worsening situation has affected males more than females. Among the explanations for this deterioration have been problems of environmental pollution, poor quality food, poor housing and poor working conditions, alcoholism and smoking, together with an inadequate health system.

One of the most significant aspects of the improvements in mortality has been the continuing improvement in life expectancy at older ages. This is a major factor contributing to the ageing of the population. Improvements in life expectancy are projected to continue and there seems no reason why life expectancies of around 80 already found in some countries for women should not be rapidly achieved for both sexes.

Some of the highest life expectancies in Europe today are found in the Mediterranean region, which raises questions about the key factors which influence mortality today in Europe. It has been suggested that Mediterranean lifestyles are important in explaining the very high life expectancies of Spain or Italy and are more important than quality of health care or income. Germany, for example, has a gross national product (GNP) almost double that of Spain (US $23,650 1991 compared with US $12,460), and 19.3 per cent of central government spending was on health between 1980 and 1990 in Germany compared with 12.8 per cent in Spain, yet Spain has a life expectancy two years longer than that of Germany.

Since about one-third of European deaths are a result of cancers and about half a result of heart disease, it is not hard to imagine that diet and other aspects of lifestyle are crucial in determining European patterns of mortality today. New diseases, however, can emerge which demonstrate that there is nothing inevitable about continuing mortality decline. AIDS (Acquired Immume Deficiency Syndrome) is a notable example of this. The future impact of AIDS on Europe's population is uncertain, but it is unlikely to reach the epidemic proportions of some parts of Africa.

3.3.2 Decline of fertility in the nineteenth and early twentieth centuries

fertility decline

Fertility has declined more erratically but ultimately even more dramatically than mortality and is of key importance in understanding why Europe's population growth is now so low. In spite of a great deal of research into *fertility decline*, we still do not fully understand the reasons for it. Economic development and income growth are certainly important but are by no means the only explanation. Broader concepts such as culture and social change are also relevant, particularly in their relation to the changing position of women in society. Education and legal changes, among others, have contributed to improvements in women's status in society, and as women's status has improved so women have chosen to have fewer children.

Activity 6 Examine the map of European fertility decline illustrated in Figure 3.6. What were the main dates between which decline occurred? Give an example of earlier decline and one of later decline.

You should be able to see that more than half the provinces on the map experienced decline between 1890 and 1920. An area of much earlier decline was most of France; Ireland and parts of Southern and Eastern Europe were areas of later decline.

Early fertility decline in France seems to be partly associated with the peasantry limiting family size from the early nineteenth century as a means of preventing the excessive sub-division of land. Under the Napoleonic code, land had to be divided equally among all heirs on the death of the father. Social and cultural factors appear to be important in explaining Ireland's late fertility decline.

Declines in fertility in large parts of Europe came towards the end of the nineteenth century as couples increasingly tried to limit the number of children born in marriage, usually by withdrawal and abstention rather than contraception in the majority of cases. Some groups in the population were reducing fertility earlier than others, especially higher socio-economic groups. Some localities experienced fertility decline before others, and some socially isolated areas (for example, with distinctive dialects or separate language) experienced later decline.

A number of varied explanations for this rapid change in reproductive behaviour have been put forward. These include industrialization, a term which includes a whole range of social and economic changes that resulted from the growth of industry. Associated with industrialization was secularization and an increasing emphasis on individualism and materialism. Smaller families became economically rational as children lost any remaining economic role they may have had, with the introduction of laws limiting the labour of children together with the introduction of compulsory education, as in England and Wales, for example, in 1880. It has also been argued that this first demographic transition can be seen in terms of a growing child-orientedness in society. Couples wanted to devote more time to children, to give them better education, and therefore could only afford a few children.

Any explanation of the decline in fertility has to include a range of factors such as rising standards of living, rising levels of literacy and falling infant and child mortality. In particular, the growing literacy of women was undoubtedly important, as was the growing feminism of the period. The movement for women's suffrage was sometimes combined with the birth control movement. In the UK, the first neo-Malthusian organization, the Malthusian League, was founded in 1877 and in the same year the trial of Charles Bradlaugh and Annie Besant for publishing a birth control tract gave the movement enormous publicity. Margaret Sanger (1932), the American birth control pioneer, certainly recognized that women being able to control their fertility was fundamental if they were to gain any real control over their lives.

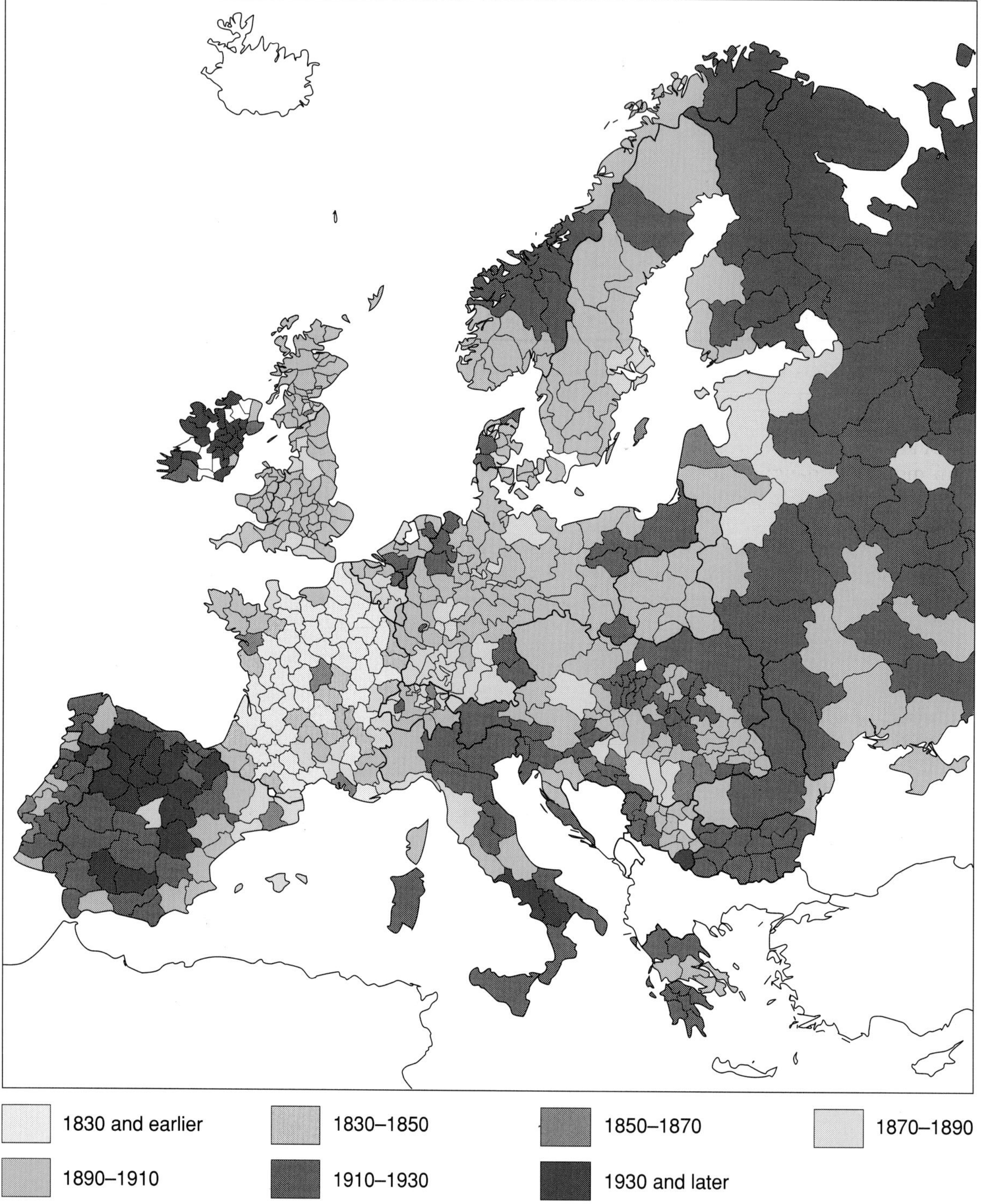

Figure 3.6 *Map of estimated date at which marital fertility has declined by 10 per cent from maximum level (and never again returned to that level), by province of Europe (Source: Coale and Watkins, 1986, Map 2.1)*

The widespread decline in fertility coincided with the development of mechanical contraceptives, although again the role of these in lowering fertility in the nineteenth and early twentieth centuries has been disputed. It has been estimated that only 16 per cent of couples married before 1910 used mechanical contraceptives. The church and the medical profession were certainly hostile to the idea of family limitation through contraception, although the middle classes, including doctors, were leaders in the decline in family size. It may be coincidence that the decline of fertility in the UK coincided with the invention of the vulcanization of rubber in the 1870s which meant that effective mechanical contraceptives were available for the first time. The conventional view is that contraceptives themselves do not produce fertility declines; people have to want to reduce fertility. Nonetheless, it is important to remember that having the means available makes it more likely that people will be able to achieve any fertility aspirations they may have. Margaret Sanger in her autobiography describes a visit to France where she remarks upon the fact that the right to knowledge of contraceptive techniques was 'almost a national right … In France information had been generally disseminated from mother to daughter for generations, since the Code Napoléon' (Sanger, 1932, pp. 72–3).

Activity 7 Have a look at Reading D, 'A wife and husband talking about parenthood and family planning in the 1930s', by Steve Humphries and Pamela Gordon. As you read, contrast the attitudes of the husband's towards his wife's desire to limit family size.

How far do explanations of fertility decline in Europe apply to other societies? Perhaps parents in all societies decide the number of children in terms of a trade-off between quality and quantity. The richer the society, or the population group, the greater the emphasis on quality rather than quantity and the greater the desire to limit family size. But other factors are clearly important, particularly the status and educational level of women. It seems clear that when women are able to assert their individual choice on family size then fertility declines. This is clearly relevant to understanding population change in other parts of the world.

3.3.3 Fertility trends from the 1930s to the present

Fertility continued to decline steadily in the twentieth century in almost all parts of Europe and by the 1930s very low levels had been reached, particularly in countries of Western and Northern Europe. Many countries had fertility rates that were *below replacement level*. Low fertility in the interwar period gave rise to great concern in many countries about the likelihood of declining population numbers and the impact this would have on the economy. Concerns other than economic ones included the strategic implications of falling numbers, especially of men for military service, and fears that some groups or social classes were not reproducing sufficiently, and that low fertility (and therefore a slow growing population) was an obstacle to progress. The result was pronatalist policies (policies to encourage couples to have more children) in a number of countries, which had very limited impact.

Fears about low fertility disappeared at the end of the Second World War when the birth rate rose almost everywhere in Europe. Partly this was a result of the end of wartime conditions, and couples who had deferred having children started to have them. At the same time, more couples were marrying and having their first child soon after marriage. Increasing numbers went on to have a second and third child. In Western Europe the post-war baby boom lasted, with variations, until the mid-1960s, longer than would be expected if it was just a readjustment to peace-time conditions. This rapid rise in the birth rate after 1945 ran counter, therefore, to the trends of the previous half century. A variety of explanations have been put forward to explain this rise in fertility. Some have related it to post-war economic prosperity, but none has been entirely convincing.

Fertility peaked in almost all Northern and Western European countries in the mid-1960s, after which birth rates began to decline, slowly at first and then, in the early 1970s, at an accelerated pace. By the early or mid-1970s most countries in the region had below replacement-level fertility. In Southern Europe fertility decline started later and with varying pace, but has continued throughout the 1980s and into the 1990s (see Figure 3.7).

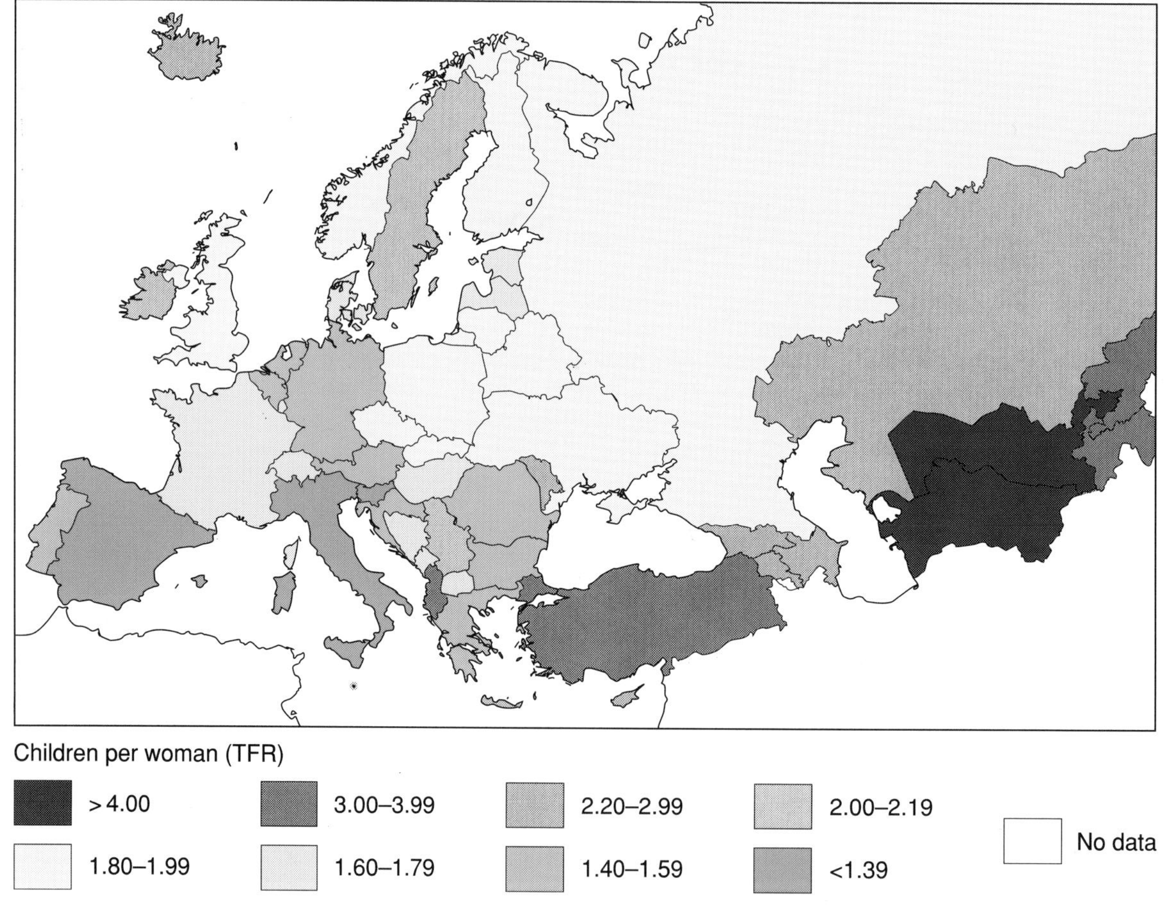

Figure 3.7 Total fertility rate in the countries of Europe around 1992
(Source: compiled from data contained in Table 3.3 of Council of Europe, 1994, pp. 49–59)

Fertility is now lowest in Southern Europe, with Italy having a TFR of only 1.25 in 1992 and Spain 1.23. Somewhat surprisingly, in Southern Europe today childlessness and one-child families are much more common than in Northern or Western Europe where two-child families are still the norm. And in Sweden, after having low fertility for much of the 1970s, there was an increase in births in the 1980s, and by 1990 above replacement-level fertility was recorded for the first time in 20 years.

Sweden's rising fertility is contrary to trends elsewhere in Europe and it might be wondered whether this is a precursor to a rise in other European countries. This is unlikely since Swedish fertility has never fallen to the extremely low levels recorded elsewhere (although 1.56 was recorded in 1983), and the likely explanation is that government social policies in Sweden encourage women to remain in the labour force even while childbearing and childrearing.

We only realize how surprising what we tend to regard as the stereotypical two-children family is when we compare it with a typical family of the nineteenth century

Such contrasts show that even in a relatively small and homogenous continent as Europe, we cannot make too many generalizations about fertility; and in detail the contrasts help us to understand some of the reasons for low fertility.

second demographic transition

Fertility trends since 1965 have therefore reverted to those current in the 1930s. But while the theory of demographic transition suggested that fertility would stabilize at around replacement level resulting in zero or stable population growth, fertility has now fallen below replacement level. If this continues, then populations will inevitably decline. As a result, this stage in population development has been described as the '*second demographic transition*', or as 'beyond the demographic transition' (Van de Kaa, 1987, pp. 4–5).

Activity 8 Construct a short family tree (of your family or a family known to you) *as far as you are able* with the information available. It is sufficient to go no further back than grandparents. How much have family sizes changed by generations, and how much did they differ between generations? Do you have any explanations for these variations?

3.3.4 Reasons for the present low rates of fertility

The remarkable similarity in European fertility suggests that Europeans have been motivated by increasingly similar reasons when making childbearing decisions. And whatever the reasons, they are choosing to have only one or two children, and in some cases none. Families of three or more children are becoming increasingly uncommon.

The desire to have children seems to have declined even as Europe continued to get richer during the 1980s. It could be argued that to achieve various materialist goals couples have to limit family size. Perhaps there is always a balance between fertility and wealth, and in order to maximize the one they need to sacrifice the other. As consumerism grows, so fertility declines? Many explanations of fertility change have concentrated on economic factors, both in terms of the broad relationships between economic conditions and fertility, as well as micro-economic explanations whereby fertility is seen as a consumer decision in much the same way as any other consumer decision. But this is too narrow an explanation.

The economic prosperity of the early post-war period enabled couples to marry earlier and it was a period when marriage and childbearing were still closely related. From the early 1970s this close relationship between marriage and childbearing began to break down, along with a variety of other closely related socio-demographic changes. This makes the late 1960s–early 1970s a real transition point and justifies the view that we are now beyond the demographic transition. European populations are changing rapidly and may indicate the way the rest of the world might follow.

To some extent these changes since the 1960s are related to changes in contraceptive technology that occurred during the 1960s, notably the contraceptive pill and the intra-uterine device (IUD) which made contraception more reliable. The decision to have a child became a much more conscious one, and it became possible safely to divorce sexual relations from marriage. At the same time, and very importantly, individual goals for women have become as important as family goals, which inevitably leads to lower fertility.

From the 1960s onwards, too, more women were entering the labour market and increasingly remaining in the labour market while rearing children. At the same time, gender inequities in childrearing responsibilities have not disappeared, and women continue to bear most of the burden. The resulting conflicts for women between paid and unpaid work are reflected in smaller family sizes. There is not, though, a straightforward relationship between fertility and female labour-force participation.

Activity 9 Examine Table 3.1 and select examples of countries where there appears to be a relationship between paid work and fertility and some where there does not. Can you suggest any explanations?

Table 3.1 Fertility and female labour-force participation (ranked by proportions of women in paid work), 1988

Country	FLFP	TFR
	(%)	*(%)*
Sweden	80.1	1.96
Denmark	78.3	1.56
Finland	73.0	1.7
Norway	72.8	1.84
UK	63.5	1.84
Portugal	59.1	1.53
France	55.7	1.82
Germany	54.4	1.42
Netherlands	51.6	1.55
Belgium	51.4	1.57
Luxembourg	47.6	1.51
Italy	43.9	1.34
Greece	43.4	1.52
Spain	39.4	1.38
Ireland	37.6	2.17

Note: FLFP = female labour-force participation; TFR = total fertility rate.

Source: adapted from Sundström and Stafford, 1992, p. 207, Table 1

It is usually assumed that greater labour-force participation lowers fertility, but this is not invariably the case, for fertility can also be influenced by child-care facilities and the wage levels of women. As Table 3.1 shows, the countries of Southern Europe, with the lowest levels of fertility, also have much lower levels of female labour-force participation than those of Northern Europe.

It has also been suggested that the population/environmental movement, with its concerns about the 'population explosion', which developed in the late 1960s may have encouraged people to have fewer children. It may have contributed to changing value systems and so helped to legitimate either the very small or even the childless family. There were certainly changing values towards contraception – as John Caldwell has said: 'By the late 1960s, the use of contraception was almost a moral priority, a very different position from the private and somewhat guilty practice which was typical in the early decades of the century and persisted until after the Second World War' (quoted in Preston, 1986, p. 181).

Organized day-care facilities are well developed in Northern Europe and enable more women to participate in the labour market

Activity 10 Think about the changes that have taken place in European fertility and some of the suggested reasons for decline in fertility. Do you think there are any lessons to be learnt from Europe which could be applied to developing world countries to help reduce fertility? Make a brief note of your response.

3.3.5 Changing lifestyles

The decline in fertility since the mid-1960s has been dramatic and unexpected. We can also see a range of related socio-demographic changes which suggest fundamental changes in what individuals expect or want from life. Lifestyles and demographic trends interrelate closely and this is perhaps most clearly seen when we look at the units in which people live: the family and household.

The traditional household of married couple and dependent children is no longer the norm, and has been replaced by a variety of household and family forms. These include people living alone, unrelated people living together, step-parents and children, lone-parent families and cohabiting couples, as well as the more traditional married couple with or without children. There is a much greater diversity of living arrangements than ever before.

Activity 11 Have a look at Figure 3.8. What is the most common household form in Britain? Consider your acquaintances: what types of living arrangement do they have?

There still tends to be the assumption among, for example, commentators or advertisers, that the two-parent, two-child family household is the norm, and yet, as Figure 3.8 shows, this is clearly not the case.

The changes which have occurred in families and households have been rapid since around 1970 and they reflect the decline in marriage rates, the rise in divorce rates, the growth in cohabitation and the rising numbers of births outside marriage which have occurred in much of Europe since this date.

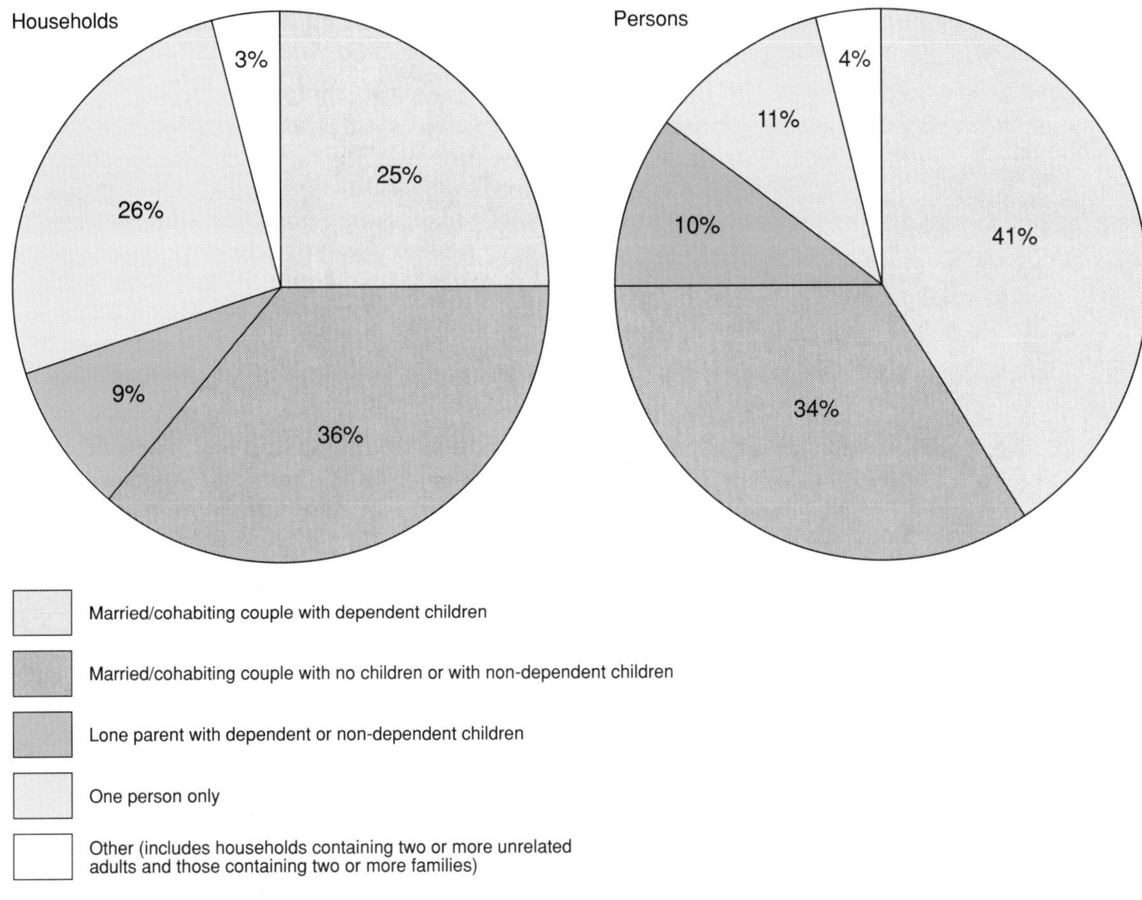

Figure 3.8 Households and people by type of household, Great Britain, 1991 (Source: OPCS, 1993, p. 13, Figure 2D)

Note: Households categorized by the type of family they contain. In the lone-parent and married-couple households, other individuals who were not family members may also have been present.

Cohabitation, for example, has become much more frequent in Northern and Western Europe since the late 1960s and appears to be part of the same process of making the family a more open system as well as reflecting changing attitudes among women. As cohabitation has increased, marriage rates have declined and age of marriage has risen – most cohabiting couples eventually marry, apart from in Sweden. But we can ask whether this will be the case in a decade's time. Will couples continue to marry? Or will relationships become more impermanent and negotiable? If women no longer see marriage and childbearing as the main source of self-fulfilment, will they decide they no longer wish to marry but maintain more control within informal relationships?

Along with the fall in marriage rates there has been a rise in divorce rates which again has been especially concentrated in the countries of the north and west of Europe. Where legislation governing divorce is restrictive, as in Southern European countries, rates are still much lower. The relaxation of legislation controlling divorce has certainly made divorce easier, but the real explanations of rising rates have to be sought in changing lifestyles and changing values. Marriage is no longer necessary to ensure economic or even reproductive survival. Women do not need men for economic security and they can and do bear and rear children on their own.

Associated with the other changes and perhaps the most interesting and dramatic illustration of changing lifestyles is the rise in births outside marriage. Marriage is no longer the essential precursor for bearing children in many European countries. Women who have inadvertently conceived may choose to have an abortion or may decide to bear the child without feeling it necessary to marry the father. Many women choose to conceive and bear a child outside marriage. The rise in extra-marital births began in the 1970s from around 5 per cent of all births in the countries of the European Community (EC) to 16 per cent in 1988 (see Figure 3.9). Rates are very high in some countries: in Sweden 52 per cent of all births are extra-marital, although most of these are to cohabiting couples.

In spite of the rapidity of socio-demographic changes, some aspects of lifestyles appear to be changing less rapidly than others. In particular, for women the burden of work has increased enormously as they have entered the paid labour force in increasing numbers. However, all studies show very little increase in the amount of domestic tasks undertaken by men, so women today face a double burden. They have to juggle competing claims as the main carers both for children and for elderly relatives and simultaneously undertake paid employment. Even so, little real attempt has been made to make it easier for men as well as women to combine their roles as parent with that of paid worker (or to encourage men to do more housework).

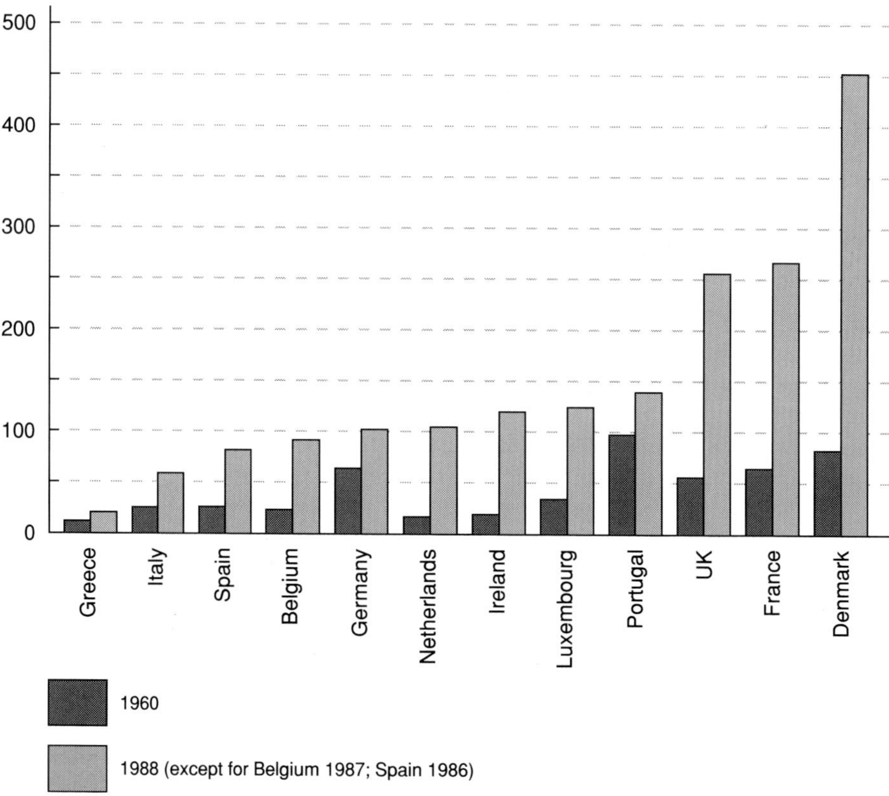

Figure 3.9 *Births outside marriage per 1,000 live births in the EC, 1960 and late 1980s (Source: Hall, 1993, p. 8, Figure 1)*

Summary of section 3.3

- Europe's population growth rate is now very low with both mortality and fertility declining, to reach very low levels today.
- The decline in fertility has been dramatic and is below replacement level everywhere in Europe. The explanations are not simply economic ones.
- Higher standards of living have resulted in couples wanting fewer children, but social factors are also important.
- The decline in fertility has been accompanied by unprecedented changes in lifestyle and living arrangements.
- The changing role of women and women's changing aspirations can be singled out as particularly important in explaining both the decline in fertility and the associated lifestyle changes.

3.4 The impact of Europe upon the world

In section 3.2 the importance of the rest of the world in containing European population growth within sustainable limits, both by providing an outlet for surplus population and providing food for the growing population, was apparent. European population growth was a product of the globalization process which today helps to explain the rapid population growth of the developing world. For Europe, the process was almost entirely beneficial and continues to be so.

In this section we look at the various ways Europeans have affected other parts of the world and the extent to which European prosperity is dependent upon the rest of the world.

3.4.1 The demographic impact

The Europeanization of the New World was of great importance in the development of European prosperity. Emigration has already been referred to as important in releasing demographic pressures, especially in Europe in the later nineteenth century, and, indeed, in stimulating the development of the New World. For example, one-third of the population growth in the USA between 1850 and 1910 came from the immigration of working-age people from Europe. But for Europeans the rest of the world was much more important than just providing areas for overflow population, and the impact of Europeans on the rest of the world was much greater than just as migrants. demographic impact

The European-settled lands of North America, Australia and New Zealand, parts of South America and South Africa, provided cheap food for Europe's growing population and other raw materials for its industries. Both were vital for European economic development in the later nineteenth century. Europe would not have been able to feed adequately its growing population from its own land, let alone keep basic food prices at a sufficiently low level to ensure there was a surplus income for consumption of other goods and services supplied by the rapidly growing economy. Even those parts of the world

Emigrants to the West, *USA, nineteenth century*

deemed unsuitable for European settlement were colonized for their raw materials while the people provided overseas markets for European goods.

The effect was much less beneficial, and, indeed, often devastating for those areas affected by European colonialization. The arrival of Europeans in many regions of the world resulted in demographic and cultural disaster for the indigenous populations. In the Americas the effects of European contact were catastrophic, particularly in Central and South America which were much more densely populated than the North. Estimates of depopulation in Mexico, for example, vary, with one estimate suggesting a decline from 25 million in 1519 (although other estimates put this figure at 10 million) to just over one million by the end of that century, recovering to just under 4 million by the end of the eighteenth century. The indigenous peoples in what is now the USA are estimated to have declined from around 5 million in 1500 to 60,000 three centuries later. They were devastated by epidemics of infectious diseases – smallpox, measles, influenza, tuberculosis, typhus and chickenpox – against which they had no immunity, as well as by genocide and the effects of forced labour (United Nations, 1973; Livi-Bacci, 1992).

Forced migration was also important for many regions as a result of European imperialism. The new colonies in the Caribbean, the southern parts of North America, and South America needed labour for the plantations and industries that were developing. In part, labour was provided by the slave trade which began in the fifteenth century, but did not reach significant proportions until the seventeenth and eighteenth centuries, and lasted until nearly the end of the nineteenth century in some regions. Although it is difficult to establish accurate figures, it has been estimated that some 10 million slaves were imported from Africa into the New World and that perhaps as many as two million more died on the sea jouney. The demographic and cultural impact on West Africa especially, was immense. Depopulation followed, not only as a result of the loss of people to slavery, but also as a result of the political and social disorganization which followed. Meanwhile, imported African slaves added consistently to the population of the Americas until around 1850 (see **Meegan, 1995***).

* A reference in emboldened type denotes a chapter in another volume of the series.

3.4.2 Colonialization

In Asia, colonialization resulted in relatively little permanent European settlement, but the exploitation of both agricultural and mineral resources was of great importance in that plentiful and cheap resources were made available for growing European industries; the ensuing economic and social impacts on regions so exploited, however, was enormous. In Malaysia, for example, both the tin mines and rubber plantations were developed by the British towards the end of the nineteenth century and the demand for labour led to the import of indentured or coolie labour from India to work on the rubber plantations, and from China to work in the tin mines, a demographic legacy with which Malaysia is still coming to terms.

Malacca, on the Straits in Malaysia, provides an example of the colonial imprint in Asia. In 1511 it was conquered by the Portuguese to gain control over the valuable spice trade, and in 1641 captured by the Dutch, again because of its importance as a trading centre. In the eighteenth century it came under the control of the British who founded the Straits Settlements, initially as a province of the East India Company from 1826 to 1858 and later as a crown colony until 1946. From there British colonial rule extended to the Malay States, and from a trading relationship the British extended their interest to the rubber plantations and tin mines.

In Bengal, there was a deliberate policy by the British to eliminate competition from the Bengal textile industry and to develop east Bengal as a supplier of agricultural raw materials for British industry.

Activity 12 Turn now to Reading E by Betsy Hartmann and James K. Boyce, entitled 'Riches to rags'. As you read, think about the following questions:

- What was the impact of the British on Bengal?
- How did Bengal contribute to British prosperity?
- What does the extract tell you about the British attitude towards their colonies?

For Europeans in the nineteenth century, the rest of the world was theirs to exploit and use as they wished and it undoubtedly contributed enormously to the development of the continent, especially in the north and west. In the later twentieth century, the relationships between Europe and the rest of the world are perhaps less easy to disentangle.

3.4.3 Trading relationships today

One way of looking at current relationships is to examine trading patterns. If we look only at the European Union (EU), then more than half of the Union's external trade is with other developed countries, and for a typical EU country more than three-quarters of its trade takes place within the wider area of Western Europe. A large proportion of trade with developing countries is with OPEC countries (Organization of Petroleum Exporting Countries) for oil.

Trading and other policies within Europe and other countries in the developed world do have an important impact on people elsewhere in the world. The Common Agricultural Policy (CAP), for example, has made it difficult for agricultural producers elsewhere to compete effectively in the

European market and, indeed, has depressed world prices in certain products: notably cereals, dairy products and sugar. So, for example, European sugar beet growers have been heavily subsidized to produce sugar to the detriment of the sugar cane industry in the Caribbean. At the same time, the lucrative European market has meant that some of the best land in some African countries, for instance, is devoted to producing luxury out-of-season fruit and vegetables for the European market rather than food for their own people. Meanwhile, surplus food produced by the EU has been exported to the rest of the world at the expense of other agricultural producers, albeit most notably those in other developed countries, and it has been argued that developing countries have benefited, at least in the short term, from subsidized imports of food (Tsoukalis, 1993). In the long run, though, it is of no real benefit if local farmers are not able to compete with cheap imported food.

Overall, in 1988 the developed countries took three-quarters of all food imports by value. The single most important import was meat, followed by dairy products and tropical beverages. This reflects the prosperity of the developed world. Cereals, however, are the single most important item in total in world agricultural trade, with North America and Australia the most important exporters. Again, the contrast between the developed and developing world is striking: developing countries import mainly food grains, while developed countries import animal feed grains which are converted into meat, which is an extravagant use of resources.

3.4.4 Consumption patterns and environmental impact

Europeans, together with the population of the rest of the developed world, do have an inordinate impact on the world's resources, and their (our) lifestyles are ultimately responsible for a large proportion of environmental degradation.

Box 3.1 Resource consumption in the developed world

25 per cent of the world's population (that is, the developed world) consume:

- 75 per cent of all energy used;
- 79 per cent of all commercial fuels;
- 85 per cent of all wood products; and
- 72 per cent of steel production;

and generate:

- 75 per cent of all carbon dioxide emissions which account for nearly half of all greenhouse gases;

but the impact of such emissions as a result of global warming, producing climatic changes affecting agriculture, is likely to fall disproportionately on developing countries.

Source: adapted from UNFPA, 1991, p. 14

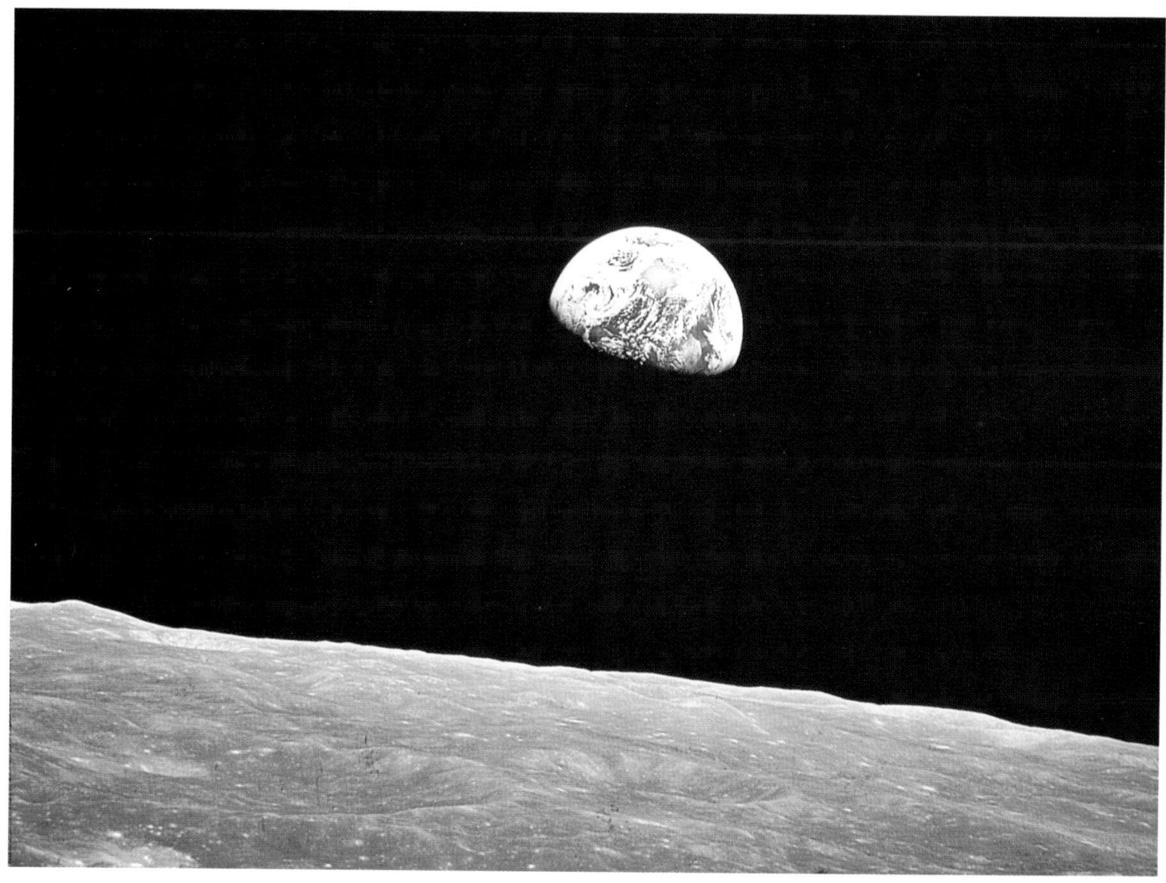

Plate 1 Earthrise as seen from above the surface of the moon, photographed by astronauts on board the Apollo 8 spacecraft as it orbited the moon in 1968. The lunar horizon is approximately 780 kilometres from the spacecraft. On earth, 384,000 kilometres away, the sunset terminator bisects Africa. The unnamed surface features in the foreground of the picture are near the eastern limb of the moon as viewed from earth. This image, locating the jewel-like earth in empty space and contrasting it with the barren surface of the moon, was a key stimulus in the development of environmentalism. By courtesy of NASA/Science Photo Library

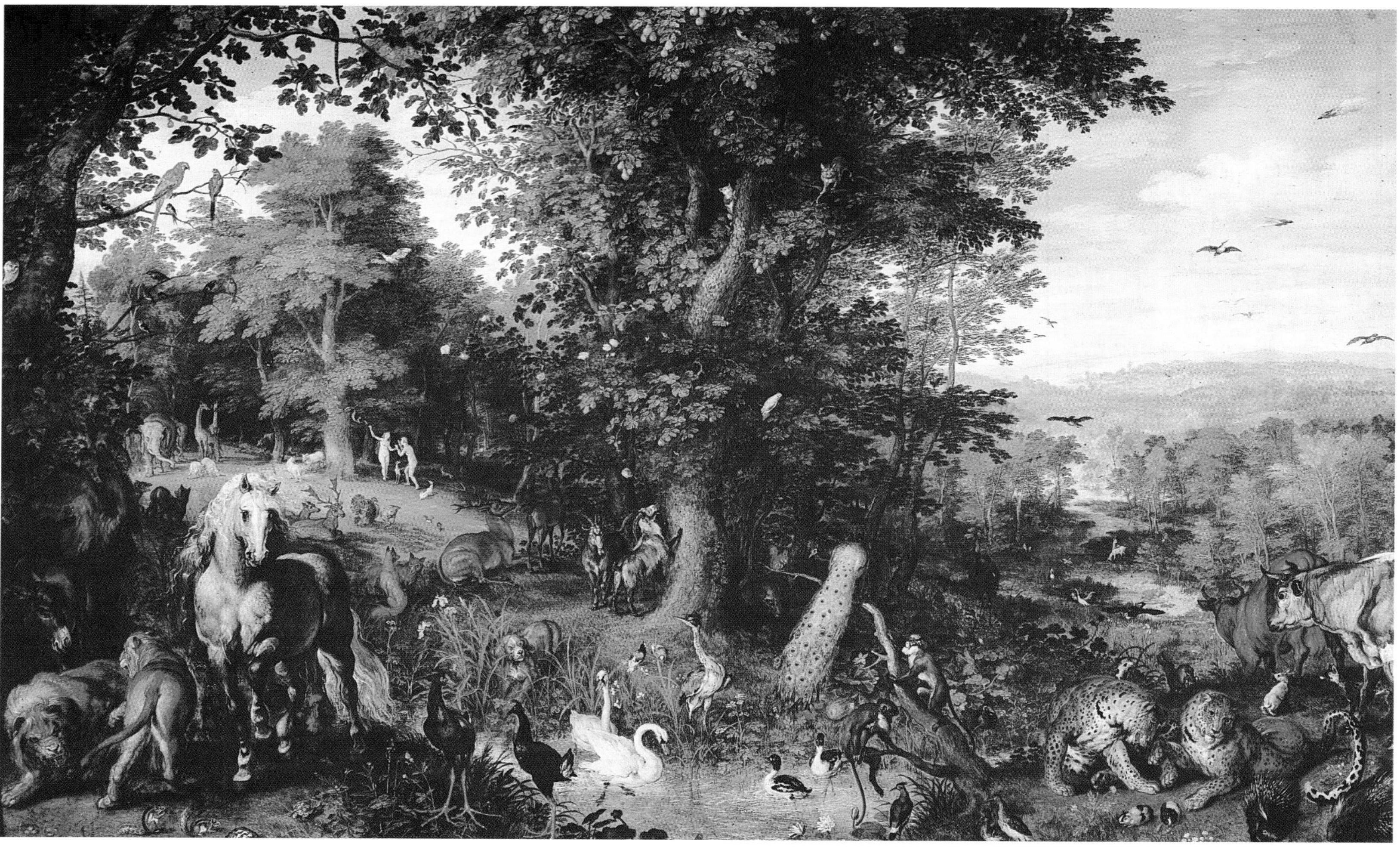

Plate 2 *Jan Brueghel the Elder,* The Garden of Eden, *copper, 50.3 x 80.1 cm. As a crucial step in Christianity's account of humanity's relation to nature, Eden is a frequent subject for Christian painters. The juxtaposition of European trees and domestic animals with exotic animals, mostly carnivores, is a very European representation of a Mesopotamian paradise. By courtesy of Galleria Doria Pamphilj/Foto Alessandro Vasari, Rome*

Plate 3 *Lorenzo Lotto,* Andrea Odoni the Venetian Banker holds Diana-Natura in his hand, *1527. Diana, Roman goddess of wild animals and the hunt, had connections through Artemis, her Greek equivalent, back to early nature goddesses. Thus, as well as his possession of antiquities, the painting symbolizes capital's acquisition of nature. By courtesy of The Royal Collection. © Her Majesty Queen Elizabeth II*

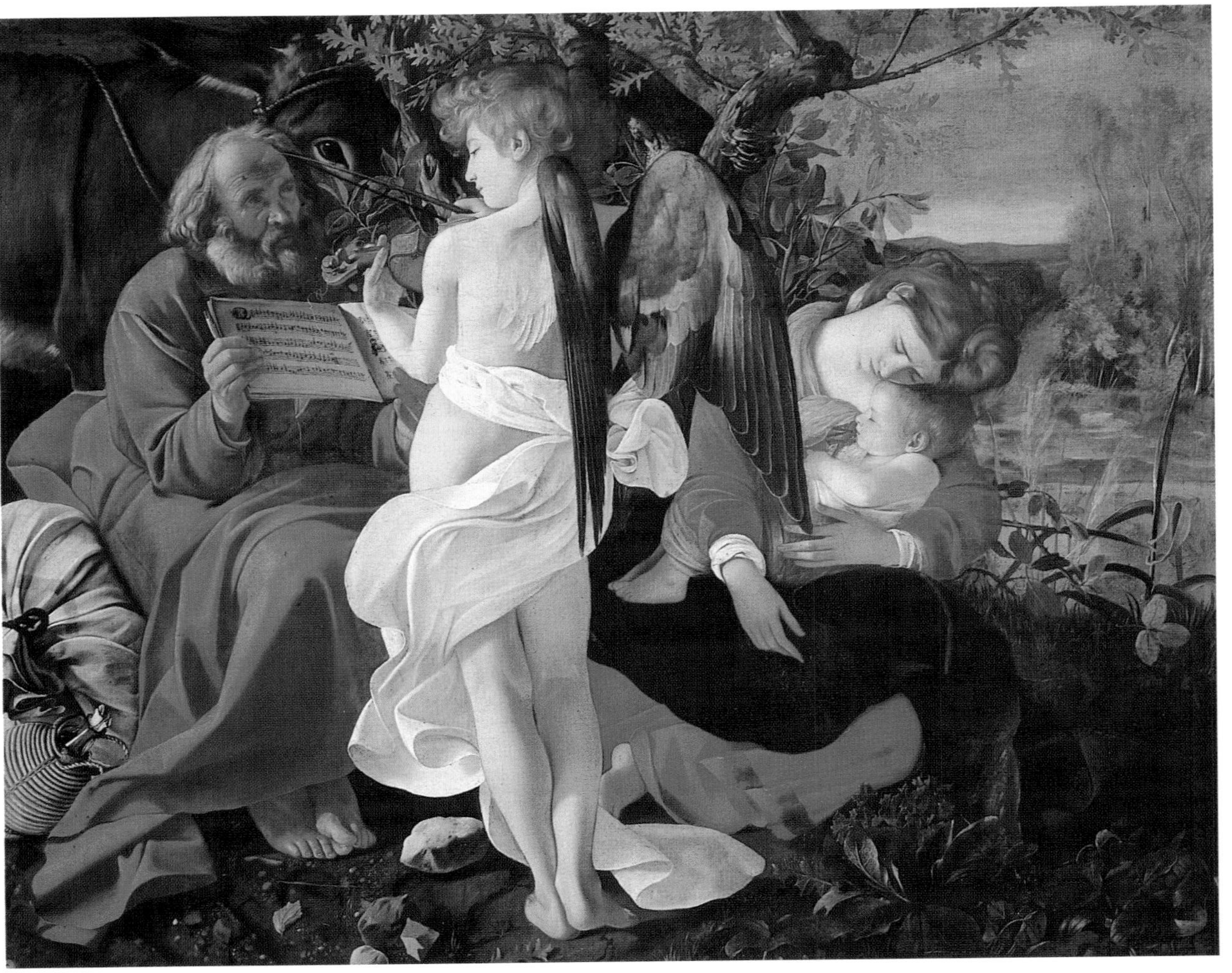

Plate 4 *Caravaggio,* The Rest on the Flight into Egypt, *canvas, 135.7 x 166.5 cm. The striking composition of this picture places the angel, symbolizing the church and playing celestial music, between man and woman. Man is associated with inorganic objects, manufactured products and the domesticated ass; whereas woman is associated with child and luxuriant nature. By courtesy of Galleria Doria Pamphilj/Foto Alessandro Vasari, Rome*

Plate 5 *Burkina Faso, West Africa. Yatenga Province, Kalsaka village. Erosion. Gullies appear as the water, which should be absorbed by the soil, runs off in channels. A single tree protects some of the soil. By courtesy of Mark Edwards/Still Pictures*

Plate 6 *Burkina Faso, West Africa. Yatenga Province, Kalsaka village. Women returning with fuelwood. They walk for an hour to an area where they can find wood. They have to make two or three visits a week. By courtesy of Mark Edwards/Still Pictures*

Plate 7 *Samuel B. Waugh,* The Bay and Harbor of New York, *1855, oil on canvas, 8 ft. 1 in. x 16 ft. 6 in. Arrival of immigrants from Europe. The chest in the right foreground carries the name Murphy. Courtesy of the Museum of the City of New York. Gift of Mrs Robert M. Littlejohn*

Plate 8 *Rubber tapper in the Amazon. This modern tapper both illustrates the way natural rubber has gone back to being a local resource of little international interest and symbolizes attempts to find sustainable uses for the rainforest. By courtesy of Mark Edwards/Still Pictures*

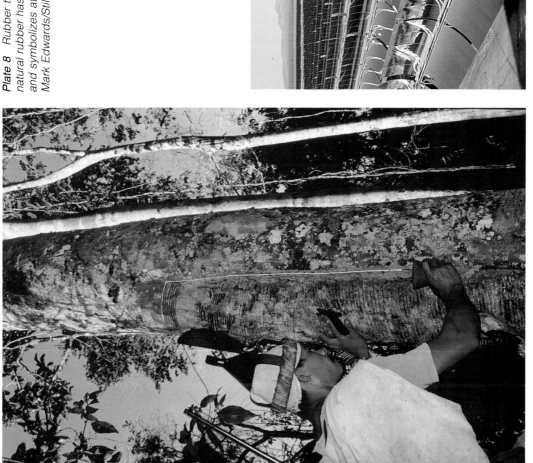

Plate 9 *Luz Solar energy plant, USA. By courtesy of Daniel Dancer/Still Pictures*

Plate 10 *Lake Nasser and the Aswan Dam, seen from the space shuttle. The Aswan Dam is the barrier near the top of the picture and separates the lighter waters of the lake from the dark blue of the River Nile. The fluctuating water level of the lake is an environmental index of drought in north-central Africa. The lake level in this photograph has decreased compared to 1981, as is shown by the exposed areas of lakebed, notably in the lower left corner of the picture. The photograph was taken from the space shuttle Discovery during the STS-26 mission in September/October 1988. By courtesy of NASA/Science Photo Library*

Plate 11 *Bingham Copper Mine, near Salt Lake City, Utah, USA. By courtesy of J. Allan Cash*

'The environmental impact of our population, now numbering 5.5 billion, has been vastly multiplied by economic and social systems that strongly favour growth and ever-increasing consumption over equity and poverty alleviation ...' (Postel, 1994, p. 5). It has been estimated that the share of global income that went to the richest 20 per cent of the world's people increased from 70 per cent in 1960 to nearly 83 per cent in 1989 (Postel, 1994). The environmental problems facing the world have their origins among the population of the developed world with their high consumption of energy, raw materials and manufactured goods. Even in apparently efficient enterprises there are environmental costs. For example, European agriculture is extremely productive and food surpluses rather than shortages are the major problem. But the costs of production are high and excessive fertilizer and other agricultural practices are resulting in land degradation: one estimate is that 23 per cent of Europe's vegetated land is degraded – a figure comparable with that of Africa (Table 3.2).

Table 3.2 Human-induced land degradation worldwide, 1945 to present

Region	Over-grazing	Deforest-ation	Agricultural misman-agement	Other *	Total	Degraded area as share of total vegetated land
			(million hectares)			(%)
Asia	197	298	204	47	746	20
Africa	243	67	121	63	494	22
South America	68	100	64	12	244	14
Europe	50	84	64	22	220	23
North and Central America	38	18	91	11	158	8
Oceania	83	12	8	0	103	13
World	679	579	552	155	1,965	17

Note: * Includes exploitation of vegetation for domestic use (133 million hectares) and bio-industrial activities, such as pollution (22 million hectares).

Source: Postel, 1994, p. 10, Table 1-3

Overall, consumption is a way of life for Europeans and North Americans. The ecological impact of the 1.1 billion people in the developed world is as great, it has been estimated, as would be the impact of 17 billion people living at Indian or Chinese standards or 77 billion living at Bangladeshi standards (Suzuki, 1993, p. 12). The following quotation highlights the contrast between Britain and Bangladesh:

In this country, we have not even asked how many people are good for Britain (or how many people are good for the World). We produce an extra 116,000 persons per year. By contrast, Bangladesh produces an extra 2.7 million persons. But because the fossil-fuel consumption of each new

British person is more than 30 times that of a Bangladeshi, Britain's population growth effectively contributes 3.9 times as much carbon dioxide to the global atmosphere, and hence to global warming, as does Bangladesh's – and Bangladesh will suffer far more through global warming than will Britain. The average British family comprises two children, but when we factor in resource consumption and pollution impacts, and then compare the British lifestyle with the global average, the 'real world' size of a British family is more like 10–15 children.

(Myers, 1994, p. 5)

How economies will develop next century is unknown, but the World Bank, for example, estimates that incomes will rise so that by the middle of the century developing countries' share of world income could have increased from less than one-quarter to almost a half (World Bank, 1992, p. 33). But even if this is so, income per head in high-income countries will still be many times higher than even in the richest of present poorer countries (see Figure 3.10). If incomes rise in high-income countries as they are projected to do by the World Bank, then it is likely that lifestyles and consumption patterns may become even more extravagant in Europe and North America.

Overall, during the next century the world should be a much richer place, but the question that arises is: will the environment be poorer?

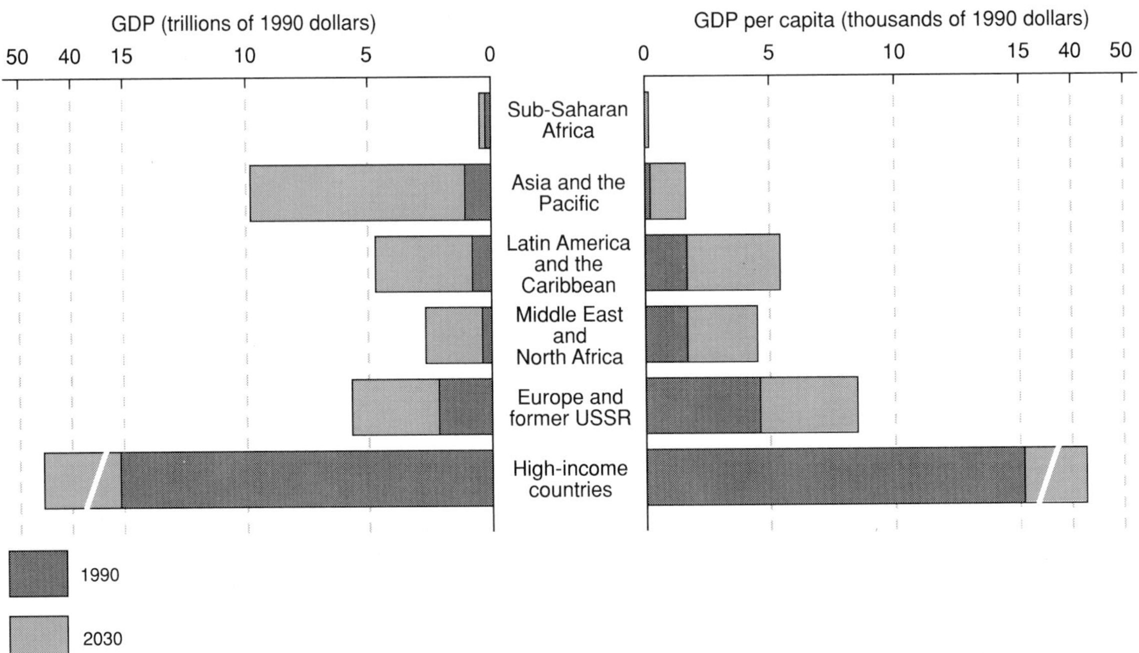

Figure 3.10 GDP and GDP per capita in developing regions and high-income countries, 1990 and 2030 (Source: World Bank, 1992, p. 33, Figure 1.3)

Note: GDP = gross domestic product. Data for 2030 are projections.

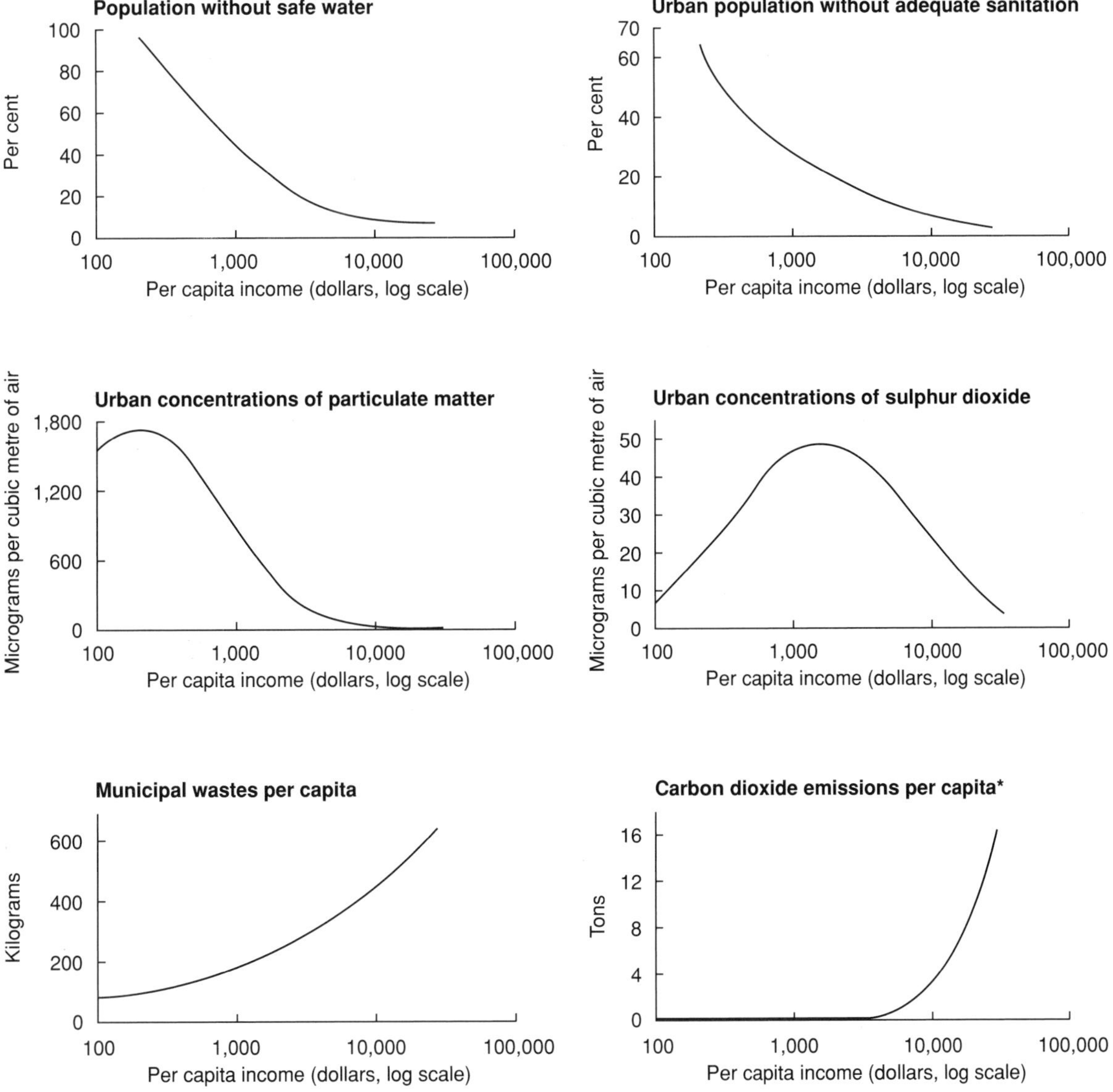

Figure 3.11 *Environmental indicators at different country income levels (Source: World Bank, 1992, p. 11, Figure 4)*

Note: Estimates are based on cross-country regression analysis of data from the 1980s. * Emissions are from fossil fuels.

Activity 13 Study the graphs in Figure 3.11. Which indicators improve with rising per capita income and which deteriorate? Suggest reasons why this should be so.

Some problems decline as income increases since increasing income provides the resources for public services. Some problems worsen to begin with then improve as income rises; examples here are for air pollution, but water pollution also fits into this category. But this only happens if countries decide to introduce policies to deal with environmental problems.

Some environmental indicators of stress worsen as income rises: municipal waste and carbon dioxide emissions are examples of these. In these cases, costs to produce lower levels are very high and the costs associated with the wastes or emissions are not yet perceived as high; only when costs are perceived as high will policies change (World Bank, 1992).

Benidorm, on Spain's Mediterranean coast

The question of conservation of rural environments is also problematic and in Europe we have seen continued degradation of natural environments; whether it be around the Mediterranean for the tourist industry or road building in areas of great scenic beauty and scientific interest in the UK, such as, for example, across Twyford Down in Hampshire. While conservation is given such low value, environmental degradation will continue, and Europe shows the way. The environmental impact of European resource demands elsewhere in the world is equally considerable, but difficult to quantify. Tropical hardwood logging is one obvious example.

Summary of section 3.4

- Europeans have had an enormous demographic, economic and cultural impact on the rest of the world as a result of emigration and colonialization.
- Today, Europeans continue to have an impact on other regions of the world through trade and economic policies.
- The levels of economic development achieved by Europeans have resulted in high levels of consumption and a market for products from all over the world.
- As a result, Europeans have an environmental impact not only on their own continent but throughout the world.
- Europe is sufficiently wealthy to deal with some at least of the environmental problems it has created at home, but it contributes little to controlling problems it produces elsewhere in the world.

3.5 A no-growth population

In this section we consider what the implications are for Europe and for the world of the European population which is no longer growing and will soon be declining. Again, the problems facing Europe are likely to be facing other regions of the world in the next century. The projected no-growth population in Europe has to be set in the context of the very high population densities found in Europe, and the already existing problems faced by crowded countries such as the UK.

In the developed world – Europe, North America and Australia – population growth rates are now very low. Europe has the lowest growth rates of all, and some European countries, including Germany and Hungary, are experiencing net natural decline. As discussed in section 3.3, death rates are very low indeed in most of Europe, but birth rates are also below *replacement level* in almost every European country. The contrasts with the developing world are sharp, and are no more evident than around the Mediterranean, as suggested in section 3.1, where, on the European fringe, some of the lowest rates of natural increase in the world are recorded: Italy 0 per cent and Spain 0.1 per cent, for example, while North Africa as a whole has a natural increase of 2.6 per cent. The Mediterranean marks one of the world's major demographic fault lines: areas of rapidly

replacement level (fertility)

growing populations, adjacent to the slowest-growing populations in the world. Compared to Africa, Europe is clearly advantaged, but does this mean that Europe's population poses no problem in a rapidly crowding world?

3.5.1 Age structures

Europe has the oldest population structure in the world as a result of declines in both fertility and mortality. Age structures are the living record of past population dynamics: past levels of fertility, mortality and migration. Today 15 per cent of the population of the countries of the EU are aged over 65. The countries of Southern Europe, where the birth rate remained high longer, have younger age structures than countries of Northern Europe. Ireland and Poland have the most youthful age structures with only 13.1 and 10 per cent of the population, respectively, aged over 65, compared with Denmark or the UK with 15.4 per cent or Norway with 16.4 per cent. Population projections assume continuing improvements in mortality and low levels of fertility which therefore result in increasing proportions of older people.

By 2025, according to UN projections, many countries will have more than 20 per cent of their population aged over 65 (United Nations, 1990). By the middle of next century 30 per cent of Europe's population could be aged over 65. How will Europe cope with this increasing proportion of older people?

3.5.2 Problems of no-growth populations

UN projections suggest that by 2005 to 2010 the population of Western Europe will be experiencing absolute decline (United Nations, 1990). In Southern Europe decline will set in by 2010–2015. Northern and Eastern Europe will continue to experience very slow growth. Europe as a whole will have a declining population of 0.01 per cent by 2015–2020 and –0.05 per cent by 2025.

What are the likely problems of a no-growth population? To a large extent they are concerned with age structures which are taking on a historically unique dimension in Europe, and which will be a challenge to health,

ageing populations

social and economic policies. At a relatively simple level, *ageing populations* imply that fewer people of working age have to support a larger dependent population, with the immediate problems of shortage of new entrants to the labour force, and social security provision for people of pensionable age.

Some would argue, therefore, that the growing proportions of older people and diminished numbers of young people present major problems for a population, which can be alleviated only by rising birth rates. This was the view in the 1930s. In the long run, though, and this applies to the world as a whole as much as to Europe, populations have to adapt to older age structures. Eventually, if birth and death rates remain constant then a stable age structure will result in which the proportions of each age group remain the same. But in the short term, populations have to adapt to fewer people entering the labour force, and perhaps, therefore, a shortage of labour unless technology intervenes. It may be too pessimistic to view those people aged over 65 as a dependent population. If life expectancies continue to improve then it is realistic to assume that length of productive, active life can also be increased, albeit in different ways from that of those aged 30 or 40. We need to take a much more flexible view of chronological age than has been the case up until now.

Policy response to ageing populations can include changing retirement ages and, indeed, reconsidering the concept of retirement. This is already on the UK agenda where the government is being required to equalize retirement age for men and women. Rather than lowering retirement age for men, the government has decided to increase the retirement age of women, with the saving of state pension payments this implies.

Other problems may lie in the perceived inherent conservatism of older age groups, if indeed this is the case. What needs to be borne in mind is that the new generations of older people will be better educated, wealthier and almost certainly more confident than any previous older generation. They will also be more numerous and need not be a force for inertia. In the future more people will live to very advanced ages: people of 80 and 90 and even 100 years old will become more numerous in the population and we have to concentrate on ways of improving the quality of life and expanding the length of active life.

One response to a no-growth population would be to encourage immigration, and this may indeed be the case in the future. In the past, immigration has been seen as a solution to labour shortages: for example, in France in the 1920s when population growth was negligible, or most of North-western Europe after 1945 when labour migration was encouraged up until the oil crisis of the early 1970s. We can only speculate about what future attitudes to immigration might be in Europe, taking into account current attitudes and laws which seek to restrict migration as much as possible in most European states. But migration is an issue facing Europe both within and from outside the continent. There is a growing movement of undocumented or illegal migrants, mainly from Africa but also from elsewhere, especially to Southern European countries. The countries of North Africa are at present the main source of migration into Europe and the pressures from this high-growth region adjacent to Europe are likely to increase, as are pressures from other regions of the world. Should Europe be isolated from the rest of the world or should it see itself as part of a wider world in which it exports expertise and welcomes newcomers? Is there really room for a 'fortress Europe' in the world of the twenty-first century? One example of contemporary European attitudes towards immigration is shown by the newspaper article reproduced overleaf.

Although Europe's population is the slowest growing in the world, Europe is still, and will continue to be, very densely populated. There is every reason to think that the material expectations of Europeans will continue to rise and consumption rates continue to increase. This does pose a problem for the rest of the planet and we must ask whether Europeans will be able to continue to enjoy the very high standards of living they do today. Will they be able to compete in a rapidly changing world where the economic and demographic centre of gravity is moving ever more rapidly towards Asia?

Pasqua in new war on immigrants

JULIAN NUNDY in Paris

CHARLES PASQUA, the French Interior Minister, has picked one of the country's toughest police chiefs to run a new agency to fight illegal immigration.

Mr Pasqua, a Gaullist, said Robert Broussard would head the new Central Directorate for Immigration Control, scheduled to start work on 15 January. Mr Broussard, a *préfet de police,* has previously been put in charge of operations against Corsican nationalists, organised crime and terrorism. Often criticised and disowned by his political bosses, he has a tendency to disappear, only to re-enter the limelight every few years. Never politically correct, he is simultaneously loathed, feared and admired for being uncompromising towards his adversaries.

As Mr Pasqua announced Mr Broussard's latest incarnation, he told a television interviewer that the government was obliged "to take coercive and administrative measures" against illegal immigration.

With six land frontiers and three coasts, France is particularly vulnerable to illegal immigration. The fears have become particularly acute since the rise of Islamic fundamentalism in Algeria, prompting talk of "boat people" from France's former North African colonies.

The conservative coalition elected last March promised a tough attitude on immigration and Mr Pasqua and his colleagues have brought in new laws on immigration, political asylum and French nationality to tighten up procedures. "Once we have sent back several planeloads or boatloads of immigrants," he said, "the world will get the message."

Mr Pasqua's use, when he was interior minister in 1986, of a charter plane to take 101 Malians home made such talk virtually taboo. It was revived in 1991 by Edith Cresson, the then Socialist Prime Minister, who suggested the use of special flights to expel illegal immigrants. Such was the outrage against her statement that the idea was again dropped.

As well as Third World immigration, France is prey to immigration from central and east Europe. Mr Pasqua said France could "not take on the misery of the whole world". If civil war developed in eastern Europe, France would suffer "a new wave of immigration". To the south, meanwhile, "by the end of the decade there will be 130 million North Africans, including nearly 60 million under 20, with no prospects and, further south, 1 billion other Africans".

Mr Pasqua acknowledged that tough measures alone were not enough and said France should lead "a development crusade" in the Third World. He suggested a figure of one per cent of gross domestic product.

Over the first 10 months of last year, 568 people were deported from France, a 21 per cent increase over the previous year. With 4.1 million foreigners living in a country of 57 million people, immigration is popularly seen as a prime cause for unemployment which is running at just over 3 million.

Source: *The Independent*, 7 January 1994

It could be argued that Europe is a continent running out of creative energy, whereas the developing parts of Asia, particularly the Pacific rim countries, appear to be imbued with the desire to grow economically and outperform the present developed world. The contrast between a country such as Malaysia, where the whole population has a target of development for the year 2020, and the UK, for example, which has no clear vision for the future, could hardly be more marked. There is still a large gap in terms of GNP between such countries and those of Europe or North America, but the pace of growth is so rapid that many are likely to achieve their targets.

Some people welcome the idea of a declining population in Europe and, indeed, would argue that numbers need to decline rapidly to reach an optimum population for the continent. But problems of defining an optimum are huge and much depends on personal views on what quality of life is desirable, how regions should interact with each other and, indeed, how resource-bases should be defined. On the other hand, there are some demographers within Europe, and indeed some policy-makers, who would argue that Europe needs a growing population and that policy steps should be taken on a Europe-wide scale to encourage reproduction. Such a policy would need acceptance of the fact that childcare and childrearing are not only the responsibilities of the parents, but should be shared by the wider society since without a renewal of the generations a society has no future.

Europe has seen enormous demographic changes over the last 200 years. It has been in the forefront of economic and social changes and it is in Europe that we can see the concluding phases of the demographic transition. The demographic, and indeed the economic, future of Europe is uncertain, and open to wide speculation. What *is* certain is that the problems relating to population which Europe faces are no less complex than anywhere else in the world.

Summary of section 3.5

- Europe's population is the first anywhere in the world to approach stability and, indeed, is likely to decline in the next century.
- Europe's population is ageing and by 2025 as much as one-fifth of the population will be aged over 60.
- The problems implied by this are both policy related and cultural.
- It is difficult to predict the future nature of European society. It would seem likely that it will continue to be consumption oriented, but such lifestyles may be difficult to sustain in an ageing, no-growth population.
- Immigration could be a solution, but it seems unlikely that barriers to incomers will be relaxed.

3.6 Conclusion

In this chapter we have seen how Europe's population grew from the eighteenth century onwards, reaching high rates of growth in the nineteenth century, and how these growth rates began to decline in the early twentieth century, reaching very low rates today. Europe was the first population growth centre in the world, growth then moving to other regions in the twentieth century. We have also examined some of the impacts of European population growth on the rest of the world.

European population growth in the eighteenth and nineteenth centuries is not comparable with that of the twentieth century for a number of reasons. Rates of growth were never as high in Europe as they have been and still are, in many regions of the world. Population growth was closely related to agricultural and industrial change and, for most of the period, economic development kept pace with population growth. At the same time, European populations had the resources of the world to call upon, which they did without compunction.

There were great pressures in European societies in the nineteenth century. Conditions in towns were appalling and there was agricultural distress, but there were some avenues of escape, via overseas migration, for some of the people at least. Fertility and mortality both declined relatively slowly. Mortality remained high until people began to understand the causes of disease – for example, the relationship of disease with water supplies, sanitation and housing. Fertility declined as couples wanted to reduce the size of families, which they often achieved without the use of contraceptives. Today in Europe, not only are fertility and mortality both at an all-time low, but the way people organize their lives is changing. They no longer rely on the traditional married unit as the only household form, but increasingly lifestyles favour individualistic modes of living: cohabitation, births outside marriage and high divorce rates are some manifestations of this process of changing lifestyles.

Perhaps the most significant change of all is the ageing of Europe's population, as population numbers reach stability and are projected to decline next century. With an ageing population Europe is entering uncharted demographic territory and we cannot predict how European society will look in the future.

Are there lessons to be learnt from the European experience that could be taken up by the rest of the world? There are probably very few. In many regions of the less developed world population growth has been independent of economic growth. Mortality rates have declined rapidly since the end of the Second World War, often as a result of the introduction of outside expertise and an understanding, gained from Europe, of the causation of diseases; and often with almost no economic growth, leading some commentators to wonder whether such low rates of mortality can be maintained without other economic improvements.

Similarly, over the last 20 or 30 years, fertility has also started to decline, often very rapidly, so decreases in average family size of five or six to two or three have been achieved in little more than a decade. Thailand, and even more so China, are examples of countries achieving very rapid declines in fertility. Such rapid declines were unknown in Europe where the decline in fertility was very slow by comparison. Populations in Europe had little access to effective contraception. Couples who now decide they want fewer children can achieve their aim relatively easily compared with Europeans in the late nineteenth or early twentieth centuries, and contraception has been adopted in many other regions of the world comparatively quickly. Yet the European experience also suggests that fertility can vary from place to place and group to group for a variety of not always easily explained reasons. Any attempt to limit population growth rates has to take into account the particular local cultures and circumstances; there is no uniform package for easy distribution.

Europe is facing an ageing population and fears the consequences. The other presently non-industrial regions of the world will also inevitably face ageing as growth rates decline and life expectancy increases, and the process will be much more rapid than has been the case in Europe, posing even more serious problems of adaptation. China is perhaps the most notable example. Fertility has been reduced from around six children in the late 1960s to around two today. Life expectancy is now around 70 years. China will be the most rapidly ageing population in the twenty-first century, which will produce a range and scale of problems unknown in Europe and at a much faster rate.

Even wealthy Europe has problems when it considers how to support an ageing population and whether its extravagant lifestyles are sustainable. Other regions of the world will see large increases in their population numbers before stabilization. At the same time, if incomes rise elsewhere, as they are projected to do, then demands on resources will inevitably increase.

One view could be that where Europe leads, the rest of the world will follow, but this can only be a very partial view. Europe had a unique set of circumstances and relationships resulting in a unique pattern of growth. Each region of the world will have its own particular patterns and associated problems. Europe can give some idea of the magnitude of the problems facing other regions of the world, but the lessons it can offer can only be limited ones. For Europe, the problems of population growth were relatively easy, the pressures never as great as those facing many other parts of the world today.

References

COALE, A.J. and WATKINS, S.C. (1986) *The Decline of Fertility in Europe: the Revised Proceedings of a Conference on the Princeton European Fertility Project*, Princeton, NJ, Princeton University Press.

COUNCIL OF EUROPE (1994) *Recent Demographic Developments in Europe, 1993*, Council of Europe Press.

FLINN, M.W. (1981) *The European Demographic System 1500–1820*, Brighton, Harvester Press.

HAJNAL, J. (1965) 'European marriage pattern in perspective', in Glass, D.V. and Eversley, D.E.C. (eds) *Population in History, Essays in Historical Demography*, London, Edward Arnold.

HALL, R. (1989) *Update. World Population Trends*, Cambridge, Cambridge University Press.

HALL, R. (1993) 'Europe's changing population', *Geography*, Vol. 78, Part 1, No. 338, pp. 3–15.

HARTMANN, B. and BOYCE, J.K. (1983) *A Quiet Violence: View From a Bangladesh Village*, London, Zed Books.

HUMPHRIES, S. and GORDON, P. (1993) *A Labour of Love: the Experience of Parenthood in Britain 1900–1950*, London, Sidgwick and Jackson.

LIVI-BACCI, M. (1992) *A Concise History of World Population*, Oxford, Blackwell.

McLAREN, A. (1990) *A History of Contraception*, Oxford, Blackwell.

MEEGAN, R. (1995) 'Local worlds', in Allen, J. and Massey, D. (eds) *Geographical Worlds*, Oxford, Oxford University Press in association with The Open University.

MYERS, N. (1994) 'Collision course', *Spa*, Spring, pp. 3–5.

OPCS (OFFICE OF POPULATION CENSUSES AND SURVEYS) (1985) *William Farr 1807–1883. The Report of the Centenary Symposium held at the Royal Society on 29 April 1983*, Occasional Paper 33, London, HMSO.

OPCS (OFFICE OF POPULATION CENSUSES AND SURVEYS) (1993) *1991 General Household Survey*, London, HMSO.

POSTEL, S. (1994) 'Carrying capacity: earth's bottom line', in Worldwatch Institute, *State of the World 1994*, London, Earthscan.

PRESTON, S.H. (1986) 'Changing values and falling birth rates', in *Below Replacement Fertility in Industrial Societies: Causes, Consequences, Policies*; *Population and Development Review*, a supplement to Vol. 12, pp. 176–95.

SANGER, M. (1932) *My Fight For Birth Control*, London, Faber and Faber.

SUNDSTRÖM, M. and STAFFORD, F.P. (1992) 'Female labour force participation', *European Journal of Population*, Vol. 8, pp. 199–215.

SUZUKI, D. (1993) 'Overpopulation is bad but overconsumption is worse', *Populi*, Vol. 20, No. 5, p. 12.

TSOUKALIS, L. (1993) *The New European Economy*, Oxford, Oxford University Press.

UNFPA (UNITED NATIONS POPULATION FUND) (1991) *Population and the Environment: the Challenges Ahead*, New York, UNFPA.

UNITED NATIONS (1973) *Determinants and Consequences of Population Trends*, Vol. 1, New York, United Nations.

UNITED NATIONS (1990) *World Population Prospects 1991*, New York, United Nations.

VAN DE KAA, D.J. (1987) 'Europe's second demographic transition', *Population Bulletin*, Vol. 42, No. 1, pp. 1–57.

WORLD BANK (1984) *World Development Report 1984*, New York, Oxford University Press.

WORLD BANK (1992) *World Development Report 1992*, New York, Oxford University Press.

WRIGLEY, E.A. and SCHOFIELD, R.S. (1981) *The Population History of England 1541–1871: a Reconstruction*, London, Edward Arnold.

Reading A: *Massimo Livi-Bacci, 'The space of growth'*

[...]

Fertility and mortality, acting in tandem, impose objective limits on the pattern of growth of human populations. If we imagine that in a certain population these remain fixed for a long period of time, then, by resorting to a few simplifying hypotheses, we can express the rate of growth as a function of the number of children per woman (TFR) and life expectancy at birth (e_o).

Figure A.1 shows several 'isogrowth' curves. Each curve is the locus of those points that combine life expectancy (the abscissa) and number of children per woman (the ordinate) to give the same rate of growth *r*. Included on this graph are points corresponding to historical and contemporary populations. For the former, life expectancy is neither below 15, as this would be incompatible with the continued survival of the population, nor above 45, as no historical population ever achieved a higher figure. For similar reasons the number of children per woman falls between eight (almost never exceeded in normally constituted populations) and four (recall that these are populations not practising birth control). For the present-day populations included in Figure A.1, control of fertility and mortality make possible e_o values of 80 and TFR of 1. Figure A.2 identifies specific examples within the more restricted boundaries of historical populations. These examples have varying degrees of precision, being in some cases based on direct and dependable observation, in others on estimates drawn from indirect and incomplete indicators, and in others on pure conjecture. Nonetheless, most of these populations fall within a band that extends from growth rates of 0 to 1 per cent, a space of growth typical of historical populations. Within this narrow band, however, the fertility and mortality combinations vary widely. Denmark at the end of the eighteenth century and India at the beginning of the twentieth, for example, have similar growth rates, but these are achieved at distant points in the strategic space described: the former example combines high life expectancy (about 40 years) and a small number of children (just over four), while in the latter case low life expectancy (about 25 years) is paired with many children (just under seven).

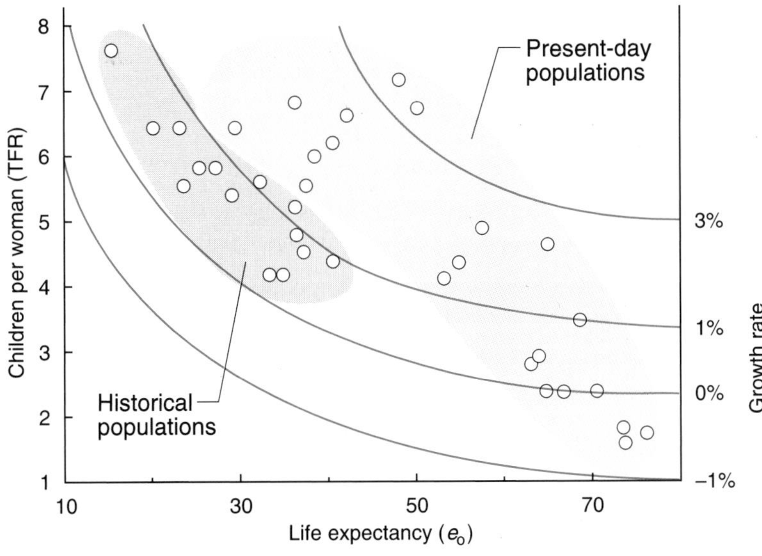

Figure A.1 Relation between the average number of children per woman (TFR) and life expectancy (e_o) in historical and present-day populations

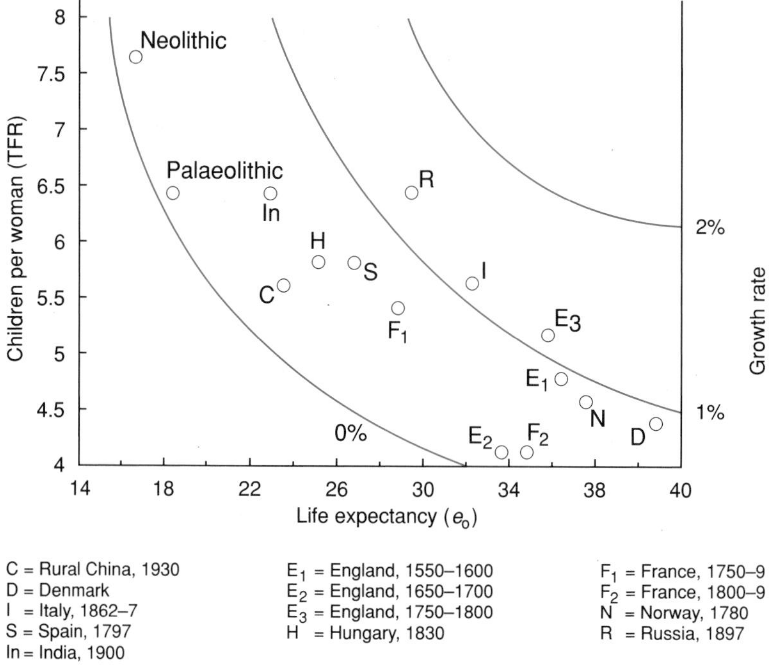

Figure A.2 *Relation between TFR and e_0 in historical populations*

Although their growth rates must have been similar, the points for Palaeolithic and Neolithic populations are assumed to have been far apart. According to a well-accepted opinion [...], the Palaeolithic, a hunting and gathering population, was characterized by lower mortality, due to its low density, a factor that prevented infectious diseases from taking hold and spreading, and moderate fertility, compatible with its nomadic behaviour. For the Neolithic, a sedentary and agricultural population, both mortality and fertility were higher as a result of higher density and lower mobility.

Figure A.3 includes points for some of the most populous countries of the world since 1950. The strategic space utilized, previously restricted to a narrow band, has expanded dramatically. Medical and sanitary progress has shifted the upper limit of life expectancy from the historical level of about 40 years to the present level of about 80, while the introduction of birth control has reduced the lower limit of fertility to a level of about one child per woman. In this much-expanded space the populations listed vary between a maximum annual potential growth rate of 4 per cent (many developing countries have a growth rate of over 3 per cent) and a minimum of –1 per cent (which will, for example, be realized by Italy should the current fertility and mortality levels remain unchanged). We are able to recognize the exceptional nature of the current situation if we keep in mind that a population growing at an annual rate of 4 per cent will double in about 18 years, while another declining by 1 per cent per year will halve in 70.[1] Two populations of equal size experiencing these different growth rates will find themselves after 28 years (barely a generation) in a numerical ratio of four-to-one!

The two situations described in Figures A.2 and A.3 differ not only in the strategic space they occupy, but also, and especially, in their permanence. The first of the two figures represents a situation of great duration, while the second is certainly unstable and destined to change rapidly, since it implies a rate of growth that cannot in the long run be sustained.

[...]

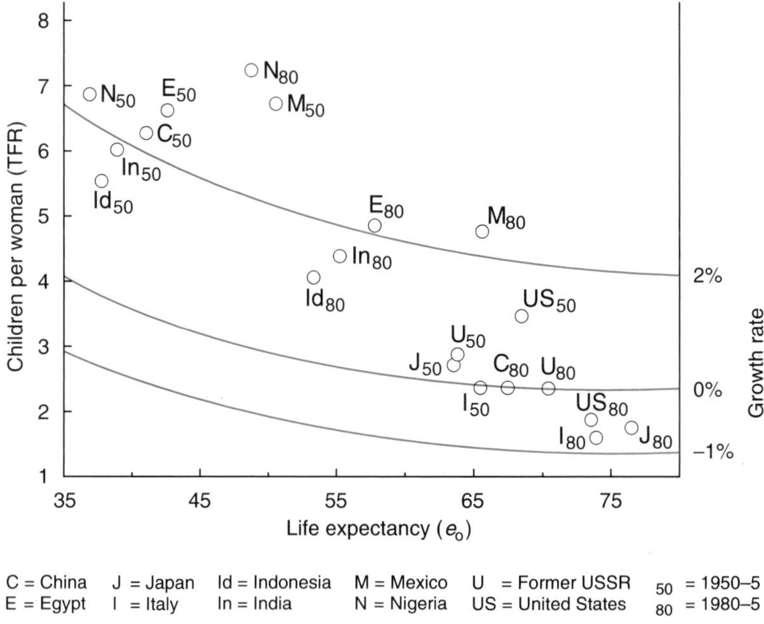

C = China J = Japan Id = Indonesia M = Mexico U = Former USSR 50 = 1950–5
E = Egypt I = Italy In = India N = Nigeria US = United States 80 = 1980–5

Figure A.3 Relation between TFR and e_o in present-day populations

Note

1 It may be useful to recall a mnemonic device for the calculation of population doubling times. These can be approximated by dividing 70 by the annual growth rate (expressed as a percentage): a growth rate of 1 per cent implies a doubling time of 70 years, of 2 per cent 35 years, of 3 per cent 23 years. Similarly, if the growth rate is negative, the population halving time is obtained by the same method: if the population is declining by 1 per cent per year, it will halve in 70 years, if by 2 per cent in 35, and so on.

Source: Livi-Bacci, 1992, pp. 21–4

Reading B: Michael W. Flinn, 'The defeat of bubonic plague'

[...]

There is a growing body of evidence to show that the defeat of bubonic plague may well have been a triumph of human organization. It is true that so long as plague was assumed to be a manifestation of God's anger at man's sinfulness there was really nothing man could do about it beyond prayer, propitiation and exhortation to a reduced level of sin: these were, of course, urged with commendable persistence by the church, with total ineffectiveness, though, miraculously, with surprisingly little loss of face. Some civil and medical authorities, however, observing the process of infection, in particular its arrival in a community after some indication of its prior existence in some other place with which the community shared a boundary or had a commercial connection, proposed various forms of isolation in an effort to break the chain of infection. It is interesting that even at the first outbreak of the Black Death in Italy in 1348 some cities, drawing on experience gained in earlier epidemics of other infectious diseases (notably leprosy), attempted to impose quarantines [Cipolla, 1976]. Though ineffective during the Black Death, the quarantine idea persisted in later epidemics.

It took a long time to develop the techniques of effective *cordons sanitaires* and quarantine management. It called, firstly, for a degree of efficiency of civil government that barely existed in much of late medieval Europe. Perhaps more important, however, was the clash between mercantile interests, often the source of urban political power, and the effects of quarantine. A quarantine involved an interruption of trade either with particular places where plague was known to exist or with the entire outside world for anywhere between a few months and a year or more. This was not good for business, and, human cupidity being what it is, there was a reluctance on the part of the local authorities (often themselves merchants) to take the drastic step of ordering a quarantine and on the part of merchants to comply with one if ordered. As plague epidemics recurred, however, and, particularly from the early seventeenth century, as epidemics were seen to be selective rather than universal, quarantines and *cordons sanitaires* were continued, and management techniques evolved. Even merchants began to see the wisdom of preferring life to profits. Local health boards were widely created in towns, which had the advantage over villages of potentially effective local government institutions. The Black Death led to their creation in a few Italian cities, and the ensuing plague epidemics of the late fourteenth and fifteenth centuries stimulated their establishment more generally. Initially these boards were ephemeral, becoming permanent later. By the end of the sixteenth century all major Italian cities had permanent health boards. Similar boards appeared in France later, mainly during the sixteenth century, while it was the seventeenth century before they appeared in Switzerland and the Low Countries. Health boards had two main functions: the management of measures taken in anticipation of and during an epidemic, which might include the establishment of a *cordon sanitaire* and quarantine lazarets; and the operation at all times of an intelligence service, the object of which was to acquire prior knowledge of the arrival of plague infection in the neighbourhood or in any place near or distant with which the town had trading links. Correspondents were appointed in all relevant foreign ports or land centres to keep the health board informed as to the state of the plague infection [Cipolla, 1976, pp. 18–31]. Cipolla [1973] has provided a dramatic example of the activities of the health board at Prato in Italy during the epidemic of 1630, while there are numerous examples of the operation of the intelligence systems at work in Italy, Switzerland and France.

Of course, these urban efforts at infection control were not always successful. But we are finding examples of success – witness the experience of Bern in 1669 [Mattmüller, 1973, pp. 138–9] and Lourmarin in Provence during the isolated Marseilles outbreak of 1720, superbly described by Sheppard [1971, pp. 116–27]. However vigorously urban governments of the seventeenth century tried, the odds were against them so long as they worked in isolation. Plague epidemics would march across Europe in the baggage or on the persons of soldiers or merchants or by sea or river in ships. An individual town might, if it were sufficiently vigorous, determined, ruthless and efficient, keep the infection out. And ruthlessness was not always wanting: when an epidemic struck the small Provence town of Digne in 1629, the inhabitants of the surrounding villages encircled the town with guards, which Biraben describes as having more the appearance of a siege than of a *cordon sanitaire.* There were even, he reported, those among the people of the surrounding villages who went so far as to advocate the destruction of the town with all its inhabitants by means of an incendiary bombardment [Biraben, 1975–76, pp. 11, 167].

But all this frightened ferocity amounted to no more than local, partial prevention. Without action on a scale wider than anything within the scope of local government, the epidemics must continue. This kind of governmental intervention was made available, for example, on a massive scale when a particularly crass example of human cupidity allowed the Marseilles *cordon sanitaire* to be breached in 1720. The French government ultimately deployed one quarter of the cavalry of the entire French Army and one third of its infantry in the erection of a new *cordon sanitaire* around the province, and the epidemic was contained [Biraben, 1968]. One hundred years earlier an epidemic of the kind experienced by Marseilles in 1720 would inevitably have spread to the rest of France and, through maritime and overland trade, to many other parts of Europe as well. In the face of a new epidemic in North Sea and Baltic ports in 1711, the English Parliament enacted, and effectively enforced, a severe quarantine for all vessels arriving from infected areas. It repeated this during the Provence outbreak of 1720–22, and on both occasions succeeded in preventing the import of infection [Mullett, 1949]. Similar rigorous action in Spain kept plague out of that country while it was raging in Marseilles – a port with which many Spanish ports had intimate trading relations – in 1720 [Chaunu, 1970, pp. 221–2].

The tendency for plague infection to spread across Europe from east to west during epidemics had not escaped attention: whatever its ultimate origin, infection entered Western Europe from Ottoman territories. When, following the Peace of Passarowitz in 1719, the boundary between the Ottoman and Habsburg empires was newly stabilized further back from the Habsburg heartland than it had been for two hundred years, and the *Militärgrenze* established, the final act of the drama of Western European plague was begun. Acting on medical advice, in a series of decrees (the *Pestpatente*) of 1728, 1737 and 1770 the Habsburg government turned the whole nineteen-hundred-kilometre military frontier between the Habsburg and Ottoman empires into a gigantic, permanent *cordon sanitaire*. Movement across the frontier was controlled by the permanent force of *Bauernsoldaten*: the peasants offered settlements in the newly conquered lands in return for a five-month frontier service each year. What may well have been the largest *cordon sanitaire* the world has ever known was underpinned by an intelligence service that alerted the Habsburg government to the state of plague infection in all parts of the Ottoman Empire. In normal times the *Militärgrenze* was permanently manned by a force of four thousand men. On receipt of information about the existence of plague in any part of the Ottoman Empire the force was immediately stepped up to seven thousand. When it was known that plague was present in any part of the southern Balkans, on the Moldau, in Wallachia, Serbia, or Bosnia, the manning was increased to eleven thousand, at which level each watch post along the nineteen-hundred-kilometre frontier was within sight of the next by day and within hearing by night. The quarantine period for any person, transport animal, or goods wishing to cross the frontier was twenty-one days in normal, plague-free times; forty-two days when there were rumours of plague in the Turkish Empire; and eighty-four days when plague was known to be present in the Balkans. There were elaborate disinfestation programmes for travellers, their animals and their goods during their quarantine in specially built lazarets. Guards were under orders to shoot to kill anyone attempting to evade the control [Lesky, 1957].

With modifications, this system was maintained until 1873. It was not 100 per cent successful in preventing the infection from crossing the frontier, but these occasional inroads were rare and relatively easily contained by the kind of local measures that had been used for centuries.

It is likely that bubonic plague was extinguished in Western Europe during the seventeenth century by vigorous local action that prevented the disease from spreading once it appeared and during the eighteenth century by national governmental action that hindered and ultimately prevented the international migration of infection. The key centres for control were the ports, above all the Mediterranean ports, and it was here that failure of control in the eighteenth century led to the last major Western European outbreaks – in parts of Northern Europe in 1710–11, at Marseilles in 1720, at Palermo in 1743, and at Cadiz in 1800. All these outbreaks were prevented from spreading far, though not without considerable governmental determination and local loss of life. But so long as plague continued to flourish in the Turkish Empire unchecked by communal action, there was a permanent risk of a renewal of epidemics. Only the great Habsburg *cordon sanitaire* of the *Militärgrenze* and the watch at the Mediterranean ports confined the Western European outbreaks after the late seventeenth century to manageable proportions.

Bubonic plague continued its periodic culls of human populations in the Turkish Empire because the Porte took no action at either the local or the imperial level of the kind that ultimately proved effective in Western Europe on grounds of indifference, religious principle, or sheer governmental incompetence. In 1840, however, almost two hundred years after local action in Western Europe had initiated the retreat of the plague, and one hundred years after the iron curtain of the Austrian *Militärgrenze* had set a limit to the age-old east–west drift of infection, the Porte, following a press campaign and the employment of European experts, decided suddenly to begin to apply vigorously and severely throughout the length and breadth of the Turkish Empire anti-plague regulations of the kind elaborated in Western Europe over the preceding centuries. Within twelve months plague was virtually eliminated from its last major human reservoir in the European and Mediterranean region [Biraben, 1975–76, pp. 11, 175].

[...]

References

BIRABEN, J-N. (1968) 'Certain demographic characteristics of the plague epidemic in France, 1720–22', *Daedalus*, Vol. 97.

BIRABEN, J-N. (1975–76) *Les Hommes et la Peste en France et dans les Pays Européens et Méditerranéens*, 2 vols, Paris and The Hague.

CHAUNU, P. (1970) *La Civilisation de l'Europe Classique*, Paris.

CIPOLLA, C.M. (1973) *Cristofano and the Plague: a Study in the History of Public Health in the Age of Galileo*, London.

CIPOLLA, C.M. (1976) *Public Health and the Medical Profession in the Renaissance*, Cambridge.

LESKY, E. (1957) 'Die Österreichische Pestfront an k.k. Militärgrenze', *Saeculum*, Vol. 8.

MATTMÜLLER, M. (1973) 'Einfürhung in die Bevölkerungsgeschichte an Hand von Problemen aus dem Schweizerischen 18. Jahrhundert', mimeographed, Basel.

MULLETT, C.F. (1949) 'A century of English quarantine', *Bulletin of the History of Medicine*, Vol. 23.

SHEPPARD, T.F. (1971) *Lourmarin in the Eighteenth Century: a Study of a French Village*, Baltimore and London.

Source: Flinn, 1981, pp. 58–61

Reading C: *Ian Sutherland, 'Farr and the public health'*

[...]

In Farr's first Letter to the Registrar General, he ranked the 32 London boroughs according to their mortality and showed that there was a general association with density of population. Farr explored this association further on later occasions and he found that the mortality varied as the sixth root of the density in London and as the twelfth root outside London. Despite this precision, however, he did not regard the density as the direct cause of the high mortality, but only as an indicator of several other aspects of the environment which he thought did contribute directly. Farr had high hopes initially that he would be able to identify these influences and thus take preventative action. But there were more theories than there was knowledge of how diseases spread. But Farr was convinced that two main needs were for improved sewerage and fresh air:

A good general system of sewers; the intersection of the dense, crowded streets by a few spacious streets; and a park in the East end of London would probably diminish the annual deaths by several thousands, prevent many years of sickness and add several years to the lives of the entire population … The poorer classess [sic] would be benefited by these measures, and the poor rates would be reduced; but all classes of the community are directly interested in their adoption, for the epidemics, whether influenza, typhus or cholera, smallpox, scarlatina or measles, which arise in the east end of the town, do not stay there; they travel to the west end, and prove fatal in wide streets and squares.

[*First Annual Report of the Registrar General of Births, Deaths and Marriages in England*, p. 113. HMSO, 1839]

Thus he expected environmental improvements to lead to a general reduction in mortality, not merely to diminish the contrast between the healthy and the unhealthy districts. Farr also stressed the value of a pure water supply:

Ely stands, with its lofty cathedral, on one of the old fen islands. It is a small city of 6,176 inhabitants (in 1851), and is in the neighbourhood of the low lands, where the great systems of modern embankments and draining were commenced by Vermuyden, one of Cromwell's colonels of horse. The Bishop of Ely in ancient times went in his boat to Cambridge. And the country around, like all our old marshes, is still imperfectly drained. The atmosphere has therefore no natural advantages. The Public Health Act was introduced in 1851. The Ely Board of Health was founded. They set on foot two great works: one for supplying the town with water, the other for carrying off that water through every house clear of the town. The public works were completed at the end of 1854; and the houses were gradually connected with the public sewers, leaving, however, at the end of 1857, 200 in 1,200 houses out of connection … In the seven years (1843–49) before the Public Health Act was in operation, the mortality was at the rate of 26 deaths annually to every 1,000 living; in the seven subsequent years (1851–57), when the sanitary measures were only partially carried out, the mortality fell down to the rate of 19 deaths annually to every 1,000 living. The mortality in the two last years (1856–57) was in the rate of 17 in 1,000. In the same periods the surrounding rural parishes underwent some improvement; but the improvement of the city has advanced so much more rapidly that its mortality was in the last two years 4 in 1,000 less than the mortality of the surrounding country … The citizens of Ely have sunk £15,000 on their sanitary works, which they appear to have conducted in something like the same determined spirit as animated Cromwell's colonel of horse. Certain ratepayers who enjoy the benefits complain of the burden of the rates.

[*Twenty-first Annual Report of the Registrar General of Births, Deaths and Marriages in England*, p. xxv. HMSO, 1860]

In addition to sanitary measures, Farr also examined the influences of climate, the seasons, and the weather upon health. He readily established an association of mortality with cold weather (already commented on by Sir Austin), but he found other association much more difficult to establish:

That smoke is irritating to the air passages, injurious to health, and one of the causes of death, to which the inhabitants of the town are more exposed than the inhabitants of the country, is probable; but if the effect were very considerable it would be most evident in the dense fogs, when the atmosphere is loaded with smoke, and is breathed for several consecutive hours by the population – men, women and children. Now we have never observed any connexion between the increase of the mortality and the London fogs.

[*Fifth Annual Report of the Registrar General of Births, Deaths and Marriages in England*, p. 416. HMSO, 1843]

The lack of association of mortality with fog persisted until the twentieth century.

[…]

Source: OPCS, 1985, p. 15

Reading D: Steve Humphries and Pamela Gordon, 'A wife and husband talking about parenthood and family planning in the 1930s'

Mary Morton Hardie

Mary was born in Lanarkshire in 1912. Her father was a civil servant. She left school at thirteen when she started work as a domestic servant. Mary met her husband, James [...] when she was sixteen, in the dairy where he was working. They were married three months later. Their first child, a son, was born in 1930. After the birth Mary decided that she didn't want any more children and, like many couples at that time, she and James decided to abstain from sex to prevent further pregnancies. Mary's second child was born in 1938, followed by three daughters in 1942, 1946 and 1948, another son in 1952, and finally a daughter in 1954. They now have eleven grandchildren and six great-grandchildren.

I didn't associate the fun that I'd had with my husband with childbirth at all. I didn't know it would ever lead to pain like this. So there and then after my first baby was born I made up my mind there would never be another baby. When I got home I told my husband I was having no more babies. 'Well,' he said, 'I can't blame you, you seem to have had a bad time.' I just wasn't going to have any more babies, it wasn't worth it. I had my baby now and I loved the baby.

My husband was really very nice about it all and I slept in a different bed with the baby. Instead of putting him in a cot I took him in bed with me. I wasn't frightened, I was terrified. I wouldn't have gone through that again for a man. I refused to sleep with him in the bed – not that he demanded it. He realized that I was terrified of having more children and he was very good, though we did have one or two sessions. But we were so careful and he always stopped before he completed. I managed seven and a half years when I didn't sleep with him to avoid being tempted.

Eventually I went down to the clinic, the doctor's surgery, and there were other young women there. I asked around to see what they did to avoid having children. So they giggled among themselves and said, 'Tell your man to get a French letter.' They call them condoms now, but I didn't know that. I'd never heard of such things. And I said to him, 'Now, look, I've been asking around at the clinic and they say you should get a French letter.' Well, my husband was so quiet and withdrawn about this. He says, 'Oh.' And he didn't say 'Yes' or 'No' and he didn't, so we didn't. As simple as that. He was really as much embarrassed as I was. Even more so.

By then my little boy was getting too big. He was about six and going to school. And so I decided I would go back and sleep with my husband because the boy needed a bed to himself. And a year later I had another baby. I didn't want to but I accepted them gratefully. I really just thought that these belong to me – I had all these children and as each one came I was so happy to have them. These were mine.

James Hardie

James was born in Motherwell in 1910. His father was an estimations clerk in the steelworks in Motherwell. When James left school at the age of fourteen he couldn't find a job so he worked in a garage for a year without pay. He then found a job as a butter scalper making butter pats in the Maple Dairy where he later became the manager. It was there that he met Mary in 1928. After the birth of their first child in 1931 James considered using condoms as a form of contraception but like many men at that time, however, found them expensive and was sometimes too embarrassed to buy them at all. He and Mary turned to the most common methods of the time, withdrawal and abstention.

We just had our family. There wasn't the same precautions or the same things as there are today and it was a case of, well, I suppose there was some kind of luck attached to it. We had to be careful, that was all. Hope for the best. I used to buy condoms if I was working away from home. When I came home I used to bring a packet of condoms with me, of course, and that was the best way. But they were difficult to get in those days. There used to be certain shops where you knew they sold them and you used to go by as if you were asking for something that you shouldn't ask for, you know, that kind of thing. It was looked down on a wee bit, when they saw you walking in with a kind of sheepish look on your face you see, knowing, 'Ah ha, here's another one.' Yes, mmm. If you went in and there was a girl behind the counter you were a bit embarrassed, you used to try and get the owner of the shop, in those days it was usually a man who prescribed the medicines, and if you could get hold of him he would slip you a packet. You just said what you wanted to him and the girl would turn her head the other way.

But it worked all right, yes. Trouble was they were quite expensive, you see. Just, I never got used to them and it didn't seem the same, it wasn't a natural thing to do, put it that way. As you know, a condom's not the same at all as the natural thing so I felt that way about it. I suppose the main way of doing it, of course, was withdrawal, yes. If you didn't have a condom that was the next best thing.

So we decided we'd be strong-willed and sleep in separate beds. Yes, we did that for a while. Well, it was just a case of either that or more kiddies, you know. It was quite difficult but we stuck it out and we did quite well. We actually had eight years between the first and second babies. But we used to cheat on it occasionally if you understand. But you just get kind of blasé about it and start off sleeping with each other again.

Nine times out of ten we were lucky – we've got a big family. That's what happens, you see, and then the odd accident happens sometimes. You're happy about it, it doesn't really matter, you're quite happy to have your family, yes. Other times it came at a time when you would not want another baby. I suppose we could have had a wee bit of an argument about it. Not terribly long, but we just accepted it. It was an expensive business – another mouth to feed. But the kiddies grew up to be healthy and we enjoyed them very much because of that. So we've got no regrets at all about the family.

Source: Humphries and Gordon, 1993, pp. 19–22

Reading E: *Betsy Hartmann and James K. Boyce, 'Riches to rags'*

[...]

We, in the industrialized nations, often view development as a straightforward historical progression: poor countries are simply further behind the rich ones on the path to development. But this view ignores the fact that the destinies of nations are linked, in ways which often benefit one nation at the expense of another. In eastern Bengal, as in most of the third world, involvement with the West began with trade, and later gave way to direct political control by a colonial power. The legacy of Bangladesh's colonial history is a variation on a familiar theme: as the region became a supplier of agricultural raw materials to the world market, local industry withered and food production stagnated. The country not only did not develop, it actually underdeveloped.

European traders – first the Portuguese in the 16th century, and later the Dutch, French and English – were lured to eastern Bengal above all by its legendary cotton textile industry, which ranked among the greatest industries of the world. After the British East India Company wrested control of Bengal from its Muslim rulers in 1757, the line between trade and outright plunder faded. In the words of an English merchant, 'Various and innumerable are the methods of oppressing the poor weavers ... such as by fines, imprisonments, floggings, forcing bonds from them, etc.' [Bolts, 1772, cited in Mukherjee, 1974, pp. 302–3]. By means of 'every conceivable form of roguery', the Company's merchants acquired the weavers' cloth for a fraction of its value.

Ironically, the profits from the lucrative trade in Bengali textiles helped to finance Britain's industrial revolution. As their own mechanized textile industry developed, the British eliminated competition from Bengali textiles through an elaborate network of restrictions and prohibitive duties. Not only were Indian textiles effectively shut out of the British market, but even within India, taxes discriminated against local cloth.[1] According to popular legend, the British cut off the thumbs of the weavers in order to destroy their craft. The decimation of local industry brought great hardship to the Bengali people. In 1835 the Governor-General of the East India Company reported to London: 'The misery hardly finds a parallel in the history of commerce. The bones of the cotton-weavers are bleaching the plains of India' [cited in Mukherjee, 1974, p. 304].

The population of eastern Bengal's cities declined as the weavers were thrown back to the land. Sir Charles Trevelyan of the East India Company filed this report in 1840:

> *The peculiar kind of silky cotton formerly grown in Bengal, from which the fine Dacca muslins used to be made, is hardly ever seen; the population of the town of Dacca has fallen from 150,000 to 30,000 or 40,000, and the jungle and malaria are fast encroaching upon the town. ... Dacca, which used to be the Manchester of India, has fallen off from a flourishing town to a very poor and small one.*
>
> *[Cited in Mukherjee, 1974, pp. 537–8]*

As Britain developed, Bengal underdeveloped.

With the decline of local industry, East Bengal assumed a new role in the emerging international division of labour as a supplier of agricultural raw materials. At first, using a contract labour system not far from slavery, European planters forced the Bengali peasants to grow indigo, a plant used to make blue dye. But in 1859 a great peasant revolt swept Bengal, and after this 'indigo mutiny' the planters moved west to Bihar. Jute, the fibre used to make rope and burlap, soon became the region's main cash crop. By the turn of the century, eastern Bengal produced over half the world's jute, but under British rule not a single mill for its processing was ever built there. Instead, the raw jute was shipped for manufacture to Calcutta, the burgeoning metropolis of West Bengal, or exported to Britain and elsewhere.[2]

The British not only promoted commercial agriculture, they also introduced a new system of land ownership to Bengal. Before their arrival, private ownership of agricultural land did not exist; land could not be bought or sold. Instead the peasants had the right to till the soil, and *zamindars*, notables appointed by the Muslim rulers, had the right to collect taxes. Land was plentiful, so if the exactions of the *zamindar* became too severe the peasants could escape simply by moving elsewhere. Hoping to create a class of loyal supporters as well as to finance their administration, the British, in the Permanent Settlement of 1793, vested land ownership in the *zamindars*, who were henceforth required to pay a yearly tax to the British rulers. In one stroke, land became private property which could be bought and sold. If a *zamindar* failed to pay his taxes, the State could auction off his land for arrears.

The architects of the Settlement set a fixed tax rate, expecting that the new landlords would then devote their energies to improving their estates. But the *zamindars* found it far easier to collect rent than to invest in farming. Instead of agricultural entrepreneurs they became absentee landlords. Numerous intermediaries – sometimes as many as 50 – each of whom subleased the land and took a share of the rent, arose between the *zamindar* and the actual tiller of the soil [Abdullah, 1976, p. 69]. Exorbitant rents had a disastrous effect on the peasants, forcing them to borrow from moneylenders whose usurious interest rates further impoverished them. As early as 1832, a British enquiry commission concluded: 'The settlement fashioned with great care and deliberation has to our painful knowledge subjected almost the whole of the lower classes to most grievous oppression' [cited by Mallick, 1961].

Little of the wealth extracted from the peasant producers by way of commercial agriculture, rent and land taxation was ever productively invested in Bengal. The budget of the colonial government clearly revealed the colonists' sense of priorities. Resources which could have financed development were instead devoted to subjugating the population. For example, in its 1935–36 budget, the Indian Government spent 703 million rupees on military services and the administration of justice, jails and the police. Another 527 million rupees were paid as interest, largely to British banks. Only 36 million were invested in agriculture and industry [Lamb, 1955, p. 490].

The British set their original tax assessment so high that many estates were soon sold for arrears, and as a result land rapidly changed hands, passing from the old Muslim aristocracy to a rising class of Hindu merchants. In eastern Bengal, where the majority of the peasants were Muslim, Hindu *zamindars* came to own three-quarters of the land. Conflicts between landlords and tenants began to take on a religious colouring.

Throughout their rule, the British consciously exploited Hindu–Muslim antagonisms in a divide-and-rule strategy. Overall, the Bengali population is about half Muslim and half Hindu, with the Muslims concentrated in the east and the Hindus in the west. At first the British favoured the Hindus, distrusting the Muslims from whom they had seized power. But as nationalism took hold among the Hindu middle classes in the late 19th century, the British tried to win the support of well-to-do Muslims by offering them more government jobs and educational opportunities. This strategy culminated in the 1905 Partition of Bengal, creating the new, predominantly Muslim province of East Bengal with Dhaka as its capital. Home Minister Sir Herbert Risley, who helped to engineer this partition, frankly revealed his motives in a memorandum. 'Bengal united is a power,' he wrote. 'Bengal divided will pull in different ways ... One of our main objects is to split up and thereby weaken a solid body of opponents to our rule' [cited in Tripathi, 1967, p. 95]. The Partition exacerbated Hindu–Muslim tensions, and, although revoked six years later, it foreshadowed events to come.

The united Bengal which Risley feared would, if independent, be the world's fifth most populous country. But as the era of British rule drew to a close, the Bengalis were unable to translate their numbers into nationhood. North Indians dominated both the Indian National Congress of Gandhi and Nehru and the rival Muslim League, overshadowing those Bengali politicians, both Hindu and Muslim, who advocated an independent, secular Bengal. So when the departing British carved the Indian subcontinent into Hindu and Muslim homelands in 1947, Bengal was again divided. West Bengal, including Calcutta, became a part of independent India; East Bengal became East Pakistan, joined in an awkward union with West Pakistan, a thousand miles away.

With the creation of Pakistan and the communal conflict that ensued, many Hindu *zamindars* fled to India, and in 1950 the oppressive *zamindari* system was formally abolished. Control of the land passed into the hands of a new rural élite, now predominantly Muslim. Although the members of this new élite lived in the villages, they were reluctant to invest in agricultural production, in part because they preferred the easier profits of trade and moneylending. Agriculture continued to stagnate.

Cut off from Calcutta, East Pakistan did experience a limited amount of industrial development. The first jute mills were at last built in the world's foremost jute-producing region. Growth remained stunted, however, by a new quasi-colonial relationship in which the West Pakistanis replaced the British. The majority of Pakistan's people lived in the eastern wing, yet those from the west dominated the military and civil service. East Pakistan's jute was the main source of the nation's foreign exchange, but development expenditures were concentrated in West Pakistan. Incomes grew in the west but not in the east, and the widening disparities fuelled political tensions between the two wings.

In 1971 these tensions culminated in civil war. The stage was set by the December 1970 elections, when Sheikh Mujib-ur-Rahman's Awami League won an overwhelming victory in East Pakistan on a platform of regional autonomy. The West Pakistani rulers, seeing the Awami League's programme as secessionist, responded by launching a vicious military crackdown, and Bangladesh's bloody birth trauma began. As the Bengalis waged a guerrilla struggle and millions of refugees poured across the border into India, Bangladesh was suddenly catapulted from relative obscurity into the headlines of the world press. The Indian Government, straining under the refugee burden and worried lest the liberation struggle assume more radical overtones, finally sent in its army in December 1971. The Pakistanis surrendered two weeks later.

Independence brought hopes that the country, freed at last from the shackles of colonial domination, could begin to address itself to the needs of its people. In January 1972, Sheikh Mujib returned from a Pakistani jail to a hero's welcome in Dhaka. Crowds surged through the streets, shouting '*Joi bangla!*' ('Victory to Bengal!'). But beneath the euphoria of independence lurked the deeply rooted problems of economic stagnation and poverty. The country's agriculture is characterized by low productivity and high underemployment: today's rice yields, which are among the lowest in the world, are roughly the same as those recorded 50 years ago.[3] Industry, constrained by both a lack of investment and by the impoverishment of the people who form the market, employs only 7 per cent of the country's work-force.

When we arrived in Dhaka in the autumn of 1974, the triumphant flush of enthusiasm had faded, giving way to anger and despair. We were greeted by the terrible spectacle of people dying in the streets. The price of rice was soaring, and soon it climbed to ten times its pre-independence level. No one knows how many people starved that autumn, but most estimates place the death toll upwards of 100,000. Although the government attributed the famine to floods, many Bengalis placed the blame on the corruption and inefficiency of the ruling party, Sheikh Mujib's Awami League. Aid officials conceded that the problem was not so much a lack of supply of food grains as inadequate distribution. Merchants hoarded grain as prices rose, and the government's distribution system broke down. A Dhaka rickshaw puller told us, 'First the English robbed us. Then the Pakistanis robbed us. Now we are being robbed by our own people.'

Despite an unprecedented influx of foreign aid – within three years of independence Bangladesh received more aid than in its previous 25 years as East Pakistan – the living conditions of the country's poor majority deteriorated. In 1975, real wages of agricultural labourers had fallen to two-thirds of their pre-independence level. Meanwhile the ranks of the landless were growing.

[...]

Notes

1 Discrimination against local textiles: see Lamb, 1955, p. 468.
2 Lack of jute mills is discussed by Ahmad, 1968, p. 222.
3 Rice yields for 1928–32 appear in Ahmad, 1968, p. 129; present yields in World Bank, 1977, p. 34.

References

ABDULLAH, A. (1976) 'Land reform and agrarian change in Bangladesh', *The Bangladesh Development Studies*, Vol. IV, No. 1, p. 69.

AHMAD, N. (1968) *An Economic Geography of East Pakistan*, London, Oxford University Press.

BOLTS, W. (1772) *Considerations on Indian Affairs*, London.

LAMB, H. (1955) 'The "state" and economic development in India', in Kuznets, S. *et al.* (eds) *Economic Growth: Brazil, India, Japan*, Durham, NC, Duke University Press.

MALLICK, A.R. (1961) *British Policy and the Muslims in Bengal*, Dhaka.

MUKHERJEE, R. (1974) *The Rise and Fall of the East India Company*, New York, Monthly Review Press.

TRIPATHI, A. (1967) *The Extremist Challenge: India Between 1890 and 1910*, Calcutta, Orient Longmans.

WORLD BANK (1977) *Bangladesh: Current Economic Situation and Development Policy Issues*, 19 May.

Source: Hartmann and Boyce, 1983, pp. 12–16

Sustainable resources?

Chapter 4

by John Blunden

4.1 Introduction

The previous chapter argued that population *per se* is not a problem, but that population growth and economic growth have produced a dramatic increase in resource use and environmental impacts. This chapter pursues the question of resource use in the context of the book's overall question 'are current lifestyles sustainable?'.

The chapter considers whether current resource use is sustainable by asking whether growing resource use will meet limits to growth, or, in the terms of Chapter 1, whether nature will set limits to growth in production through shortages of resources.

These questions immediately raise a prior question: what do we mean by resources?

Activity 1 Read Boxes 4.1 and 4.2 and write down your own definition of a resource. Does the contrast between the two examples given suggest to you any key differences between types of resource?

Box 4.1 Energy for the millions

In the mid-nineteenth century an event took place at Titusville in north-western Pennsylvania that was to transform the industrial and social base of the developed world. The Drake Well was drilled for the specific purpose of finding oil. Only a few years before this event oil was often considered a nuisance since it could contaminate salt wells. Nevertheless, in spite of the fact that the initial purpose-drilled well was no more than an entrepreneurial effort to provide a steady supply of lamp oil, it had started its existence as a valued commodity and a 'new era in world fuel resources was underway' (Haggett, 1975, p. 192). However, this was just a beginning, for only 20 years later its potential as a motive fuel had been realized.

By then 30 million barrels were being filled from wells around the world, although four-fifths of output still came from Pennsylvania. First of all, its advantages as a more flexible and more portable substitute for coal as a power source in boilers, in industrial plant, in locomotives and in ships was recognized. But the discovery of a lighter range of oils led to new uses in motor cars and aircraft and to the rapid expansion of two industries that have remained dominant in the world context ever since. At the same time, a petrochemicals industry was starting up, based entirely upon oil as a feed-stock. This has since developed 'end products that range from plastics and synthetic fibres to pesticides and drugs' (Dunham *et al.*, 1979, p. 66). All these developments caused the world demand for oil to soar, especially after the Second World War, with total consumption in 1992 approaching three billion tonnes.

Box 4.2 The rise and fall of natural rubber

According to Zimmermann (1951), latex from the Amazon area had been known for centuries by the Western world, but it was of little consequence until 1839 when Charles Goodyear discovered how to vulcanize rubber, thus greatly enhancing its strength and elasticity. This meant that rubber could now be used to satisfy a whole range of human needs, and factories were set up to produce such articles as motor tyres. The difference between manufacturing costs and prices to consumers largely determined the price commodity dealers would pay for latex at its point of production in the forests of Brazil. This in turn decided the number of local people who could be hired to gather it from the trees (see Plate 8, for example) and thus the extent of the areas worth tapping. 'In short', Zimmermann concluded, 'the demand of people throughout the world for vulcanized rubber goods governed the process by which *neutral stuff* in the wilds of the Amazon could be converted into the rubber resources of Brazil' (Zimmermann, 1951, p. 14).

But, as he goes on to point out: 'not only do modern science and technology backed by wants and needs create resources; they also destroy them and reconvert them into *neutral stuff* (p. 15). When the possibilities for rubber were fully recognized it was obvious that the supply of wild latex would be too small and too costly and a bottleneck would form in the rubber goods industry, particularly in that section of it making tyres. So it was that botanists set about breeding trees with a high latex output and these were grown as plantation crops in Sri Lanka and the Malay Peninsula. Costs were reduced to a fraction of those of the Amazon Basin and commercial latex production was squeezed out of the area where it had first begun.

But this was not the end of the story. The Second World War virtually cut off supplies of latex to industrialized consumers and led to the rapid development of a synthetic equivalent. This new activity was at first protected by wartime conditions, but the increasing efficiency of its manufacture meant that, with the return of peace, its products were cheaper than those derived from plantation latex, thus further reducing the extent of cultivated sources.

Source: based on Zimmermann, 1951, pp. 14–15

4.1.1 Some basic considerations

natural resource

A *natural resource* is a natural product which is deemed by humans to be useful. Most of the discussion of resources in this chapter is about materials or energy sources used in agricultural or industrial production, but, as Chapter 1 showed for wilderness, genetic materials and attractive scenery can also be defined as resources.

Because definition as a resource depends on usefulness to human society, natural materials may be regarded as resources by societies in some times and places but not in others. Box 4.1 shows that petroleum only became a resource a century and a half ago, and Box 4.2 that latex moved from a local resource to an international one and then back to a local one.

The examples in these two boxes also indicate that a number of different factors enter into the definition of natural resources. First, the existence of the natural material must be recognized, both in general terms and as regards availability in particular places. This depends on the second factor – the existence of the technology to detect and use the resource. However, these two factors are not sufficient: particular resources are rarely used unless the cost is competitive with alternative technologies or other sources of the same materials. Finally, knowledge, technology and cost are all affected by political factors: wars, international disputes and ideological differences may work to prevent resource exploitation in one area or shield an otherwise uncompetitive resource use, as was initially the case with synthetic rubber. All these aspects of resource use are influenced by uneven development, so the definition of resources is inherently geographical.

Current resource appraisals often identify the currently exploitable resources as *reserves*. These are the supplies of materials or energy which are exploitable in political, economic and technical terms in current conditions. Other occurrences may not be currently exploitable but may become so if technology improves or demand is high enough to justify exploitation of more difficult or more costly sources. In practice, new reserves are often identified as demand grows because exploration becomes more effective, costs rise or technologies improve. Since the collapse of the Soviet bloc in 1989–1990, most world resources have tended to be appraised in economic terms, though environmental controls increasingly affect costs, especially in the developed countries.

reserves

The two examples in Boxes 4.1 and 4.2 – petroleum and latex – were also chosen to highlight a crucial distinction between resources. Some, such as metal ores and petroleum, exist as finite *stocks*, which are non-renewable in the short term. Others, like latex, timber, animals, river water and solar energy are renewable so that, potentially, the *flow* can be harvested indefinitely. The distinction between stock and flow, non-renewable and renewable, is a crucial one in relation to long-term sustainability.

stock/non-renewable resource

flow/renewable resource

The distinction is not absolute, however, as the only totally renewable resources are solar radiation and the winds and tides. Many renewable resources depend on the existence of soils and vegetation which have taken long periods to reach their current state and which are more or less fragile. Resources such as whales, timber and even water can be over-harvested or mismanaged and degraded, making renewability far from guaranteed.

So to examine the sustainability of current resource use, this chapter will look at a range of resources – metal ores, fossil fuels, renewable energy sources and water. The aim is to assess the sustainability of current and foreseeable levels of resource use and whether a move from stock to flow resources is needed to improve this.

Activity 2 Before turning to sections 4.2, 4.3 and 4.4, write down your own expectations about the sustainability of stock and flow resources. Which would you expect to present the greatest problems in future?

4.2 A future for industrial minerals?

4.2.1 Consumption in the recent past

Any attempt to look into the future availability of what we now know as stock resources, in order to ask whether our current use of these minerals is sustainable, has to look at estimates of the likely availability of supplies of those finite resources, and consider them against forecasts of consumption.

Probable future trends in the quantities of minerals likely to be used can best be extrapolated from evidence of what has occurred in the recent past. In this respect it is worth looking back over the last 40 or so years at the consumption of the major non-precious metals, since, for the sake of simplicity of argument, these may be taken as reasonably representative of industrial minerals and metals as a whole.

However, such a review is revealing in other ways. It not only underlines the dynamic within the industrial base of nations over time, whether they are from the developed or the developing world, but it also illustrates the fact that, in spite of rising levels of consumption in some of the developing countries, the gap between them and their developed counterparts remains wide. For example, even in 1950 the USA, with 5 per cent of the population of the world, consumed more than half of the world's aluminium output, 44 per cent of its copper, 43 per cent of its lead and 44 per cent of its zinc. The industrial capacity of the USA, untouched by the Second World War, had enabled it to maintain a high standard of living well above any other part of the world. Western Europe, with a population larger than that of the USA, accounted for 20 to 25 per cent of non-ferrous metal consumption, whilst the remaining key world market areas accounted for 85 per cent of world population, but commanded only 25 per cent of the market for metals.

By 1974, as statistical sources indicate (USBM, 1950, 1974), the change had been dramatic. The economies of Europe had been reconstructed and living standards had risen markedly in the industrialized countries outside the USA. Moreover, in this period Japan's consumption of all these metals rose more sharply than those of the other market areas as that nation began to develop its world leadership in the production of consumer durables.

For aluminium, copper and zinc the second greatest increases in consumption were recorded by the developing countries. However, while the gains were in percentage terms, the actual volume of this market remained comparatively small against the enormous size of its population, at a mere one-tenth of that consumed by the developed nations.

The communist bloc showed fast gains between 1950 and 1974, but with an aggregate population several times that of Western Europe, it consumed only 70 to 85 per cent in actual volume compared with those nations, although its percentage increase was well above that of Western Europe. Nevertheless, Western Europe, when compared with the USA, was making greater overall use of the metals in question by 1974, in stark contrast to the situation as it had been two decades earlier.

Since neither the North American countries nor Australia suffered the need for infrastructural reconstruction after the Second World War, their demand for these minerals remained lower than those of Western Europe and Japan. Nevertheless, in an era of economic growth, consumption did rise, particularly for aluminium. Indeed, in the period from 1950 to 1974 aluminium consumption worldwide rose 781 per cent, compared with 242 per cent for copper, 189 per cent for zinc and 137 per cent for lead.

The boom period up to 1974 was, however, ended by the energy crisis. As statistical information from the United States Bureau of Mines (USBM) shows, during the 16 years to 1990 world consumption rose to new records for all the base metals, but the rate of increase was markedly below that of the earlier years (USBM, 1974, 1990). Whilst the overall gains in consumption were 28 per cent for aluminium, 30 per cent for copper, 25 per cent for lead and 16 per cent for zinc, the data for individual markets show strong regional variations. Much of the expansion of the period was the consequence of rapid economic growth in South East Asia with many countries in the area making enormous strides towards industrialization.

Consumption in the rest of the developing countries rose by 113 per cent for aluminium, 198 per cent for copper, 107 per cent for lead and 104 per cent for zinc between 1974 and 1990. Although the percentage gains proved less than in the earlier period, the actual increases in tonnes were greater, quite a remarkable performance given the general slow-down in economic growth.

In terms of tonnages, minerals consumption for these developing countries in the period from 1974 to 1990 rose by 27 per cent for aluminium to 1.5 million tonnes, by 61 per cent for copper to 1.06 million tonnes, by 32 per cent for lead to 0.403 million tonnes, and by 15 per cent for zinc to 0.65 million tonnes. No other market had comparable gains. Japan was second fastest, followed by Western Europe. Progress in the then communist bloc was very slight. The USA, Australia and Canada increased their lead consumption but, apart from the USA, lost ground with copper. For all three countries aluminium consumption fell, having been hit hard by the fact that its manufacture uses a great deal of energy.

The modest performance of zinc during the period 1974 to 1990 was largely due to the loss of the zinc die castings market in the car industry. To save weight in vehicles and reduce petrol consumption plastics were used instead. As for lead, towards the end of the period concern over its environmental impact reduced its use in paints and in petrol.

From this discussion of changing patterns of minerals consumption since 1950, what trends can be identified in the years up to the end of this decade and beyond into the twenty-first century? Strauss (1993) and other analysts certainly suggest that the greatest advances in consumption will be amongst the developing countries. They argue that if population levels are to be contained and living standards raised then the creation of industrial infrastructure will be required, and this will continue to show through in the use of minerals. If South East Asia appeared to be the engine of growth in the 1970s and 1980s, then it seems likely that Latin America has this potential for the rest of this decade and beyond. There is also a potential for very considerable increases in minerals consumption from the former Comecon countries. However, this assumes that the transition from state ownership to free enterprise can be accomplished without social disruption.

Since at the beginning of the 1990s the developed countries accounted for two-thirds of minerals consumption, further increases by them in the more immediate future seem likely to be of a modest order; indeed, their industrial infrastructures are already well established and living standards are comparatively high. Looking at the demand situation as a whole, minerals analysts therefore suggest that the tonnages consumed in the year 2000 may exceed those consumed in 1990 by about the same amounts as 1990 exceeded 1974, with a disproportionate share of the increase recorded by the developing nations. However, demand forecasts need to be seen against a capacity for them to be met.

4.2.2 Are there future constraints?

Questions regarding the constraints upon stock resources have exercised the developed countries for some time. At the beginning of the 1950s President Truman appointed his Materials Policy Commission under William S. Paley to survey the world resource position. Reporting in 1952, the Paley Commission, as it became popularly known, concluded that consumption would rise sharply by 1975 as a result of rising living standards and increased population. More significantly, it also suggested that resources would not be adequate to meet the demand and that, in consequence, prolonged shortages of minerals would result (Paley *et al.*, 1952). Twenty years later a similar study was published by the influential Club of Rome (Meadows *et al.*, 1972) in which it was forecast that inadequate resources – specifically mineral resources – would create 'limits to growth' at a global level towards the year 2000 and total depletion for certain minerals mid-way through the twenty-first century.

As the previous section indicates, neither forecast has proved to be accurate. Indeed, the last 40 years have provided ample evidence that the dangers of mineral resource exhaustion are more remote than either report could have envisaged. Dealing specifically with the major industrial metals, whose consumption was looked at in section 4.2.1, one of these, aluminium, in the guise of its raw material bauxite, is the third most common element on the planet. Estimates are that it represents 8 per cent of the earth's crust, outranked only by silicon and oxygen. Some of the other metals – copper, lead and zinc – are by comparison much rarer. It is estimated that their shares of the earth's crust are 0.0058 per cent for copper, 0.001 per cent for lead and 0.0082 per cent for zinc. It is unlikely, of course, that attempts could ever be made to work minerals at these percentages because of the environmental damage such operations would cause. For copper alone, vast open pits would be required, along with space to store gigantic quantities of waste rock. Moreover, the amounts of energy required for mining and basic processing would be very great indeed.

Activity 3 Look at Figure 4.1 which shows energy requirements for mining and processing per tonne of copper against the copper content of the ore (rock). If the leanest ores currently extracted are around 0.275 per cent copper content, by what order of magnitude would energy use have to rise to exploit copper at its average value in the earth's crust?

However, in spite of the prognostications of Paley and then the Club of Rome, the prospects their analyses suggested have remained far from realized. Despite substantially increased consumption of copper, lead and zinc, in the 40 or more years since the first report appeared, known reserves of all these metals are greater now than the amounts suggested either in 1950 or in 1970. As for aluminium, the increase in identified bauxite reserves have been staggering. As an example of how the reserve base (never mind the resource base) has changed, the Paley Commission concluded that the mine production of copper in the USA would at best be 800,000 tonnes in a year, reaching this peak in 1975 and falling away rapidly thereafter. However, US mines in 1990 produced 1,739,000 tonnes of copper – more than twice the Paley prediction of maximum production.

The question that remains, therefore, is how is it that these forecasts of imminent minerals exhaustion have been so wide of the mark?

Summary of section 4.2

- Natural resources are human constructs which are time and space dependent.
- They can be subdivided into renewable (flow) and non-renewable (stock) resources.
- Analysis of the post-Second World War consumption of industrial mineral resources shows marked inequalities, especially between the high levels of the developed countries and those of the developing world, but future substantial increases in demand seem most likely to occur in the latter.
- Forecasts made in 1952 and 1972 of 'limits to growth' as a result of depletion of stock industrial minerals have proved wide of the mark. Why?

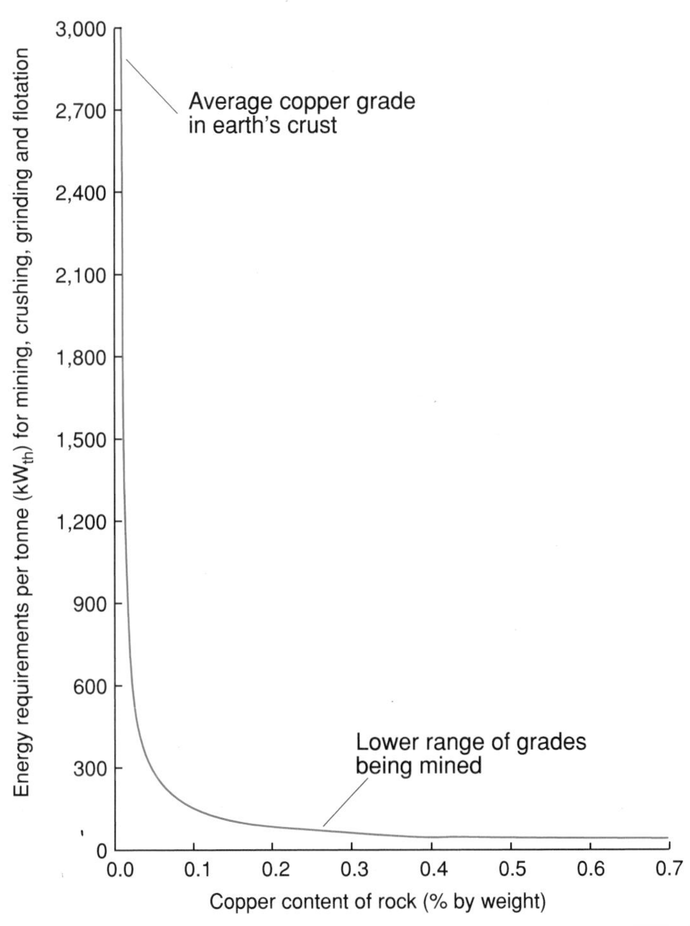

Figure 4.1 *Energy requirements for mining and milling different grades of copper sulphide-bearing rock from an open pit. The fact that lower grades of rock could entail finer grinding is not considered. Also excluded are the likely increases in energy needed to support efforts to dispose of the very large quantities of waste rock (Source: Singer, 1977, p. 128, Figure 1)*

Note: kW_{th} = thousands of kilowatts.

4.3 Minerals and the technological fix

4.3.1 Innovations in exploration

To begin with, the investigations of both Paley and the Club of Rome failed to recognize how much of the earth's land surface had not been thoroughly explored at the time of their investigations and then failed to anticipate technological advances in exploration for exploitable mineral deposits. According to Andrews (1992), these advances have been revolutionary in the last 40 years and most notably include the development of remote sensing techniques operated from space satellites. Using digital sensors, radar imaging systems and stereo cameras, these feed data back to receiving stations where they can be analysed by computers. Maps can then be produced which identify structures that may contain high concentrations of minerals of economic value, including those of oil and gas. These can then be proved and the scale of the reserve more accurately assessed by on-site drilling.

Such techniques have been used worldwide and have been responsible for the identification of many of the more recent viable finds, including those in Latin America, the territories of the former USSR (particularly those bordering the Caspian Sea where massive oil deposits have been located), and the Republic of China. In Latin America alone, whilst it had long been known that a metalliferous province existed along the Andes mountains from northern Peru to the southern tip of Latin America, new exploratory techniques recently identified the so-called 'Venezuelan Arch' which extends across Venezuela, Guyana and Surinam and has considerable base metal and gold potential. These techniques have also been applied to its geologically related neighbour, Antarctica, with results which are perhaps not surprising. However, the immensity of its wealth, in terms of all those mineral resources that characterize Latin America, seems unlikely to be realized because of an international agreement, the Antarctic Treaty, which came into effect in June 1961. This accepts that the fragility of the ecosystems of Antarctica would be jeopardized by extractive processes. Moreover, even the most modest of oil spills in the waters immediately surrounding the continent would devastate marine life and would be very slow to disperse.

Activity 4 Pause for a moment and ask yourself why it appears that Antarctica is better able to resist mining development in favour of environmental protection than many undeveloped areas of Latin America.

What other 'costs', apart from those relating to the environment, might be incurred if mining development proceeds in many parts of Latin America?

However, just as new technology has permitted a fuller exploration of the minerals potential of the land, so entirely new methods of examining the sea-bed have revealed a hitherto untapped source of mineral wealth – that of the so-called manganese nodule. It has been estimated by Enzer (1980) that 1.5 trillion tonnes of these lie on the floors of the major ocean basins, anywhere between 30 and 3,000 metres deep. These contain not only 24 per cent manganese, but also 1.6 per cent nickel, 1.2 per cent copper, 0.21 per cent cobalt and 0.005 per cent molybdenum. Until recently, only international agreement about the division of this rich source of minerals and the development of appropriate cost-effective systems for harvesting the nodules using techniques unlikely to harm the sea-bed, stood in the way of these being added to the reserve base. Now, according to Hoagland (1993), even these difficulties have been resolved. Such a rich source of minerals was, of course, unimagined by the Paley Commission or the Club of Rome.

4.3.2 The role of recycling

Innovation and technological development have not been restricted to exploration alone. Comment has already been made about their role in enabling what seemed part of the stock of a particular mineral to be transformed to a workable reserve, whether this involves methods of extraction or processing, or transport. In addition, the *recycling* of metals has become increasingly practicable and may well have a substantial role to play in securing the availability of some minerals well into the future, irrespective of our ability to continue to extract these from the earth. Moreover, the environmental advantages of recycling, as opposed to extraction, increase notably as the ores extracted grow leaner.

recycling

The industrial nations already re-use between 25 and 30 per cent of materials previously fabricated into consumer goods, or industrial plant, or in the form of wastes from their manufacture. However, this still falls far short of what might be achieved, in spite of the fact that in some instances the quantities recycled are quite high. Where lead is used in batteries, for example, in the USA some 90 per cent is recycled. As for the substantial market in dry batteries, most of which contain a range of metals including iron, cadmium, manganese, lead, copper, aluminium and mercury, plant capable of individually separating these from any type of cell now exists and operations have begun in Japan and Switzerland (Rodger, 1992). The recycling of mercury is particularly important because supplies are limited and the metal expensive. This is why, as with all precious metals, recycling rates are very high. Apart from batteries, nearly all of the mercury in boilers, instruments and electrical apparatus is retrieved when these are scrapped. The same is true of gold and silver used in jewellery, optical frames, photographic laboratories and chemicals. However, where the platinum minerals are concerned it is only when these are used as metals or in the chemicals industry that they are retrieved in large amounts. Where they are used as an agent for pollution control in exhaust systems, little has so far been achieved.

As for the recycling of steel, the most commonly used of all metal products, although quantities of available scrap rose through the 1980s, the amount recycled began to fall because of changes in production techniques which favoured the basic oxygen method. Whilst the earlier open-hearth approach had been able to accommodate pig iron (which is the basic ingredient of steel) and scrap in a fifty-fifty mix, the more environmentally friendly oxygen method could only utilize 30 per cent scrap. However, the latest development, an electric arc furnace, which is also available as a low-cost minimill, is able to cope with a 100 per cent scrap, even if it finds it difficult to handle impurities this may contain. Steel with high percentages of chromium, cobalt, molybdenum, nickel, tungsten and vanadium, currently presents a problem. Ultimately, when these technical difficulties are overcome, it is expected that this will restore the upward trend in steel recycling.

However, if steel alloys offer problems for recycling, those relating to the metallurgical complexity of many items of redundant household and business equipment have been largely overcome. For example, copper and high silica steel components are frequently closely associated in electric motors and transformers. Here technology developed by the United States Bureau of Mines can dissociate the metals. This involves the use of preferential melting in molten salt baths, a technique which seems likely to assist in the separation of the constituent metals in steel alloys.

Where aluminium is concerned, it may seem strange that the most abundant metallic mineral should enjoy a high level of recycling. The reasons for this have already been hinted at and are essentially ones of cost. The mining of the bauxite ore and its subsequent processing to form aluminium is a much more expensive means of accessing the metal than reclaiming it in its used

Aluminium can collection in Milton Keynes. Recycling aluminium cans is simple since no other metals are involved and yet it reduces the need to mine and process bauxite, which is one of the least environmentally friendly of all metallic ores because of the toxic wastes produced. The added bonus is that recycling aluminium uses only a twentieth of the energy required to produce virgin metal from bauxite

form. These additional costs arise in a number of ways. To begin with, the processing of the ore involves the utilization of a number of other minerals, including soda, lime, oil derivatives and cryolite. Moreover, expensive pollution controls have to be exercised at primary production sites to avoid the emission of toxic gases and liquids. In addition, the treatment of waste materials from bauxite processing poses a difficult and, until recently, an intractable problem. If recently discovered technical solutions regarding the rehabilitation of such disposal sites represent an important step forward in the interests of an improved environment, they nevertheless remain costly. Although there are also other reasons for recycling, more than anything else it is the energy used in the extraction and processing of bauxite that makes this practice so attractive, as Box 4.3 indicates.

Box 4.3 Record levels for aluminium can recycling

Aluminium can recycling rates in Europe and the USA reached record levels in 1992. In Europe the rate rose from 21 per cent in 1991 to 25 per cent, while the US rate jumped from 63 per cent to 68 per cent.

The main reasons why cans provide cheaper metal are that (i) they can be recycled on a 'closed loop' system (from cans to scrap and back to can again); (ii) they save as much as 95 per cent of the energy needed to produce new aluminium; and (iii) the capital cost of a recycling plant is only one-tenth of that for a smelter.

In Europe at least 39,595 tonnes of aluminium were recycled in 1992, up from 32,350 tonnes. The European industry has set itself a target of recycling 50 per cent of all aluminium cans by the mid-1990s at a time when its share of the beverage market is also growing rapidly – it reached 51 per cent in 1992. Sweden, which has a mandatory can deposit scheme, had the highest recycling rate in 1992 – 86 per cent. In Switzerland, where 'cash for aluminium cans' machines have been installed even high in the Alps next to ski huts, the rate was 68 per cent.

While the recycling rate in Austria jumped from 25 to 40 per cent in 1992, and steady increases occurred in Greece (26 to 29 per cent) and Italy (15 to 18 per cent), the UK managed a 5 per cent increase to 16 per cent. In the meantime, the chief manufacturers of aluminium cans in Europe and the USA are providing co-ordinated support for recycling in all of their domestic markets.

Source: based on Gooding, 1993

Energy requirements for recycled metals are generally less than for primary production. This has the effect, not only of reducing energy demand, itself often produced as a result of using stock resources such as coal, oil or natural gas, but also of diminishing the potential for environmental pollution from power generators. However, the extraction of minerals also has other adverse impacts on landscape and ecosystems which, once again, can be diminished by recycling.

Activity 5 Pause again and consider why it is that reduced output from power generators might benefit the environment.

What other environmental benefits do you think can emerge as a result of recycling? (You will be able to check your response when you come to Reading A associated with Chapter 5.)

4.3.3 Substituting new materials

Recycling is just one way of extending the life of metals that are produced from ores obtained from the earth's crust and that are ultimately finite. Equally significant, and perhaps even more important in the long run, is the notion of *substitution*. This can be looked at in two ways. First, there is the substitution of one metal for another. Where this is possible technically, and it is not always the case, it might be envisaged that a mineral in short supply might be used for another that was more plentiful. However, there are drawbacks. While aluminium is an excellent substitute for copper in its electrical applications, it is less conductive. Thus it can be used in overhead power cables, but not where they are housed in underground ducts or other enclosed spaces since over-heating becomes a problem. This also means that there is an energy cost too – in the heat given off by the cable. Moreover, as we have noted, the primary production and processing of the aluminium metal is likely to be much more expensive than that of copper, as well as less environmentally acceptable.

substitution

The other form of substitution is where a metallic mineral is replaced by an entirely different material. Among the more obvious examples of this is the replacement of lead and copper piping by their plastic equivalents. So far, this has largely occurred because of the ease with which plastic can be handled and its non-toxic qualities. However, plastics, particularly those made from polymers toughened with high-strength graphite or glass fibres, are also being used to replace metal.

Activity 6 Can you think of some of the industries whose products might immediately benefit from such a substitution?

Clearly, the automotive and aerospace industries can benefit from the strength and lightness of these plastics, as well as their lower costs. Indeed, plastics are now widely employed in the bodywork of cars, replacing the metal in those areas most vulnerable to impact. Their lightness also improves fuel performance, thus extending the life of hydrocarbon reserves. It is difficult to be precise about the in-roads plastics have made here, but the most recent estimates suggest that, in an average four-seater saloon car, about 460 kilograms of metals have been replaced by about 220 kilograms of plastics. This far from exhausts the possibilities of the further use of a material that uses less energy in its production and is readily recycled, an operation which saves over 90 per cent of the energy used to make it in the first place. However, as Griffiths (1993) has pointed out, a drawback exists and where different sorts of reclaimed plastic become mixed, their re-use becomes confined to plastic timber, park benches and building materials such as piping and sound-proofing. This, together with the relative youth of plastics recycling – it only began in the mid-1980s – may account for the fact that in the USA only 7 per cent is recycled and in the UK one per cent. A further drawback is that derivatives of oil are used in the production of plastics.

Substitution also occurs as a by-product of technological change. Video-tape (made from plastic and the relatively ubiquitous iron oxide) can now be used in movie cameras, replacing photographic film with its heavy reliance on the precious metal silver. Copper, formerly used in cables for telecommunications, has been replaced by glass fibre-optics derived from common sands. Because of the message-carrying capacity of the latter they are also hugely more efficient. Common clays in the form of ceramics are also being developed with the long-term aim of using them as a substitute for metals in a wide range of purposes, as Box 4.4 shows.

Box 4.4 Ceramics

Ceramic materials are produced from the non-metallic minerals of clay and silicon and fired at high temperatures. Their production is rooted, therefore, in one of the most ancient of technologies – the making of pottery and bricks.

Advanced ceramics have quickly established a global market worth US $12 billion in 1990. Following on from their early success in fibre-optics, applications now include automotive heat engine parts, where their resistance to high temperatures and corrosion, their hardness and strength, their low friction qualities and comparative lightness are all valuable assets. However, many of these same attributes mean that ceramics are now being used in the construction industry, in the manufacture of machine tools and in medical implants such as joint replacements.

The first ceramic-intensive car and lorry engine containing at least six ceramic components is scheduled for Japan's market in 1995 and that of the USA in 1999. The Japanese dominate the world market for advanced ceramics – with some 58 per cent of it in 1988, compared with 31 per cent in the USA and 12 per cent in Europe.

It is widely thought that ceramics are destined in the medium term to relieve much of the pressure on conventional metallic elements, especially the scarcer alloying metals such as manganese that are used to give steel particular properties of hardness and strength. The fact that the raw materials are so widespread and cheap is an obvious and immediate boon to Japan and other industrial countries of the Pacific rim already poor in the traditional forms of mineral resource.

Source: based on Ghandi and Thompson, 1992

The advantage of using clay and similar minerals as industrial materials is that, for the longer term, they represent a virtually limitless resource which is readily available in most parts of the world to meet whatever level of demand that might occur. Their growing use also helps to underline a key concept that pervades these chapters and that is the dynamic nature of what we socially define as resources.

Undoubtedly one of the mistakes made by both the Paley Commission and the Club of Rome was to ignore the significance of changing technology and to think of industrial resources as fixed for all time.

4.3.4 What governments can do

Whatever technical solutions may be produced to assist in recycling or substitution, some of which may be longer term, it is clear that governments can play a major role in the shorter term to conserve stock resources and at the same time reduce damage to the environment. Local authorities might, for example, forbid the tipping of wastes containing metals at land-fill sites or insist on their sorting at the household or the factory in order to reduce recycling costs. National governments might take steps to reduce the consumption of newly won minerals by the levying of a tax on their use in manufactured goods.

A number of variations on the operation of such a scheme have been put forward, but commonly they would impose the tax according to the weight or value of the new minerals contained in the item. Some, however, favour the inclusion of a negative tax or subsidy according to the proportion of recycled materials used. This can be achieved by exercising a uniform rate across a range of stock resources used from new, or by penalizing those resources that are in short supply or in most immediate danger of exhaustion.

A further suggestion favours the inclusion in the selling price of a consumer durable, such as a car or a refrigerator, a sum which would be refundable when the item is no longer usable and is sent to a collection centre for recycling. At a most modest scale, this can certainly work, as was noted in Box 4.3. In the USA deposits on aluminium cans, equivalent to 15 pence per can, have resulted in a recovery rate of 80 to 90 per cent.

Other actions might certainly involve, as Cairncross (1991) has suggested, the abandonment of all positive activities in support of primary metals production of the kind that has been prevalent in recent years – for example, depletion allowances for mining so as to hold down minerals prices, and capital tax relief on mining exploration and development. Such reliefs in the USA amounted to $7 billion in 1982, although this has declined dramatically in the years since then. Equally important would be the ending of favourable treatment given by many countries to underwriting the production of aluminium from bauxite by subsidizing their heavy consumption of electricity.

Governments can also play a major role in the conservation of traditional energy resources, a role that may have other beneficial environmental effects in terms of reducing the discharge of pollutant gases such as carbon and sulphur dioxides into the atmosphere. Such a duality of purpose lies behind attempts by the European Union (EU) to reduce carbon dioxide emissions by the end of the century by the imposition of energy taxes. In the UK the government claims that the imposition of value added tax (VAT) on energy bills from April 1994 is its first move in this direction.

However, schemes which are more directly aimed at energy conservation are to be found. In the northern European countries such as Sweden and Denmark, as well as in Canada where between 25 and 50 per cent of all energy consumed is used to heat buildings, governments in the 1970s introduced strict construction insulation standards aimed at energy savings. Moreover, in Scandinavia, Germany and Austria, instead of dissipating heat through the use of cooling towers at power stations, local authorities have sponsored the use of this in neighbourhood heating schemes.

Since in industrialized countries between 30 and 40 per cent of all electricity consumed powers electrical appliances, efforts have been made to make these more energy efficient. In California regulations have been made concerning the maximum of the amounts of energy to be consumed by lighting, air conditioning and domestic appliances, although in general terms improved energy-efficient designs have been produced by manufacturers seeking to gain market advantage through products which can be advertised as cheaper to run and environmentally more acceptable.

Governments have also begun to promote lifestyles which seek to diminish energy use, but which have added attractions usually of an environmental type. The local authority in Zurich in Switzerland has planned much of the newer sectors of the city so as to closely associate places of employment with housing and leisure facilities, all integrated by an effective public transport system. The EU, taking a different approach to a similar problem, is sponsoring telecommuting schemes. These, with the aid of advanced telematic systems, including digital telephones, modems, fax machines and computers, encourage both individuals and communities to undertake work in the services sector from where they live, rather than travel to urban centres to do much the same thing (European Commission, 1993). Envisaged as a means of maintaining the viability of rural areas in a post-Common Agricultural Policy (CAP) environment, its implications for the conservation of non-renewable resources are considerable. In the UK alone, telecommuting is expected to account for between 2.5 and 3.3 million workers, or more than 15 per cent of all employment, by 1995. This alone could save £2 billion in fuel costs and substantially reduce the level of vehicle exhaust fumes that currently run at 2.4 billion gallons a day. A basic reorientation of the distribution of work-place activities first predicted by the geographer Brian Berry in the late 1960s would appear at last set to become a reality.

Telecommuting. Over 2.5 million people in the UK currently enjoy the advantages of working from home, while their companies enjoy the financial benefits this helps to create. All of us, however, enjoy the reduced levels of environmental pollution that follow from a fall in the use of private and public transport

Activity 7 As a means of summarizing many of the ideas in sections 4.3.1 to 4.3.4 let us take one new element in the supply situation where metallic ores are concerned, that of sea-bed manganese nodules, and consider the factors likely to stimulate or retard their harvesting.

What you should do is draw up two columns for your positive responses (covering the long and the short term) and two columns for your negative responses (again covering the long and the short term).

If you are unsure about how to make a start, quickly look at the layout of the suggested response at the end of this chapter. By the way, this answer is not meant to be definitive!

4.4 A future for fossil energy?

In the previous section we noted that sea-bed mining, recycling and the substitution of new materials such as ceramics could all lead to a diminution in the physical impacts caused by traditional mining on the landscape. However, in the case of recycling and substitution, energy savings can accrue, leading to a fall in energy production and thus atmospheric pollution. Such savings, however, seem unlikely to have a profound effect in terms of the conservation of stock resources of energy.

Self-evidently, unlike metallic minerals, energy stock resources cannot be recycled. Moreover, the supply of energy is not only fundamental to industrial advance in the developing countries, but it underlies any attempts to raise the basic standards of domestic welfare. Thus it is here that resource vulnerability in the future seems most apparent.

At the beginning of this decade oil reserves at the then rates of world extraction were estimated to be around 44 years, but with considerable variation in spatial terms. Oilfield reserves ranged from under 10 years for those of the UK, to over 100 for those of several Middle Eastern states. For natural gas the picture was not much brighter with reserves forecast to last 56 years. The UK and North America had reserves likely to last about 13 years, but for some Middle Eastern states these could last for more than another century. Coal had a forecast life of 200 to 400 years.

However, since then, in spite of further consumption, the situation has not worsened as far as oil is concerned, with many additional new fields located in the former USSR, Latin America and the Middle East. Even in Europe the North Sea has yielded about a dozen small new oilfields each year since 1990. In 1993, Department of Trade and Industry (DTI) estimates put reserves at 1,960 million tonnes, enough for more than 20 years at the then annual level of production, 91 million tonnes (Lascelles, 1993). Also in this period fresh reserves were found in the Yemen. These were first tapped in January 1991. By December 1991, 14 wells had been drilled, giving access to an estimated 235 million barrels of oil. By September 1993, another 37 wells had been sunk and reserves increased to 460 million barrels.

These examples, if nothing else, give added emphasis to the need to treat depletion forecasts with considerable caution. Apart from the fact that the prospect of new reserves has been far from exhausted, better ways of extracting oil and gas are continually being developed. Estimating the life of reserves is therefore a notoriously inexact science. However, evidence has begun to emerge that oil companies have now embarked on a 'frontier strategy' which involves examining the likely production profiles of the most remote or most environmentally inhospitable fields and finding ways of lengthening the life of those already in production. In other words, they see their future only in terms of extending the limits of the known oil possibilities (Hargreaves, 1992). This means that it must be sensible to take a cautious view of the future as far as the hydrocarbons are concerned and to look for ways of conserving these if only to avoid their continuing use in ways which pollute the atmosphere. This is particularly so, given the commitments of the developed countries made at the Rio Earth Summit in 1992.

In this respect, some comment has to be made about the role of nuclear energy. Although another non-renewable resource, it was, in the 1950s, considered to be capable of providing cheap power in abundance far into the future. Unfortunately, it is now unlikely to prove helpful on two counts. First, efforts to increase nuclear power production sharply would run into problems of declining uranium ore grades as world reserves of this, its basic raw material, are depleted. Secondly, the safety of nuclear operations cannot be guaranteed, as accidents at Three Mile Island (1979) and Chernobyl (1986) readily testify. Moreover, the decommissioning of nuclear plants and the disposal of radioactive wastes are now seen to be formidable environmental problems with very long-term implications, as well as being extremely expensive. Where energy is concerned, therefore, the longer term must involve a combination of greater efficiency in terms of energy consumption combined with the use of renewable energy sources such as solar, wind and water power, and that to be derived from biomass and geothermal sources. The extent to which these offer a substitute for stock resources is the subject of section 4.5.

4.5 Renewable energy – an alternative to fossil fuels?

4.5.1 Sustainable energy resources

At present the contribution to world energy of renewable resources amounts to about 2 per cent, illustrating the immensely strong hold which the non-renewables – oil, natural gas, coal and, to a lesser extent, nuclear power – have on the energy market. But, in the light of the inevitable constraints which may well apply to their future availability, can renewable resources, almost certainly in association with energy conservation schemes, provide a viable and sustainable alternative? Before attempting to answer that question we need to consider the individual elements that constitute those resources, most of which have always been available as a source of energy. Even if technology did not permit their exploitation, as was the case in some instances, many of them have been used traditionally as energy sources. Their importance was particularly significant before the exploitation of fossil fuels. This began with coal at the time of the industrial revolution and reached its peak in the period after the oil crisis of the early 1970s when the worldwide search for new reserves outside those of the established oil-producing countries had come to full fruition.

Apart from the fact that the source of all forms of energy is radiation from the sun – oil, natural gas and coal merely represent stored forms of it – the sun's direct rays can also be used to produce heat or electricity. Solar panels may be employed to collect heat from the sun to produce hot water.

However, at a more commercial level large numbers of photo-voltaic panels containing semi-conductor materials resembling those used in the electronics industry, can be used together to produce commercial quantities of electricity by converting it from heat. Experimental solar power plants using these are to be found in California and Japan (see, for example, Plate 9).

Activity 8 Make a note of what you think might be the drawbacks to utilizing the direct rays of the sun as a source of power, particularly in the UK.

There are, however, indirect ways of using solar energy. Although the burning of wood has a long tradition here, the use of other plant materials to produce energy is now beyond the experimental stage. Indeed, 'energy crops' or fast-growing plant materials which can be converted to gaseous, liquid or solid fuels are now available. Energy crops such as these used to be seen as competing with food crop production, but at a time when the industrialized countries are over-producing conventional foodstuffs to the extent that farmers are being paid to keep their land uncultivated through the scheme known as set-aside, the production of what is known as biomass seems attractive and not without long-term implications, especially in terms of meeting the needs of transport, as Box 4.5 indicates.

Box 4.5 From farm to fuel tank

As long ago as 1912 the inventor of the diesel engine demonstrated the use of vegetable oil as a source of fuel for it. So, why is it that crude oil is still being used for tractors, lorries and cars when they could be run on fuel derived from crops grown afresh each year on farms? Moreover, if these low carbon/low sulphur fuels were being used, air pollution could be reduced. The reason is cost. The price of oil needs to rise by a factor of three, or, alternatively, the price of farm-based fuels (for example, oil seed rape) produced inside the European Union needs to come down by two-thirds for the relative costs of the two products to correspond.

Some downward pressure on the cost of producing such crops is evident following the successful completion of the Uruguay Round of the General Agreement on Tariffs and Trade (GATT) in December 1993. Moreover, the European Commission has already proposed the imposition of a carbon tax designed to encourage consumers to use fuels that cause the least air pollution by raising the cost of fossil fuels. Both may be seen as moves in the right direction. In addition, it has agreed that the levels of excise duty to be charged on bio-fuels should be limited across the European Union to only 10 per cent of those charged on fossil fuels.

In Austria, where the use of oil derived from oil seed rape is well advanced, the 5 per cent replacement of its total fuel requirement so far achieved is being concentrated in areas where clean air is important. Its low carbon and sulphur emissions make it ideal for powering snow machines at ski resorts and public transport in large towns where a reduction in pollution is particularly desirable.

Diesel fuel is not the only fuel that can be replaced from farm-based sources. In Brazil many cars are run on fuel derived from alcohol made from sugar cane. This can also be made from maize or wheat starch and during the 1970s oil crisis US fuel companies set up plants to make 'gasahol'. Now that the oil price has fallen these have been mothballed, but the techniques exist and, given a favourable cost comparison, they could be reactivated.

Source: based on Fisher, 1993

Other means of using the sun's energy by indirect methods derive from the hydrological cycle and weather systems. Hydroelectricity, which currently supplies about 4 per cent of the world's energy needs, depends on a sufficient supply of water, usually accumulated behind a dam, which can be allowed a rapid descent into a turbine. This is used to generate electricity. But unlike other forms of power generation, its potential is largely tapped in Europe and North America where few sites suitable for such development remain to be exploited. Although the costs of developing hydroelectric schemes can be very large and their realization controversial, there is considerable further potential for energy production through this method with some estimates putting possible total capacity at around one-fifth of current energy needs.

The use of wind as a source of energy has a long history in the windmill, though the use of wind turbines is recent and growing. Suitable sites at locations where there is a reasonable chance of wind blowing on a consistent basis show a great deal of spatial variation. Since most wind turbines have a rating between 200 and 1,000 kilowatts, they have a useful potential for serving small communities on an individual basis. However, small towns largely dependent on such sources would require large arrays of these, particularly when it is realized that a conventional power station can have a rating of 1.3 gigawatts, according to Rand (1990). Proposals for quite modest 'wind farms' have, however, generated opposition on environmental grounds in that they damage the appearance of the landscape. At the close of the last decade total wind power capacity in the USA only amounted to 1.5 gigawatts.

Windmill Park wind farm, Ebeltoft, Jutland, Denmark. Small wind farms can be a useful source of energy for isolated communities, but in many areas these are opposed on aesthetic grounds, particularly in upland areas of outstanding natural beauty

However, a rapid expansion in wind farms is contemplated in Denmark and Holland by the end of the century.

Like wind power, the potential of wave power, produced by the wind blowing over the sea, is very considerable as a world resource with the possibility of 5 gigawatts being available from around the British coasts. However, the appropriate technology remains in the experimental stage, although machines producing up to a megawatt have been successfully tested in the UK, Norway, Japan and the USA.

Tidal sources of energy can be derived in the same way as the power generated from a dam. A high tide can be trapped behind a barrage and then at low tide the water released through turbines. However, rising tides may also be used since it is the differential in the height of the water either side of the barrage that is the key factor. One of the best examples of this form of energy production is to be found on the Rance estuary in Brittany, France. Operating since the early 1970s, it has a capacity of 240 megawatts. A similar scheme has been looked at for the Severn estuary in England where it seems likely that an annual output of between 1,300 and 20,000 gigawatt hours might be generated, representing about 7 per cent of total UK demand. The large capital expenditure involved, together with the strong hostility of environmentalists to the project, has successfully prevented any further action upon it for the present. There are, however, 15 other sites around the coast of Britain that may well prove feasible for generating power in this way with an estimated total annual energy output, excluding the Severn, of about 26,660 gigawatt hours.

Lastly, there is some potential in the utilization of geothermal energy emanating from the interior of the earth. In the UK, for example, experiments show that hot rock material lies at a depth of 5 kilometres in south-west England. Techniques have been developed for fracturing these and pumping water through them in order to make use of the heat. It has been suggested that the total energy available in the UK from such a source is at least equal to the total coal reserve as it stood at the end of the last decade, whilst at a global level the total potential for geothermal energy may stand at around 360 gigawatts. Such a figure, however, remains highly speculative.

Activity 9 Consider the environmental effects that have been identified from the use of fossil fuels and critically compare these with the likely environmental impact of renewable energy sources. What conclusions do you draw? Bear these in mind as you study the next section.

4.5.2 Sustainable energy policies

If all of these renewable energy sources represent feasible opportunities, what then are the constraints on their use, environmental considerations apart? One undoubted drawback, at the time of writing, is the plentiful supply of alternative sources available at low price, particularly those derived from the hydrocarbons. In February 1994 the price of oil reached a low of US $13 a barrel. This, plus a degree of short-termism amongst governments, has resulted, in the immediate past, in inadequate investment in research and development except where specific countries have an obvious short-term energy problem, as Table 4.1 makes clear.

Table 4.1 Research and development expenditure on renewable energy (1986)

Country	Expenditure per capita (US $)	Total outlay (millions US $)	Percentage of total energy outlay
Sweden	2.06	17.3	21.8
Switzerland	1.57	10.2	14.7
Netherlands	1.17	17.0	10.6
Germany	1.09	65.9	11.6
Greece	0.97	9.7	63.2
Japan	0.82	99.2	4.3
USA	0.73	177.2	7.8
Italy	0.52	29.5	3.9
Denmark	0.51	2.6	17.8
Spain	0.50	19.4	27.6
UK	0.29	16.6	4.4

Source: Shea, 1988, p. 42, Table 6

In the UK, for example, the government per capita investment performance on renewables was relatively poor in the mid-1980s.

Activity 10 Can you speculate about why this should have been so?

At the end of the 1980s, if the then current trends had been maintained, by 2030 renewable sources may have contributed a mere 18.5 per cent of likely electricity demand. Of this it was expected that wind power would supply 6.5 per cent, tidal 5 per cent, geothermal 2.5 per cent, biomass 3.5 per cent and the remaining 1 per cent would be made up from wave and hydroelectric power. These figures were produced by the Central Electricity Generating Board and endorsed by the Department of Energy.

Without any form of subsidy, costs will certainly play a crucial element in whatever is achieved in terms of renewables. In the UK in 1987 the Watt Committee on Energy produced its own set of pence per kilowatt costings for major renewable contenders. These may be compared with what this committee reckoned would be the 'break-even' figure for private conventional generators of power: 2.49 pence per kilowatt.

Activity 11 Compare the Watt Committee's break-even price for conventional power generation with Table 4.2 below (also derived from the Watt Committee). What do you conclude for at least the last three of the renewables listed in that table?

Table 4.2 The cost of renewables (pence per kilowatt)

Hydroelectric schemes (small scale)	1.35
Wind	2.26
Geothermal	3.01
Onshore waves	3.17
Tidal	3.23

Source: adapted from Blunden and Reddish, 1991, p. 216, Table 5.4

Comparisons between the 'conventional' Watt price per kilowatt and the figures in Table 4.2 do not look that encouraging for some renewables. However, if it were possible to add to the 2.49 pence the environmental costs of using fossil or nuclear fuels a different picture might emerge. Certainly the disparity between government support for nuclear and renewable energy is still all too evident in one respect since the cost of nuclear energy per unit does not contain any component which will reflect the outlay on decommissioning nuclear power stations as they reach the end of their normal lives.

However, since these forecasts for renewables and their competitive costs were produced, the UK government has enacted a series of Non-Fossil Obligation Renewables Orders which provide subsidies for renewable energy forms. The two of these in 1990 and 1991 underwrote 58 wind farms and were in support of its now avowed general objective of achieving 1,500 megawatts of renewable electricity capacity by the year 2000. The important bonus to this objective, if not one of its spurs, is that this will help to reduce carbon dioxide emissions in line with the Earth Summit commitment of 1992.

Similar motives are behind the European Commission's new alternative energy programme (known as ALTENER), details of which became known in 1993. This, along with EU policies to encourage energy savings, and to provide financial support for specific renewables projects, aims at trebling the use of renewables for energy requirements by 2005, with overall renewable capacity rising from 8 gigawatts to 27 gigawatts. Table 4.3 shows the ALTENER targets, with the major increases coming through small hydroelectric schemes (thus doubling capacity to produce energy from this source), geothermal (with trebled capacity) and wind (with capacity rising by a factor of sixteen). Biomass would also make a major contribution to the production of energy from heat, rising from a figure of 20 million tonnes of oil equivalent.

Table 4.3 ALTENER targets for 2005

	Gigawatts (GW)	Millions of tonnes oil equivalent
Electricity		
Small hydro (under 10 megawatts)	10.0	2.6
Geothermal	1.5	5.4
Biomass	7.0	8.6
Wind	8.0	1.7
Photo-voltaic	0.5	0.1
Large hydro	86.6	17.1
Thermal		
Biomass (fuelwood)		50.0
Geothermal		0.4
Solar		0.2
Bio-fuels		11.0
Others		2.7

Source: ALTENER, 1993, p. 10

Direct subsidization in the short term is undoubtedly important in stimulating these alternative energy supplies. But as programmes gather momentum, notable reductions in costs will certainly be achieved as a result of technological improvements. To cite but one example, since 1990 the costs of the silicon wafers used in photo-voltaic cells to produce power from the sun have halved following the development of techniques to slice the material much thinner without losing generating capacity. At the same time considerable reductions have been achieved in the costs of installing these solar panels using lighter weight, more flexible materials. Indeed, improvements in the costings related to each renewable can be identified back into the 1980s, according to a series of papers on renewable sources of energy which appeared in 1991 in *Energy Policy* and which were subsequently summarized by Jackson (see Jackson, 1993). These also made forecasts to the year 2030 based on projections arrived at from an examination of a wide range of relevant evidence, as Table 4.4 indicates. In examining these cost reductions (given in US cents per kilowatt), it is as well to remember that they are likely to be accompanied by increases in the costs of fossil fuels as the availability of some of these, especially oil derivatives, begins to decline.

Table 4.4 *Economic costs of renewables (costs in US cents per kilowatt, 1992 prices)*

Renewable	1980	1990	2000	2030
Solar (photo-voltaic)	100–400	30–100	10–15	4–6
Solar (thermal)	25–85	10–40	6–10	5–8
Wind	30–40	13–30	4–5	3–4
Wave	40–80	10–20	8–10	5–8
Hydroelectric	5–20	5–15	5–15	5–10
Tidal	15–30	10–20	10–15	8–10
Biomass	5–15	5–5	5	4

Source: based on Jackson, 1993, p. 28, Table 3

Activity 12 Briefly summarize the reasons why the costs of renewables are likely to fall as programmes supporting their utilization gain momentum.

4.5.3 Alternative scenarios for renewables

The arguments put forward in *Energy Policy* undoubtedly feed the views of those for whom the transfer to renewable energy sources is desirable and entirely feasible. A report from the Overseas Development Institute (ODI), for example, suggests that renewables could take over from fossil fuels over a 50- to 75-year period, provided this is alongside a firm energy conservation commitment. Even using a conservative estimate of costs of such an achievement, the report concludes that there is little reason or evidence to suggest that economic growth prospects in either developing or industrial countries would be seriously affected by the use of renewable energy fuels (Anderson, 1992). While concerned that no instant solutions are feasible and the conversion to renewables will take time and that for some sectors (for example, non-electric end uses) the problems and costs could be significant, the report nevertheless offers further support for the view that the falling cost of the technology of renewables will play a vital role. The long-term view is optimistic in stating that renewables offer prospects of meeting rising energy demands in developing countries and maintaining supplies in industrial countries while reducing carbon emissions (Anderson, 1992). Only moderate levels of financial aid, it argues, would be needed and much of this would be beneficial anyway, quite apart from the global warming issue. A fairly Panglossian view perhaps, since it suggests growth without pain.

However, the World Energy Council's interim report, *Renewable Energy Resources* (1992), is even more optimistic concerning renewables and although its analysis considers a more restricted period, it postulates the notion of a 21 per cent contribution to world energy by 2020 using what it calls a conventional 'business as usual/current policies' scenario, but a more ambitious 30 per cent using an 'ecologically driven' scenario.

The alternative and far less optimistic view for renewables, widely shared in energy establishment circles, was well summarized by Professor Peter O'Dell in a seminar run by the Organization for Economic Co-operation and Development (OECD) in 1989. Based on the evidence of relatively cheap and globally abundant supplies of oil and, more recently, gas – including the huge Russian reserves now available on international markets for the first time – he concluded that, with current technology and given present and expected levels of energy prices, there seems to be virtually no chance of any significant increase in non-nuclear alternative energies' contribution to total energy supply over the rest of the twentieth century (OECD, 1989). He also stated that even the longer-term (post-2000) 'prospects still depend on the very near future investment of much increased research and development funds': and 'there are few signs to date that these will be forthcoming from either the public or private sectors'. To some extent this investment has begun to materialize, as we have noted. However, perhaps even more significant was the caveat that O'Dell added at the time, that: 'this pessimistic conclusion could be undermined by the impact of rapid build-up of political and policy concern for the global warming effect' (OECD, 1989). While many believe that this provides the key to increasing the part played by renewables in the short to medium term because 'in the longer term there is no alternative but the use of natural energy sources', it is not a universally held view. Some analysts contend that although this is the direction of the developed countries, for the 4,700 million people in the developing world the situation could well be different. These countries, driven by population growth and the need for higher per capita incomes, will attempt rapid economic growth which cannot happen without major increases in energy use. At present they consume less than a fifth of the per capita consumption of the 870 million people living in the OECD nations. It is suggested that they will seek the least expensive energy forms to achieve such ends, rather than the immediately more expensive renewables. This argument is advanced in much more detail in Reading A by Norman Duncan. However, in addition he also remains unconvinced about the role of renewables even in the developed world.

Activity 13 Turn now to Reading A, 'A declining future for renewables' by Norman Duncan, which you will find at the end of the chapter. After you have read this article, explain the premise on which Duncan builds his notion of 'a declining future for renewables' in relation to the OECD countries, and think about the following questions:

- In what respect is his argument about the role of a particular renewable misplaced?
- How important does he rate improvement in energy efficiency in the debate about meeting our future energy needs?
- What is his solution to the problem of rising carbon emissions in non-OECD countries?
- What view does he take of the relationship between nature and society?

If the non-OECD nations do take the route suggested by Duncan in the short to medium term then perhaps the challenge may be the extent to which the developed nations are prepared to underwrite the required technology transfer to those of the developing world in order to contain carbon emissions. However, the fact remains that if it is the case that 'energy growth will occur in the 80 per cent of the world that is poor', then the life of fossil fuels in meeting energy needs is likely to be even shorter lived, notwithstanding improving efficiency in the use of energy. Thereafter there is no alternative, as O'Dell concludes, 'but the use of natural energy resources' (OECD, 1989)..

Summary of section 4.5

- Renewable energy resources only contribute about 2 per cent of world energy consumption at the time of writing (1994).
- Of the technologically feasible sources of renewable energy, their capacity to be of value is spatially variable and, in some instances, environmentally contentious.
- In the short to medium term, governments need to subsidize renewable energy schemes if they are to compete with low-cost fuels, or otherwise recognize the real costs of using fossil fuels by including in their unit price those accruing to the environment.
- The economic costs of renewables will only fall as programmes supporting their utilization gain momentum and, in the longer term, shortages of non-renewable fossil fuels begin to show in rising prices.
- Although developed countries may increasingly invest in renewable energy sources, driven by the need to meet air pollution targets, the danger lies with the developing nations which may see cheap fossil fuels as their means of underpinning rapid economic growth.

4.6 Fresh water as a renewable resource

4.6.1 Global abundance?

the water cycle

Fresh water is a renewable resource because it is continuously recreated through *the water cycle*, as illustrated in Figure 4.2. However, the diagram also shows that the amounts of water available for human use are small in comparison with amounts involved in other parts of the cycle. Most water (1,400 million cubic kilometres) exists as salt water in the world's seas and oceans. 434,000 cubic kilometres of fresh water are evaporated from the surface of the sea each year, but 398,000 km^3 are returned to the oceans as rain, leaving a net movement of 36,000 km^3 from sea to land. Over land, 107,000 km^3 of water falls as precipitation, but 71,000 km^3 is lost by evaporation, leaving 36,000 km^3 to run off to the sea, balancing the transfer from the sea in the form of clouds. Over time, the balance has not been perfect and 43.4 million km^3 have been stored as ice. About 15,300 km^3 is stored underground with very much smaller quantities in surface waters such as lakes.

The key message of the water cycle is that only the throughput of the system is renewable, that is the 107,000 km^3 that falls as rain and especially the 36,000 km^3 that flows through major river systems. This averages about 75 centimetres per annum over the land surface, but ranges from less than 25 centimetres in hot deserts, interior Asia and the Arctic regions, up to 250 centimetres in tropical forests and temperate west coasts. Supply is poorly related to demand.

Because of the unevenness of supply, a keynote paper delivered to an international conference on water and the environment held in Dublin in 1992, which attempted to survey the future of the world water environment for the United Nations Environment Programme (UNEP), commented: 'Water, like energy in the 1970s, will probably become the most critical natural resource issue of the next century' (Koudstaal *et al.*, 1992, p. 278). Any attempt to answer the question, 'why should this be so?' needs to address a number of central issues, including the availability of fresh water in relation to present levels of utilization; water quality; and population growth.

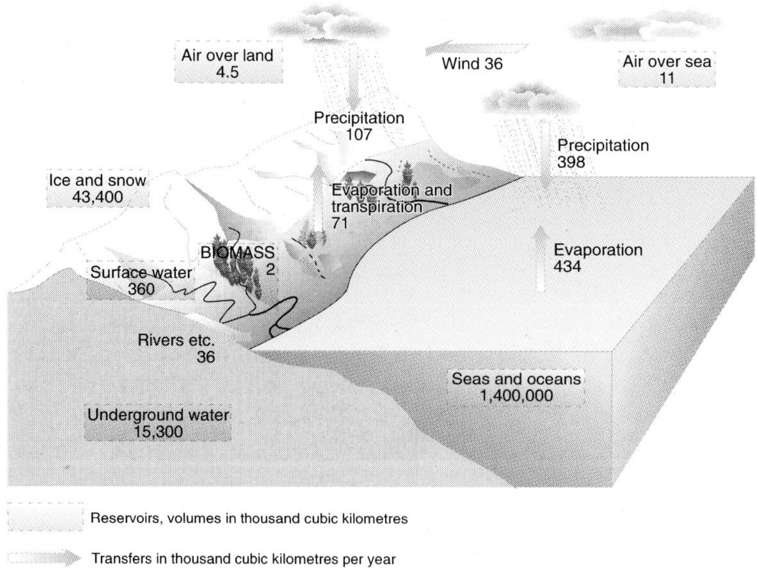

Figure 4.2 The global water cycle, indicating estimates of the many reservoirs in which water is stored on earth, and the transfers of water each year between them (Source: Nisbet, 1991, p. 24, Figure 2.13)

4.6.2 Regional disparities in availability and use

Perhaps the key issue is that of availability, in so far as the main source of fresh water, rainfall, does not always occur where it is needed. The largest body of fresh water, containing one-fifth of the world's available supplies in an area roughly the size of Belgium, is Lake Baikal. But this is in Siberia, remote from centres of population and 'off the map'. Moreover, while Iceland gets enough rain every year to fill a small reservoir for each member of its 250,000 inhabitants, Kuwait, with seven times the number of people, scarcely receives any at all. Its only redress is to employ a plant which desalinates sea-water, even though operations of this type can cost up to ten times the cost of fresh water from conventional sources. Nevertheless, the Kuwait operation is now one of the 7,500 desalination plants in operation around the world, although they only supply one-thousandth of its consumption of fresh water.

Self-evidently, the desire for water in sufficient quantities, at least for the purposes of agriculture, is greatest where its availability cannot be assured by adequate rainfall.at the right season. Here crop production has to be supported by irrigation which, in fact, makes use of around 75 per cent of the world's fresh water resources. The reason for this is not merely the prevalence of irrigated agriculture, but the notorious inefficiency of most surface irrigation systems with around half the water lost by seepage or evaporation even before it reaches the fields. Thus, whilst the production of one kilogram of wheat when fed by normal rainfall requires about 500 litres of water to reach maturity, one kilogram of irrigated rice uses over 1,800 litres.

By contrast, all the developed countries are also prime users of water, not least because of the demands made by industry. Where the concentration of industry is particularly apparent, regional water shortages, even in countries which are generally amply supplied, have led to major catchment developments including inter-basin transfers of water, as Figure 4.3 indicates. The magnitude of such schemes is perhaps less surprising when the needs of particular manufacturing

Chinese cabbage production on a Thai farm, Bangkok. Wetland cultivation tends to be synonymous with the growing of rice, but other vegetables are also produced in this way where the crops are adapted to their roots growing in an oxygen-deficient soil

processes are appreciated. For example, the manufacture of one tonne of cement requires 3,600 litres of water, one tonne of steel 8,000 to 12,000 litres, one tonne of paper over 27,000 litres, and a car 38,000 litres. Even 'a tonne of bread produced industrially consumes 2,100 to 4,200 litres' (Simmons, 1989, p. 274). However, 'in the UK 1,000 litres of beer has needed up to 4,200 litres of water for its production. In the USA the figure for beer appears to be larger at 15,000 litres of water. American beer is either cleaner or weaker than its British equivalent we may suppose' (Simmons, 1989, p. 274). Notwithstanding this minor puzzle, the evidence certainly suggests that, amongst the industrial countries, the USA has the greatest per capita consumption of water. Thus, although Phoenix, Arizona, has the same amount of rainfall as Lodwar in the north of Kenya, it uses 20 times as much water per head. Even its domestic consumption of water for washing, cooking and the flushing away of wastes is four times greater than that of the UK at over 600 litres a day. Only in the luxury hotels of the EU is such a figure reached in Europe.

However, in the 1990s, even the USA has recognized that further growth in levels of fresh water consumption is not sustainable and that the use made of this flow resource, at least in certain regions, must be brought under control. The time is well past when the profligate use of river water could be supplemented by pumping supplies from deep-lying aquifers such as the Ogallala which stretches under eight Great Plains states from South Dakota

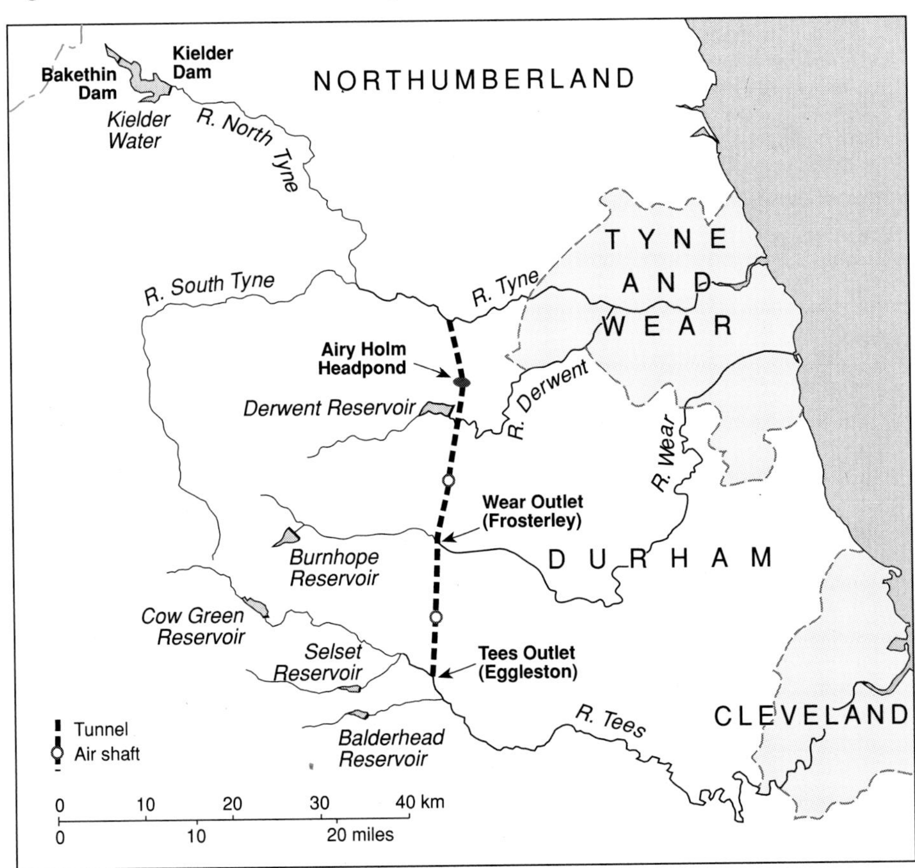

Figure 4.3 Manipulating water supplies in north-east England. Inter-basin transfers by tunnel from Kielder, the biggest reservoir in Europe, to industrial development in Cleveland (Source: Simmons, 1989, p. 238, Figure 5.18)

to Texas. Once containing as much water as Lake Superior, it is said that one out of every six ears of grain grown in the USA depended on it. But this stock source of fossil water, like others, is now seriously depleted and at present levels of use the Ogallala can last no more than a further 20 years. The replenishment of aquifers like this can take many centuries. Soon, the USA will be totally dependent on flow water resources, which, as Box 4.6 shows in a regional context, will need careful management in the interests of meeting both basic human needs and those of the environment.

Box 4.6 Reaching the bottom of the barrel?

For 90 years US government engineering works, through the construction of dams, reservoirs and water delivery systems, wrested the rivers of the West from their beds to bring water to the fast-growing cities of the region or to provide the farm irrigation that accounts for 80 to 90 per cent of consumption. But as the limits to ever increasing water provision are being reached, it has been realized that water, in future, will have to be used primarily to meet the most basic needs of the population. At the same time it will be needed to conserve the environment and endangered ecosystems. As Craig Bell, Executive Director of the Western States Water Council, has said: 'The challenge is that most of the water was originally allocated under a system that did not realize the ecological value of allowing water to remain in the river beds' (Graham, 1993).

Under pressure from vociferous and powerful environmental interest groups, the reallocation process has, in fact, already begun. Dams in the Pacific northwest must now control their outflows in the interests of local salmon species which their construction put in peril, while the Glen Canyon Dam on the Colorado River has been forced to stop the water releases that were eroding the beaches of the Grand Canyon.

The Central Valley Project, California's largest water delivery system, decided in 1993 to allow nearly one-fifth of its throughput, some 600,000 acre feet, to flow straight into the Sacramento–San Joaquin Delta. The idea was to help restore the habitat of such species as the endangered delta smelt, even though this leaves farmers in the San Joaquin valley, the project's biggest beneficiaries, with barely a quarter of the water they believe they require.

The corollary of all these actions is the need to curb the use of water for irrigation and domestic consumption. Farmers working the irrigated areas of the Los Angeles and Imperial Valley have been persuaded, for example, to adopt more economical irrigation methods which involve the use of sprinklers or drip methods. These have efficiency rates of 65 to 75 per cent and 90 per cent respectively, compared with 40 to 50 per cent for surface methods. Alternatively, farmers can be encouraged to grow less water-intensive crops.

As for domestic consumers, the approach here has been one of demand-side management, thus moving away from the notion of water as a cheap commodity, or a 'free gift of nature'. As a result, average water bills in Los Angeles have been increased by 100 per cent in the six years up to the end of 1992. The Metropolitan Water District, southern California's biggest supplier, authorized a further rise of 22 per cent in 1993. Indeed, many city authorities are now into water saving by demanding smaller toilet cisterns and the increased use of showers with controlled flow systems. Some local governments are now allowing the use of 'grey' water (waste water from baths) to be used for flushing toilets or watering lawns.

Source: based on US National Research Council, 1992

4.6.3 Industry and water quality

Unfortunately, some problems of water distribution and use are further aggravated by pollution since, for every litre of dirty waste water that is discharged into a lake or river, many more become, to some extent, contaminated. Indeed, the World Health Organization in a report suggests that as many as 10 per cent of the world's rivers that have been monitored are polluted (World Health Organization, 1992–1993). Taking the USA as an example of a developed country with a mature industrial manufacturing base which features prominently in this survey, all of its water-using factories discharge considerable quantities of liquid waste. Most of this effluent passes into rivers or lakes where, in the course of the breakdown of its organic content through natural biological processes, it consumes the oxygen present in the receiving water body. This is sometimes to the detriment of its biota or sometimes its complete destruction. In addition, however, much of the waste is in the form of chemicals which, when discharged to other water bodies, can create a hostile environment of considerable local toxicity to both plants and animals unless dilution is very rapid.

Activity 14 Consider Table 4.5. Although some fresh water is consumed by industry in its manufactured products, much process water remains available for beneficial re-use if its impurities are removed. What overall percentage is currently reclaimed to this end? Think about the regions of the West in relation to the water supply problems outlined in Box 4.6. How satisfactory is the performance of their industries in reclaiming a scarce resource and improving water quality?

Table 4.5 Industrial fresh water use and reclamation by US water-resources region, 1990 (figures may not add up as independently rounded)

	Deliveries	Reclaimed waste-water	Consumptive use
	(millions of US gallons per day)		
New England	698	0.0	71
Mid Atlantic	2,400	63	256
South Atlantic Gulf	3,570	0.5	470
Great Lakes	5,040	0.0	458
Ohio	2,990	0.0	297
Tennessee	1,290	0.0	163
Upper Mississippi	1,430	0.0	214
Lower Mississippi	2,690	0.0	286
Souris-Red-Rainy	52	0.0	9.5
Missouri Basin	282	0.0	87
Arkansas-White-Red	625	1.9	113
Texas Gulf	884	20	359
Rio Grande	31	0.5	16
Upper Colorado	9.7	0.0	4.7
Lower Colorado	254	2.3	227
Great Basin	127	0.0	55
Pacific Northwest	1,200	1.6	125
California	625	0.8	102
Others	203	0.0	21.7
Total	24,500	90	3,330

Source: US Department of the Interior, 1991

These high levels of effluent discharge prevail in spite of the 1972 Federal Water Pollution Control Amendments Act which took over responsibility for water quality from the individual states and provided a comprehensive set of powers to attack both industrial and municipal polluters. Although 1985 was a target date for the achievement of clean effluent discharges, this has not been realized. Amending legislation in 1977, weakening the powers of the original Act; a series of legal challenges brought by industrial interest groups; and cuts in the budgets and staffing of the Environmental Protection Agency, during the Reagan presidency, have meant that the present standards fall well short of those originally envisaged.

Other developed countries have enacted legislation to control the quality of effluent discharged to water bodies – the European Union, for example, operates a system of uniform emission limits which it is attempting to tighten. However, only rarely do industries either totally recycle or purify their process waters (see Table 4.5). As they have successfully claimed in the USA, the cost is too great.

As for Europe at large, one of the worst records for the discharge of liquid industrial effluent is to be found in Poland where about 75 per cent of its rivers are now too contaminated even for further industrial use. Where Russia is concerned, it was revealed in a report published in April 1993 that three-quarters of its surface water was unfit to drink and a third of its underground water supplies contaminated. As the report in question put it: 'water is a prime victim of decades of the abuse of Russia's environment' (Boulton, 1993). Equally discouraging is the situation in the developing world where industrialization has occurred. More than two-thirds of China's rivers are seriously polluted, while 40 of Malaysia's rivers are reported to be biologically dead. In the meantime, the European Commission remains frustrated with progress over pollution control by the responsible administrative body Directorate General XI. Indeed, it has expressed 'disappointment that the state of the aquatic environment in the European Community has not improved to the extent expected', according to a survey of the supply and use of fresh water undertaken in 1992 on its behalf (Ecotec, 1992). This also argues that future economic growth in the Union will be constrained by the contamination of fresh water, not only from industrial, but also from agricultural sources.

4.6.4 Agricultural and domestic impacts

Certainly, both in the developed and developing countries, the contamination of water supplies by agricultural chemicals used as fertilizers – such as nitrogen, phosphate and potassium – is becoming increasingly serious. The problem is that these are, more often than not, delivered in such quantities that the plants that make up the field crop can rarely use all of them. This situation particularly prevails in those regions of the developed countries where large-scale grain production is practised and in the developing world in those areas where the 'green revolution' has brought about substantial increases in rice production.

Applications of artificial fertilizer to field crops grown by high output farmers in Europe frequently result in 40 to 50 per cent of the nitrates entering the run-off waters. The same is true of phosphorus, but to a much lesser extent since most of it found in fresh water arises from the discharge of sewage and detergents. The addition of these nutrients can produce a water body that rapidly suffers eutrophication. This is characterized by the presence of algal blooms whose decay robs the water of oxygen. Evidence of the growing environmental impact of nitrates during the 1970s and 1980s has been found around the world. Concentrations in the River Tomo in Japan rose during these two decades by 14 per cent, in the Rhine by 27 per cent, and in the River Wear in England by nearly 50 per cent. Nitrogenous agricultural fertilizers also appear to be the main source of nitrates present in the ground waters used to supply regional domestic consumers. At concentrations above 10 to 12 milligrams a litre, they can cause health problems, especially to bottle-fed babies.

In addition, United Nations (UN) agencies have reported the presence of organochlorine pesticides in rivers in agricultural areas of developing countries at levels considerably higher than in European rivers. In Colombia, Malaysia and Tanzania, levels have been such that their ingestion by fish and their concentration within the fish presents a health hazard to those who consume these as a major part of their diet. However, in the developing countries water is often additionally affected by organisms, some of them derived from sewage, which give rise to a range of diseases. These include typhoid, food poisoning and hepatitis. Water diseases present a particular problem for women. Because of their domestic role which includes their function as water collectors and the part they play in producing the staple foods for their families, women are, as a group, the most vulnerable to water-related illness. This was recognized by the United Nations Committee for the Implementation of the International Drinking Water Supply and Sanitation Decade which not only gave particular attention to the vulnerability of women in this respect, but also incorporated the UN International Research and Training Institute for the Advancement of Women (INSTRAW) in the membership of the Committee. According to Pietilä and Vickers (1990), a task force on women, water supply and sanitation has since been established to provide guidelines on future policies in this field with regard to their full participation in attempts to resolve these problems.

Activity 15 Turn to Reading B, 'Women and water' by Borjana Bulajich who is Social Affairs Officer at INSTRAW. In her article, Bulajich recognizes the need to include women in the resolution of the broad problems attached to water supply and sanitation.

As you read, briefly note down Bulajich's view of the current problems related to the design of water projects and what needs to be done for the future.

Not surprisingly, the problems of water contamination remain high on the agenda in spite of UN initiatives. It is estimated that contamination of this kind, which still affects 25 per cent of the world's population, is a major factor in more than five million deaths a year. Thus the provision of clean drinking water is a major task for many developing countries – in Tanzania alone this involves 20 million rural inhabitants – and any attempts to achieve such an end can mean competition with the provision of water for irrigated agriculture. Certainly the magnitude of the contaminated water problem is brought home by the realization that if all of this water could be purified, it would be equal to half the current demands of the industrial nations.

Meanwhile, demand for fresh water is rising, linked as it is with population growth in developing countries and the spread of agriculture. UNEP has been reported as stating that the world's use of fresh water has 'increased nearly four fold in the last 50 years to 4,130 cubic km a year' (Maddox, 1993). In using three-quarters of this total, the area of irrigated land has increased more than a third in the last two decades. The growth in Asian demand for water is the fastest. In 1993 it used just over half the world's water, but by the year 2000 UNEP expects this to rise to nearly two-thirds, mostly to support irrigation.

Activity 16 Think about your study of water so far. Can you identify the different ways in which society and nature interact? Look back at the last two sentences of the second paragraph following Table 4.5. What view of water does this suggest? Who would benefit and who would lose out from this view?

4.7 Fresh water as a source of conflict

4.7.1 Fresh water resource development – a contested domain

Developing countries faced with substantial increases in the demand for fresh water, spurred by increasing populations, frequently resort to water management schemes. These, more often than not, involve the construction of dams and water storage facilities along river systems, which, apart from rainfall itself, are best able to meet regional or national needs. But what appears to be desirable in the interests of a nation can run counter to the interests of local populations most affected by such schemes. For example, in Lesotho in southern Africa the construction of the Katse Dam, the first of five to be built as part of the Lesotho Highlands Water Project, will result in the displacement of 20,000 people by 1996. In this instance, however, this development, which is the largest of its kind in Africa, will not only provide water for Lesotho itself, as well as hydroelectricity, but it will also allow the valuable export of water to meet the growing needs of South African industry. Nevertheless, the scheme remains controversial, not merely because of the damage done to the environment as the result of the flooding of thousands of hectares, but also because of the destruction of the way of life of the indigenous peoples for which a Rural Development Plan to promote alternative sources of income appears, in their eyes, to offer no meaningful compensation.

A similar situation is to be found in India. The construction of the Sarovar Dam, one of a complex of 30 dams on a river system that crosses three states – Gujarat, Maharashtra and Madhya Pradesh – may mean the removal 250,000 tribal people from the Narmada valley. This has caused considerable unrest in the area and the arrest of many protesters, including India's leading environmentalist, Medha Paktar. However, local objections, even though deeply felt, are unlikely to do more than temporarily delay what is seen to be a project of overwhelming national importance.

Neither of the projects cited could have been countenanced without the support of international loans from the World Bank. Thus, as Rich (1994) suggests in his analysis of environmental and social record of this organization, it does call into question its support for what must remain controversial forms of development in the light of their negative impact on local ecologies and the dislocation of the traditional lifestyles of indigenous peoples. Whatever may be seen as the wider benefits of such schemes, fresh water management in such instances offers opportunities for conflict.

4.7.2 Fresh water management – sharing the resource

Against this problematic background regarding water resources, there are, however, further difficulties relating to the availability of supplies which will almost certainly be exacerbated by the rise of nationalism in the 1990s. This is because whole drainage basins, natural physiographic regions, which may cover many thousands of square kilometres, are often not fully contained within the national boundaries of a single state. The tensions between nations to which this gives rise offer a good example of the problematization of political relationships stretched over space and boundaries referred to in the introduction to **Allen and Hamnett, 1995***. In extreme cases of the problem of the shared resource very little of the surface water available to a country originates inside its own territory. For Hungary, the figure is only 4 per cent, the other 96 per cent entering it as a cross-boundary flow in the form of the Danube and its tributaries. A similar example in Africa is Mozambique. In the south less than 2 per cent of its available surface water originates in its territory.

Activity 17 Rogers (1993) has identified a number of major river systems shared by more than one country – for example, the Colorado (Mexico and the USA) and the Ganges (India and Bangladesh).

Using a world atlas, identify the number of national territories which are covered by the Danube, Congo, Zambesi, Amazon, Rhine and Brahmaputra.

Whilst in the case of the Rhine the sharing of a river water resource, as well as agreement about its standard of quality, can be considered much less problematic since most of the nations involved are embraced by the European Union, elsewhere problems may arise. This is particularly so in areas where agriculture cannot be adequately practised without the aid of irrigation, or the demands on water for domestic and industrial purposes are rapidly rising in an effort to provide better living standards for increasing populations. Perhaps this may be better understood by reference to Table 4.6 where the impact of upstream activities in one country on the downstream activities of another lays emphasis on the negative aspects of the supply of water to the latter.

* A reference in emboldened type denotes another volume in the series.

Table 4.6 Downstream effects of upstream water use

Upstream water use		Downstream effects
Direct use	Hydroelectric power	Helps regulate river (+) but can cause surging when working at peak load (–)
	Irrigation	Removes water from system (–)
	Flood storage	Offers downstream protection (+)
	Industrial/municipal usage	Removes water from system (–)
	Wastewater treatment	Adds pollution to river (–)
	Navigation	Keeps water in river (+)
	Ecological maintenance	Keeps water in river (+)
	Aquifer recharging	Reduces water in river and stream flow (–)
Indirect use	Agriculture	Adds sediment and agricultural chemicals (–)
	Forestry	Adds sediment and chemicals, increases run-off (–)
	Animal husbandry	Adds sediment and nutrients (–)
	Wetland drainage	Reduces ecological carrying capacity, increases floods (–)
	Urban development	Induces flooding, adds pollutants (–)
	Minerals exploitation	Adds chemicals to surface and groundwater (–)

Source: Rogers, 1993, p. 119, Table 1

4.7.3 Shared waters – international conflict and conflict resolution

In such circumstances where no agreement between the countries exists regarding the quality of the shared water and the amount to be taken, it is easy to see how this may become a major source of international conflict. Nowhere is this more true than in some of the river basins of the semi-arid or arid developing countries of the Middle East, as the short case studies in Boxes 4.7 and 4.8 indicate. There, although there has been a steady decline in the importance of farming to the national economies concerned, irrigated agriculture remains important. Against a background of rapidly rising populations, food security and self-reliance are certainly seen as key national economic goals, in spite of the tremendous drain such policies place on limited water budgets.

Box 4.7 Water – a source of international conflict?

The Nile Basin

'In the next few years the demographic explosion in Egypt, Sudan, Ethiopia and Uganda will lead to all those countries using more water; unless we can agree on the management of water resources we may have major international or inter-African disputes on our hands' (Boutros Boutros Ghali, UN Secretary General, in Tucker, 1993).

Notwithstanding this comment from the Secretary General of the United Nations, a treaty relating to the allocation of the Nile waters was drawn up in 1959, but the so-called Nile Waters Agreement binds only Egypt and Sudan. Both countries agreed to adhere to an annual allocation of 55.5 billion cubic metres in the case of Egypt, and 18.5 billion cubic metres for the Sudan. This was based on a mean discharge from the river of 84 billion cubic metres and an assumption that about 10 billion cubic metres would be lost by evaporation and seepage from Lake Nasser behind the Aswan Dam in Upper Egypt (Plate 10). However, the head waters of the river consist primarily of the Blue Nile which flows from Lake Tana in the Ethiopian Highlands, and contributes to 75 per cent of the main flow of the Nile. The rest comes from the White Nile which flows through Uganda and southern Sudan (see Figure 4.4).

In 1990 Ethiopia presented plans for an extensive irrigation and hydroelectric power generation scheme which would involve damming the Blue Nile and a take of vast quantities of water from the river. It argued that if Egypt improved the efficiency of its irrigation schemes it should not be unduly disadvantaged. However, Egypt saw this proposal as a major threat to its well-being and blocked a loan to Ethiopia from the African Development Bank which would underwrite its proposed development. In return, Ethiopia stopped the provision of basic hydrological data to its downstream neighbours. Although a so-called African water summit was called, it failed to resolve the dispute. In the meantime, further problems have been averted only because of the civil war in Ethiopia which has meant the temporary abandonment of the scheme.

However, with the population of Egypt, Ethiopia and Sudan reaching an expected total of 170 million by the year 2000, the need for agricultural expansion and increased food production which in turn will demand more water, will be inevitable and the imperative for agreement amongst the countries along the Nile overwhelming if conflict is to be avoided.

Source: based on Beschorner, 1992, pp. 45–61

Box 4.8 Water – a source of international conflict

The Tigris–Euphrates Basin

Both Syria and Iraq, which share the area between the Black Sea and the Persian Gulf, believe that the region could ultimately be developed to be agriculturally self-sufficient using irrigation systems. However, the problem is Turkey in whose eastern territory both rivers rise.

Turkey, through its South East Anatolian Project (GAP), plans to encumber the head waters with a number of dams which would cut the flow of the Euphrates into Syria by more than a third in dry years. Moreover, Iraq is worried that Syria, too, will increase its share of water (see Figure 4.5).

But problems over the use of water from these shared rivers are not new. As far back as 1975 war was narrowly avoided when Turkey temporarily plugged the Euphrates to fill the reservoir behind its recently completed Keban Dam, while downstream Syria dammed the river to fill a reservoir of its own. By the time the Euphrates reached Iraq the flow was down to one-fifth of the average, jeopardizing the livelihoods of some three million farmers. Iraq blamed Syria, which blamed Turkey. Baghdad threatened to bomb the Syrian dam and it was only after intervention by Saudi Arabia which persuaded Syria to make a goodwill release of water, that the crisis was defused.

Then in 1989 Turkey threatened to use its control of the headwaters of the Euphrates as a political weapon by cutting the flow of the river because Syria was supporting Kurdish dissidents. In 1990 further controversy arose when Turkey stopped the Euphrates for a month while it filled up its new Ataturk Dam. This, combined with its Keban and Karakaya Dams, is designed to store three years' flow of the Euphrates.

Turkey's latest GAP scheme, with its 22 dams and 17 hydroelectric plants in the area of the head-waters of the Euphrates, will generate half Turkey's electricity and provide irrigation for thousands of hectares of otherwise arid land. The official expectation is that agricultural production will double and bring prosperity to the local Kurdish population. However, the ultimate realization of this scheme can only worsen relations with Syria and Iraq.

Source: based on Beschorner, 1992, pp. 27–44

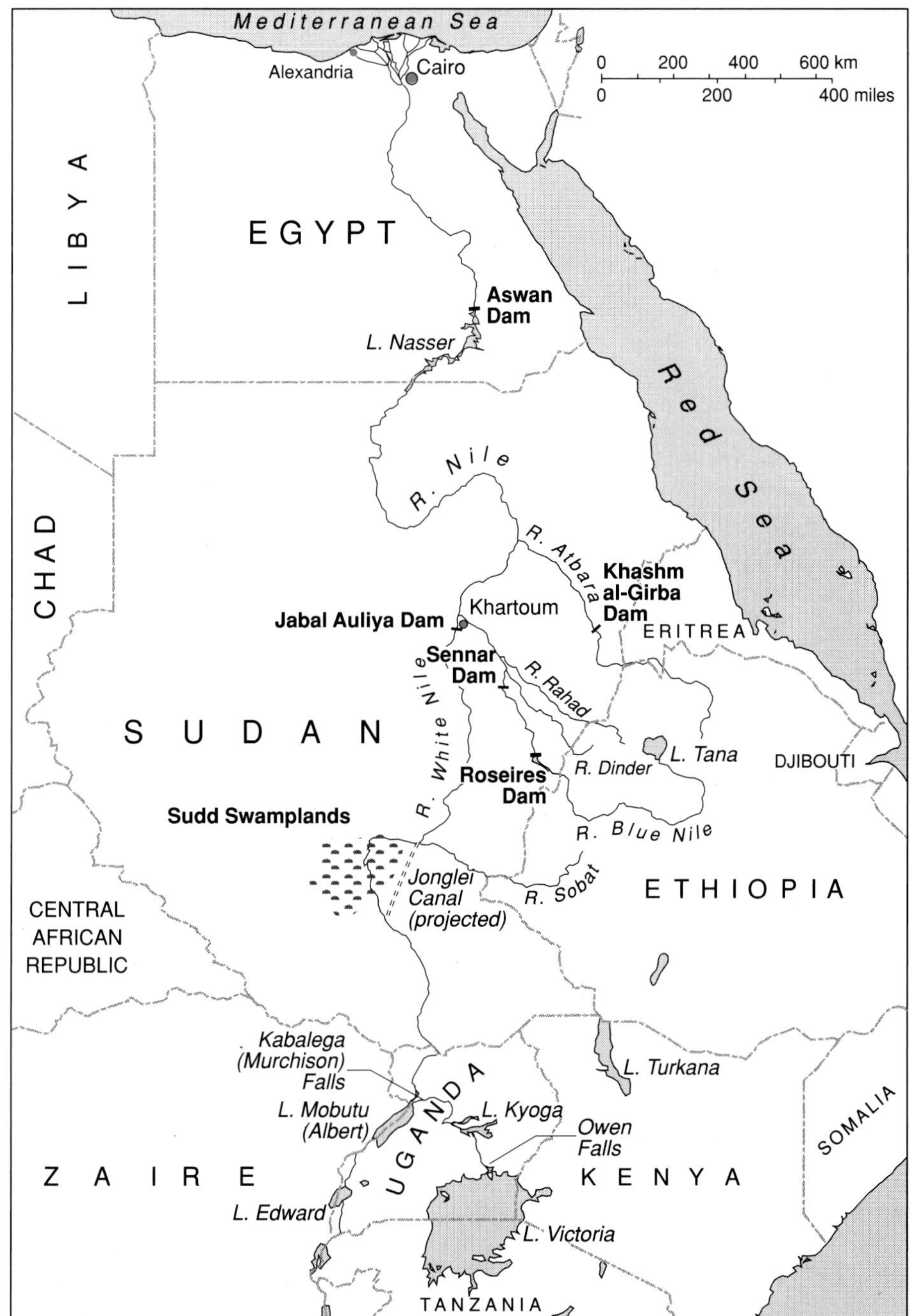

Figure 4.4 *The River Nile and its tributaries (Source: Beschorner, 1992, p. 46)*

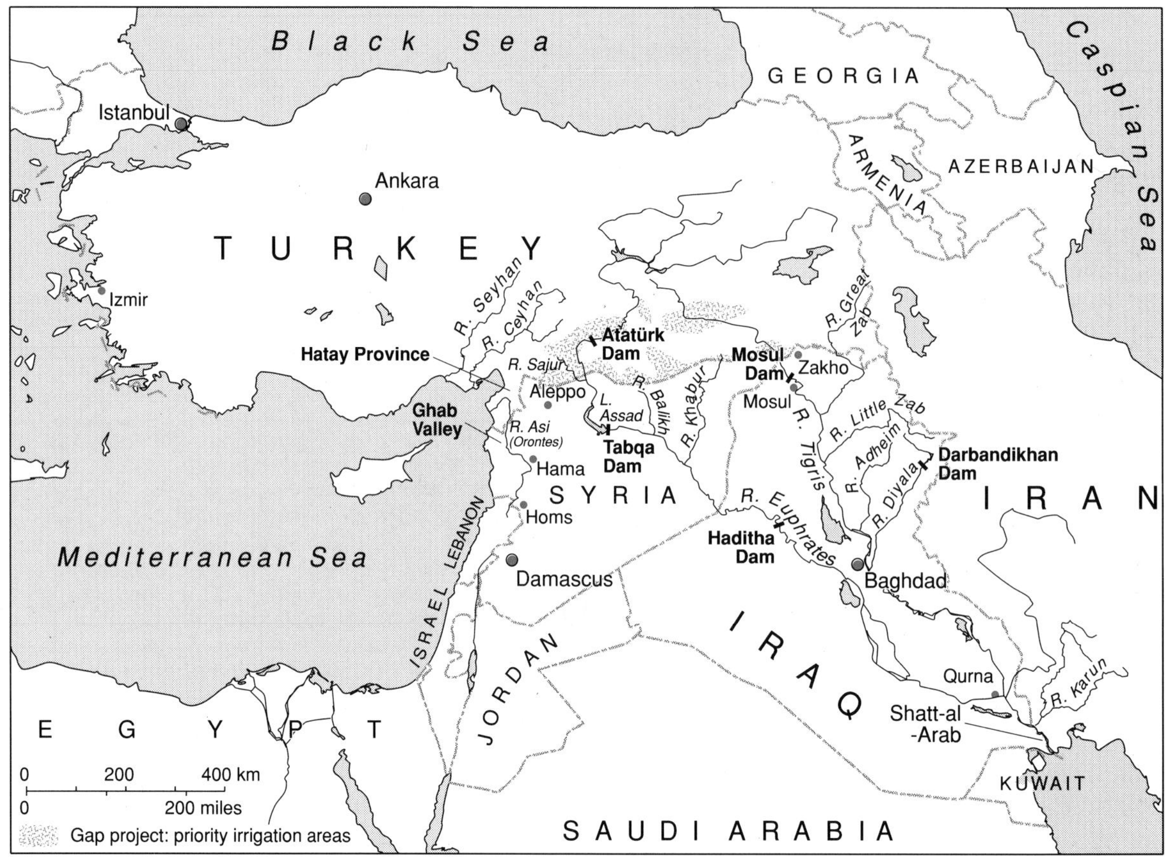

Figure 4.5 The Tigris–Euphrates Basin (Source: Beschorner, 1992, p. 28)

Since nearly half the world's peoples live in the more than 200 countries that share rivers or lakes with their neighbours, political tensions are likely to grow over the control of water supplies in many other parts of the world. In the central Asian republics of the former Soviet Union, for example, the five countries Uzbekistan, Tajikistan, Tirkmenistan, Kazakhstan and Kirgizstan share two rivers, the Amu Darya and the Syr Darya, and the Aral Sea, once the world's fourth largest inland body of fresh water. In the past three decades the sea has lost two-thirds of its volume as the rivers that feed it have been drained for growing cotton, which, under the former Soviet regime was sold overseas to raise hard currency. Recriminations regarding responsibility for the damage to the Aral Sea are hardly likely to make the reasonable division of access to these water resources any easier and although meetings were started in April 1993 to this end, commentators regard any attempts at their eventual resolution as a likely cause of major political instability. Some unresolved disputes are, though, of considerable long-standing. Perhaps the most significant of these is the quarrel between India and Bangladesh over the diversion of the Ganges at the Farakka Barrage in India in order to support a massive irrigation scheme.

However, not all nations which share access to such water resources are in a state of tension with their neighbours regarding their use. Some 280 nations have taken successful steps to produce workable agreements in the form of treaties. This is particularly so in Europe and North America whose water

The shrinking Aral Sea. Formerly the size of Tanzania, the Aral Sea had lost more than a third of its area by the mid-1980s, leaving beached vessels and once flourishing ports far inland. Its feeder rivers have been diverted for agricultural purposes. Pessimists forecast the complete disappearance of the Aral Sea early in the twenty-first century

problems were the first to become acute. Indeed, two-thirds of all the agreements relate to these areas. However, others include India and Pakistan who, in 1960, after a 13-year wrangle, finally settled their conflict over the use of the Indus under pressure from the World Bank; Mali, Mauritania and Senegal, who have co-operated for the last 20 years to manage the Senegal River; the states surrounding the Mekong River which have reached a series of multi-lateral agreements on its use and have set up a powerful Mekong Committee to ensure these are adhered to; and finally the eight countries of the Zambesi Basin which, at the time of writing, are hammering out a joint development plan under the auspices of the United Nations Environmental Programme.

4.7.4 Fresh water management – is there a global crisis?

But whilst agreements to share and manage water resources in a rational way will assist some regions or countries, fresh initiatives over the use of water will be needed as well. The considerable quantities of water used to support surface irrigation with its current dismal record of efficiency will require to be reduced by ensuring water reaches the plants without waste and in just the right quantities via sprinkler or drip systems. This will be essential for those countries driven to maintain or expand irrigation under pressure from rising populations.

Other nations which permit irrigation for social, economic or security purposes may need to think again. Is it really necessary for the Israelis to export citrus fruit; is rice growing in southern California essential; are US national irrigation programmes, such as the Newlands Project in the Nevada Desert, still relevant; are there not other ways to slow the trend of rural–urban migration; indeed, is the notion of food security really worthwhile where water budgets are limited? However, the quantitative significance of water used for irrigation should not, perhaps, overshadow the need for the demand-side management of water which can be effectively employed both in relation to manufacturing industry and in the domestic context, as was noted earlier. Such measures may have considerable regional significance, especially those which enjoy a high degree of economic development. Similarly, the problems of water pollution need to be addressed. As the Director of the UN Natural Resources and Energy Division made clear at the United Nations Conference on Environment and Development in June 1992, it is also necessary that 'far reaching measures ... be taken to preserve the quality of water bodies ... involving new approaches, such as integrated pollution control, inside an international framework for effective collaboration' (Pastizzi-Ferenic, 1992, pp. 6–7).

Without such progress in fresh water management, and given the likely increases in population discussed in earlier chapters, by 2020 it is expected that another 40 nations will have joined the 26 nations, mainly in Africa and the Middle East, that already have inadequate fresh water resources. The number of people thus affected would be expected to grow ten-fold from the present 300 million to 3 billion – one-third of the projected population of the world.

Activity 18 Look back over sections 4.6 and 4.7 and pick out the social factors that affect water supplies. How do they operate differently in the developed and the developing countries? Why is it the latter have most of the inadequate supplies? To help you prepare an answer look again at Reading B – what does it suggest? Why have the developed countries been able to ensure adequate water supplies?

Summary of sections 4.6 and 4.7

- If the world's needs for water are to be met adequately in the next century, both water resources and water demand will need to be managed.
- At present there is no single water crisis operating at a global level, only a series of looming crises scattered amongst different regions, some in the developed world, but predominantly and most significantly in the developing world.
- In the Middle East and North Africa, for example, water shortage is already constraining the socio-economic development needed to raise living standards and to control population growth.
- Attempts to come to terms with water shortages have often resulted in over-pumping and mining of ground water resources, or the misuse of waters shared with other nations.
- Pollution from industries, cities and agriculture is prevalent and remains outside attempts to make the necessary link between economic development, properly managed water resources, and social and environmental well-being.
- In the developed world there is now some evidence of attempts at a more equitable approach to the disposition of scarce water

resources amongst different interest groups, and indications of holistic attempts to manage water resources in the interests of a sustainable future.

- Evidence of attempts to manage water which are inappropriate still exist, the problems of the Aral Sea representing a prime example.
- Probably the single most significant danger for many regions will be that of resource management failure with 'political stalemate and economic grid-lock in countries sharing international river basins' (Edwards, 1993, p. 60), since this could push water towards a scarcity value and an emotional intensity resembling that of oil in the period after 1974.

4.8 Conclusion

The analysis of the sustainability of a variety of resources has reached a surprising conclusion. Contrary to the commonsense expectation that 'the limits to growth' would involve a global shortage of industrial minerals, it appears that a combination of improved exploration, product substitution and recycling can overcome material shortages at least in the developed world. The fossil fuels will pose problems, but over a long time period and with the potential resort to renewable energy ameliorating the situation. The most intractable problems in the next few decades seem likely to involve a renewable resource: water. The prospects are for growing supply and distribution problems in semi-arid areas with political and economic repercussions which could add up to a global crisis. The only remedy seems to lie in less profligate lifestyles.

The existence of serious problems with water as a renewable resource in particular geographical areas is indicative that use of renewables may be problematic. This message is emphasized in Reading A, which argues that the less developed countries will be obliged to use finite and polluting fossil fuels by the cost advantages over renewables. Here again, uneven development enters as a barrier to the adoption of sustainable technologies. Geographical differences and inequalities may affect resource use in large parts of the world in such a way as to reduce overall sustainability.

The next chapter moves on to reassess the interplay between global integration and uneven development, asking a new question: is the solution to global problems of supply of industrial minerals being achieved at the expense of local environments and lifestyles?

References

ALLEN, J. and HAMNETT, C. (1995) *A Shrinking World? Global Unevenness and Inequality*, Oxford, Oxford University Press in association with The Open University.

ALTENER (1993) *Targets for 2005*, Brussels, European Commission.

ANDERSON, D. (1992) *The Energy Industry and Global Warming*, London, Overseas Development Institute.

ANDREWS, C.B. (1992) 'Mineral sector technologies; policy implications for developing countries', *Natural Resources Forum*, Vol. 16, No. 3, pp. 212–20.

BESCHORNER, N. (1992) 'Water and instability in the Middle East', Adelphi Paper 273, London, The International Institute for Strategic Studies.

BLUNDEN, J. and REDDISH, A. (EDS) (1991) *Energy, Resources and Environment*, London, Hodder and Stoughton in association with The Open University.

BOULTON, L. (1993) 'An unhealthy drink for a nation', *Financial Times*, 7 April.

BULAJICH, B. (1992) 'Women and water', *Waterlines*, Vol. 11, No. 2, pp. 2–4.

CAIRNCROSS, F. (1991) *Costing the Earth*, London, The Economist Books.

DUNCAN, N. (1993) 'A declining future for renewables', *Oxford Energy Forum*, No. 14, pp. 6–8.

DUNHAM, K. *ET AL.* (1979) *Atlas of Earth Resources*, London, Mitchell Beazley.

ECOTEC (1992) *Research and Technological Development for the Supply and Use of Freshwater Resources*, ref. no. EUR 14723, Luxembourg, EC Office of Publications.

EDWARDS, K.A. (1993) 'Water, environment and development – a global agenda', *Natural Resources Forum*, Vol. 17, No. 1, pp. 59–64.

ENZER, H. (1980) *Economic Assessment of Ocean Mining*, Institute of Mining and Metallurgy Conference: National and International Management of Mineral Resources, May, London.

EUROPEAN COMMISSION (1993) *Employment Trends Related to the Use of Advanced Telecommunications*, Report to DG XIII/B, Brussels, Commission of the European Communities.

FISHER, A. (1993) 'Business and the environment – rapeseed popularity grown among "greens"', *Financial Times*, 8 September.

GHANDI, M.V. and THOMPSON, B.S. (1992) *Smart Materials and Structures*, London, Chapman and Hall.

GOODING, K. (1993) 'Aluminium can recycling reaches record levels', *Financial Times*, 23 April.

GRAHAM, G. (1993) 'America reaches the bottom of the barrel', *Financial Times*, 24 March.

GRIFFITHS, J. (1993) 'Energy efficiency', *Financial Times Survey*, 7 December.

HAGGETT, P. (1975) *Geography: a Modern Synthesis* (2nd edn), London, Harper Collins.

HARGREAVES, D. (1992) 'Oil and gas industry', *Financial Times Survey*, 3 November.

HOAGLAND, P. (1993) 'Manganese nodule price trends', *Resources Policy*, Vol. 19, No. 4, pp. 287–98.

JACKSON, T. (1993) in 'Reviews: Renewables are go!', *RENEW: Technology for a Sustainable Future* (Energy and Environment Research Unit, Open University), No. 81, p. 28.

KOUDSTAAL, R., RIJSBERMAN, F. and SAVENIJE, H. (1992) 'Water and sustainable development', abridged from a report for the International Conference on Water and the Environment – Development Issues in the 21st Century, Dublin, January 1992, *Natural Resources Forum*, Vol. 16, No. 4, pp. 277–90.

LASCELLES, D. (1993) 'Oil and gas industry', *Financial Times Survey*, 6 February.

MADDOX, B. (1993) 'The world's tap seizes up', *Financial Times*, 17 March.

MEADOWS, D.H., MEADOWS, D.L., RANDERS, J. and BEHRENS, W.W. III (1972) *The Limits to Growth: a Report for the Club of Rome's Project on the Predicament of Mankind*, London, Earth Island.

NISBET, E.G. (1991) *Leaving Eden – To Protect and Manage the Earth*, Cambridge, Cambridge University Press.

OECD (ORGANIZATION FOR ECONOMIC CO-OPERATION AND DEVELOPMENT) (1989) *Energy in the Twenty-First Century*, OECD seminar, 20 April, London.

PALEY, W.S. (ED.) (1952) *Resources for Freedom*, A Report to the President by the President's Materials Policy Commission, Washington, DC, US Government Printing Office.

PASTIZZI-FERENIC, D. (1992) 'Natural resources and environmentally sustainable development', *Natural Resources Forum*, Vol. 16, No. 1, pp. 3–10.

PIETILÄ, H. and VICKERS, J. (1990) *Making Women Matter: the Role of the United Nations*, London, Zed Books.

RAND, M. (1990) *Developing Wind Energy for the UK*, London, Friends of the Earth.

RICH, B. (1994) *Mortgaging the Earth – the World Bank, Environmental Impoverishment and the Crisis of Development*, London, Earthscan.

RODGER, I. (1992) 'Business and the environment – charged up in the Alps', *Financial Times*, 18 November.

ROGERS, P. (1993) 'The value of co-operation in resolving international river basin disputes', *Natural Resources Forum*, Vol. 17, No. 2, pp. 117–31.

SHEA, C.P. (1988) 'Renewable energy: today's contribution, tomorrow's promise', Worldwatch Paper 81, Washington, DC, Worldwatch Institute.

SIMMONS, I.G. (1989) *Changing the Face of the Earth – Culture, Environment, History*, Oxford, Blackwell.

SINGER, D.A. (1977) 'Long-term adequacy of metal resource', *Resources Policy*, Vol. 3, No. 2, pp. 127–33.

STRAUSS, S.D. (1993) 'Prospects for the mining industry in the year 2000', *Resources Policy*, March.

TUCKER, E. (1993) 'A source of conflict on the Nile', *Financial Times*, 21 April.

USBM (UNITED STATES BUREAU OF MINES) (1950, 1974, 1990) *Mineral Commodity Summaries*, Washington, DC, USBM.

US DEPARTMENT OF THE INTERIOR (1991) *Estimated Use of Water in the US in 1990*, US Ecological Survey Circular 1081, Reston, VA.

US NATIONAL RESEARCH COUNCIL (1992) *Water Transfers in the West: Efficiency, Equity and Environment*, Washington, DC, National Academy Press.

WORLD ENERGY COUNCIL (1992) *Renewable Energy Resources: Opportunities and Constraints 1990–2020: Energy for Tomorrow's World*, London, WEC.

WORLD HEALTH ORGANIZATION (1992–1993) *Bi-annual Report of the Director General to the World Health Assembly and the United Nations*, Geneva, WHO.

ZIMMERMANN, E.W. (1951) *World Resources and Industries*, New York, Harper and Row.

Response to Activity 7

Factors affecting the development of sea-bed mining

Positive long term	Short term	Negative long term	Short term
World population expansion implies expanded demand for natural resources	Regional flare-ups (Middle East; Bosnia, etc.) imply continuing need for defence material	Post-Cold War era; defence industry retrenchment; industrial conversion	Low-capacity utilization (manganese, nickel, copper) implies land-based sources come on line first (if needed)
World economic growth	Asian economic growth (expanded demand for copper); world emerging from recession	Exploration and discovery of on-shore deposits	Current slow pace of economic growth in the industrialized world
Industrialization of developing economies (implies expanded demand for copper; stainless steels)	Applications for stainless steel expected to grow (implies expanded demand for nickel)	Materials conservation (doing without); intensity of use declines with increasing per capita gross domestic product (GDP)	Supply of low-grade Russian cobalt to the world market increasing (hard currency needs)
Cost-reducing technological advances	Lower exploration, and development costs for sea-bed mining	Recycling becoming more commonplace (especially copper and steels)	Sales of primary metals and stainless steel scrap increasing from the Commonwealth of Independent States
Real energy costs declining; OPEC unable to maintain pricing discipline	Zaire in turmoil (might have an effect on supply of copper and cobalt – driving prices up)	Introduction of material-saving technologies (e.g. fibre-optics); newer steels require less manganese	Price trends for nodule metals well below estimated required price levels
Depletion of on-shore deposits (nickel, cobalt especially)	Sea-bed mining patents expiring	Substitution with other metals and new materials (ceramics)	Increase in strategic stockpile sales from developed nations
Introduction of new transport technologies (maglev trains; nickel alloy automobile batteries)	Prospects for revised sea-bed mining legal regime	More effective competition amongst transnational mining corporations	Demand for superalloys developed (related to restructuring in the commercial airline industry)
Continuing slow-paced deep sea-bed mining technological development efforts	Governments continue to invest in deep sea-bed mining as a source of metals for the long term	Increased attention to environmental effects of large-scale industrial activities in the ocean	Perception in industrialized world of oppressive deep sea-bed mining legal regime
Environmental effects of land-based mining operations enhances outlook for ocean mining	Strategic behaviour related to maintenance of deep sea-bed exploration claims	Risk due to effect of deep sea-bed mining on metals prices	Risk due to inexperience with deep sea-bed mining technologies

Source: based on Hoagland, 1993, p. 296, Table 3

***Reading A**: Norman Duncan, 'A declining future for renewables'*

Let me start with an assertion: 20 years from now, renewable energies will contribute about the same percentage to the energy consumption mix as they do today. How could things possibly turn out that way when the environmental benefits of renewable energy forms appear to many to be so large?

The answer is that from now on, all energy growth will occur in the 80 per cent of the world that is poor. That energy growth will fuel these poorer countries' economic growth. And these countries, while paying some heed to environmental concern, will inevitably choose the least expensive forms of energy. These will be fossil fuels. Renewable energies will be cheapest only under very special and limited circumstances.

Those who champion renewable energy forms argue that their present-day higher costs will be lowered. Undoubtedly they will be, but not to less than that of today's fossil fuels and, certainly, not within 20 years. When confronted with this, the argument is shifted to, 'Well, the cost of fossil fuels must go up substantially because of resource depletion, or cartel action, or environmental tax.' Wishful thinking, that!

I choose to tell my story from the context and perspective of one future for energy consumption, but a future – Figure A.1 – that is optimistic as to forward energy conservation. (That future is explored in somewhat greater detail than possible here in my article 'The energy dimension of sustainable development' in *The Colombia Journal of World Business*, 1992.) I have divided, for simplicity, the nations of the world into two groups: the 24 richest and industrialized, the OECD; and the group of all other nations, the non-OECD. (This future, if projected to 2020, is very close in total amount to the ecologically driven case of the World Energy Council's projections at Madrid last year once correction is made to eliminate 'traditional fuels' from the WEC figure.)

The energy history is well-known to this readership. Energy consumption for the 870 million people of the OECD has about plateaued. Energy consumption for the other 4,700 million people of the world is just now reaching, in total amount, that consumed by the OECD. (An OECD person, on average, thus consumes some 5.4 times as much energy as does a non-OECD person, on average!) Energy consumption in the OECD, from now on, will diminish in total amount; energy consumption in the non-OECD grouping, from now on, will increase.

It is the projections which tell the story and I will stress these. However, there is some housekeeping to tend to first. I define renewable energy forms to exclude nuclear energy and the so-called 'traditional' fuels such as peat, dung and firewood gathered for cooking, and so on. The only significant renewable form today is hydropower, providing about 7 per cent of world energy consumption. Most, but not all, OECD potential hydro sites have been developed. The potential hydro resource in the non-OECD is

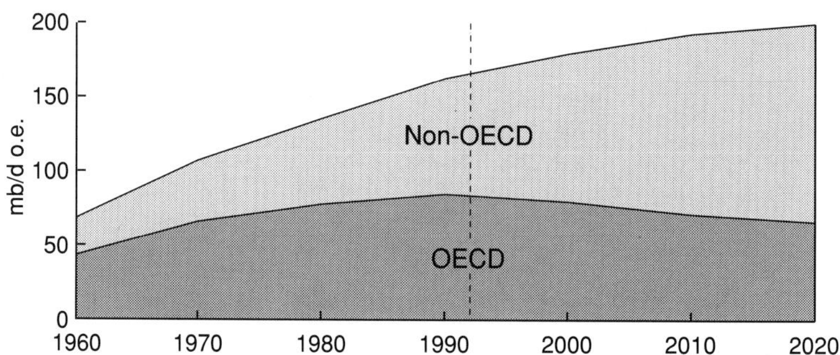

Figure A.1 World energy consumption (Source: BP Statistical Energy Series)

large, but environmental and social concerns already impede many large projects. Then too, large hydro projects are very capital intensive. That suggests that smaller hydro projects will be favoured. The 'new renewables' – solar power including photovoltaics, geothermal, biomass and windpower – in total provide less than 1 per cent of world electric power production.

The non-OECD projection

There is no secret to the expected energy consumption increases in the non-OECD grouping. They are driven by population growth and by these populations seeking higher living standards. Figure A.2 shows world population history and projections. Growth will be about one billion people in each of the next decades and almost all this population growth will be in the non-OECD grouping.

People in these poorer countries, of course, aspire to the things that have been gained in the industrialized nations – potable water sewage systems, health care, education for their children, mobility, and a lot of other things we group under the term 'quality of life'. These do not happen without economic growth. And economic growth in poorer countries does not happen without energy growth. But how much?

Energy consumption in the non-OECD averaged 6.37 barrels oil equivalent (o.e) per person in 1990 up from 5.90 in 1980 and 5.15 in 1970. The 1990 level of energy consumption was able to achieve an average GDP per person of about $1,900, up from about $1,550 in 1980 in today's dollars. If GDP per person is to continue to grow, energy consumption per person will need to grow, albeit at slower rates of growth than in the past since more energy-efficient technologies can now be applied. However, there is no reservoir of potential energy conservation to be tapped for further economic growth as there is in OECD countries.

The processes of economic growth and economic development are of enormous complexity and diversity. Figure A.3 presents a 1990 snapshot (and some history) of how countries struggle up the economic development path to higher income – each tortuous step of the way using more energy. Stages of economic growth are evident in the 1990 snapshot. African nations and China, with incomes less than $500 per person per year, are still largely agricultural. Latin American nations, averaging about $1,850 per person per year, are moving toward manufacturing, while Taiwan and South Korea, at $4,000–$6,000, are well advanced in building industrialized economies. Somewhere above the $6,000 per person per year income level, a plateau in energy consumption is reached, as, perhaps, will soon be evident in the economies of Taiwan and South Korea.

The non-OECD countries will seek the least expensive energy forms applied as efficiently as can be mustered. Their energy consumption per person per year will be about 6.90 barrels (o.e) in 2000

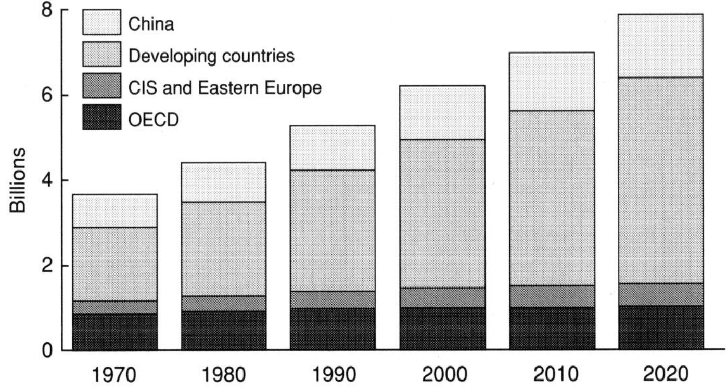

Figure A.2 World population (Source: World Bank)

[Note: CIS = Commonwealth of Independent States.]

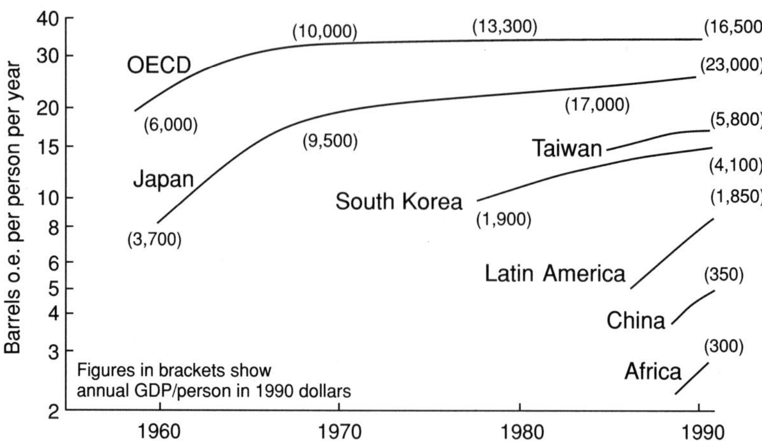

Figure A.3 The economic development process in energy terms

and about 7.15 barrels (o.e) in 2010 allowing average real GDP growth in the non-OECD grouping to be +2 per cent per year. In very few cases, in my opinion, will any of this extra energy be the 'new renewables'. The non-OECD countries cannot afford them.

The OECD projection

Twenty years ago, in the industrialized countries, economic growth and energy growth went hand-in-hand: a unit of economic growth required a unit of energy growth. By the early 1980s, the unit of economic growth required only 0.4 units of energy growth. Today, industrial economies are achieving economic growth with no increase in their overall energy consumption. And, they are challenged tomorrow to continue economic growth while lowering their total energy consumption.

This is a story of continually increasing efficiency of energy use. The reservoir that is being tapped is the high level of energy use that has typified OECD economies – 35.7 barrels o.e. per person in 1990. How quickly can this be lowered? What role will renewables play?

The recent history of OECD energy intensities is shown in Figure A.4. For the past 20 years, we have witnessed a march-down in intensity at a pace of 2.1 per cent per year. Whereas it took the OECD 3.24 barrels o.e. to produce $1,000 ($1990) of GDP in 1970, it required only 2.15 barrels o.e. to do this in 1990 – a gain of 33 per cent. That trend of improvement, as a minimum, will continue. However, note that the –2.1 per cent trend means that, if GDP growth in the OECD averages +2.1 per cent per year, then, total energy consumption in the OECD will stay at about today's 84 mb/d o.e. [million barrels per day of oil equivalent] level.

If OECD total energy consumption is to be cut, say by 15 per cent by 2010 (the decline shown in Figure A.1), and average economic growth of 2.1 per cent per year retained, then the pace of improvement in energy intensity will need to be upped to about 2.6 per cent per year. I conclude that this can be done since the technologies are in hand and there are many OECD governments ready to give energy conservation efforts that extra push.

What will be the role for 'new renewables' in an OECD situation which would shrink 15 per cent and 13 mb/d o.e. of its energy consumption in the next 20 years? It will not be as a substitute for transportation fuels since this is not the field of renewables. Nor will it likely be in the residential/commercial sector; rearrangements, there, are likely to be taken by natural gas. Power generation, especially peak-power generation, seems the only possibility. Here, solar thermal as practised by Luz International under contract with Southern California Edison, would seem to have some chance. However, cogeneration using a natural

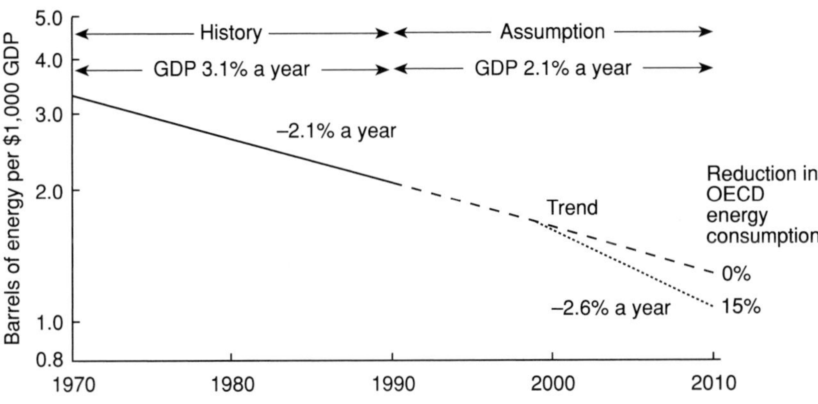

Figure A.4 OECD energy intensity

gas-fired turbine would seem the overwhelming approach of choice. It is now providing upwards of 40 per cent of new generation capacity in the United States, allowing idling of older facilities.

A final note

The cinder in the eye that is possibly at the root of the differences between advocates of renewable energy forms and those of us that are sceptical about major additions soon is more carbon emissions and the further increase of atmospheric CO_2 [carbon dioxide] concentration. Further increase in carbon emissions seems inevitable to me. Without entering the debate of whether these will cause climate change, it would seem a more productive approach than trying to deny energy to the poorer countries of the world for us to undertake more research into economic ways to recover carbon emissions from, say, power plants, and to chemically fix such carbon in ways that it does not re-enter the atmosphere.

Source: Duncan, 1993, pp. 6–8

Reading B: Borjana Bulajich, 'Women and water'

Until the way in which water supply and sanitation policies are decided changes to include women in every step of the process, the huge improvements that are necessary for healthy communities will not take place.

Water is a prerequisite for the survival of human beings and for their development. Current and projected problems with freshwater resources arise from the pressure to meet the agricultural, human settlement, food and industrial needs of a fast-growing global population. Inadequate water and sanitation facilities are some of the most critical problems faced by the developing world today.

Statistics tell the story. In the 1990s, over one billion people in the developing world lack safe and adequate drinking-water, while those lacking sanitation number almost two billion. The lack of a healthy environment and the lack of safe drinking-water are the cause of 900 million cases of diarrhoeal disease every year, which cause the deaths of more than three million children: two million of these deaths could be prevented if adequate sanitation and clean water were available. At any time 200 million people have schistosomiasis or bilharzia, and 900 million more have hookworm, cholera, typhoid, or paratyphoid.

The ever-increasing scarcity of water, combined with environmental degradation, continues to have serious impacts on primary water carriers, managers, end-users and family health educators. The economic and social costs resulting from the lack of safe drinking-water are high. The average proportion of their working time spent on water collection by women in East Africa varies from about 12 per cent in humid areas to 27 per cent or more in dry or mountainous areas. By virtue of their domestic functions, women are in constant contact with polluted water and are therefore the group most vulnerable to water-related diseases.

Power and responsibility

In the area of water resource development and environmental protection, women are at present more often victims than beneficiaries. As women are responsible for the domestic use of water as well as for the provision of fuel and for the production of food crops, they are the ones most keenly aware of and the most adversely affected by the current negative developments. At present, women are still often excluded from both environmental and river basin development projects. For example, planners of projects related to soil conservation, agricultural extension, and credit for water conservation activities seldom include women or women's groups in the planning stage. In most irrigation projects, land and water rights are vested in the male head of the household, leaving women without the land or water to grow the staple foods essential for family health.

Women are more than target groups; they are active agents who can contribute constructively and knowledgeably to policy decisions, and who can also mobilize labour, provide resources, and disseminate and implement innovations. By involving women, particularly in the planning, design, operation and maintenance stages as well as in health education programmes, water and sanitation projects could more effectively achieve the ultimate goals of more and safer water, resulting in better health.

The multi-sectoral nature of water supply and sanitation activities among women requires appropriate co-ordination among the national institutions and authorities concerned with water, health, sanitation, and agriculture and rural development, as well as among agencies in charge of education and training, including international organizations. Appropriate agencies at the national level should be instrumental in the co-ordination between responsible ministries and women's organizations. The strategy for women's participation needs to include water and sanitation if it is to become an integral part of the whole development process. Improved water supply and sanitation facilities could have many direct benefits, such as the reduction of the drudgery of water collection, an improvement in health, nutrition, and food supply, and environmental protection. Moreover, there are indirect benefits in the form of improved potential for economic and social development, such as the rise in productivity and incomes, and improved standards of living.

The management of local water sources and watersheds, forestation and the prevention of local pollution are typical areas where the interests of women and development planners go hand in hand, and in which women have already played constructive roles at the neighbourhood and community levels. So far, many countries have adopted policies of community involvement and women's participation and many *ad hoc* examples can be found of how this involvement makes a difference to local support, use and maintenance. Development planners and engineers put policies into practice, and the responsibility ultimately lies with them as to whether these policies surface visibly in projects and programmes. Women's and non-governmental organizations can support this endeavour by establishing co-operative structures in engineering programmes and by educating their own colleagues about how women can play a more prominent role.

Despite their increasingly important and multiple roles, there is still not enough attention paid to women's roles in water supply and sanitation. They are the primary resource in water collection, and they are the greatest users of water. Their water-related work is taken for granted and denied its economic and social value. Women are often excluded from both the early planning and the final implementation of water projects. Projects concerning women lack the elements of communication and information on the relation between water, sanitation and health. Local women's customs, preferences and traditions are not taken into consideration during the selection of the technical design and the location of the projects. Training programmes on water activities rarely include women, and evaluations seldom consider the impact of water projects on the lives of women specifically.

Positive action

The role of women working in the field could be greatly increased through education, training and the inclusion of gender issues in water policies, programmes and projects. Equally, the recognition and enhancement of women in water supply and sanitation depends on a firm commitment at the national level. For example, in low-income urban areas women play a prominent role in innovative approaches to more sustainable water and sanitation services. They are (co-) managers of communal waterpoints, latrines and vending stations, they run local water supply and wastewater treatment systems, promote domestic sanitation systems, and manage and collect domestic waste for recycling and re-use. In rural areas women, as managers of communal waterpoints, are concerned with drainage and hygiene, the proper use of taps and pumps, and the prevention of damage by children and livestock.

There are many examples of successful solutions to water problems that have also had a positive impact on the environment and the socio-economic well-being of women and their communities. Still, a lot remains to be done to involve them effectively in environmentally sustainable water programmes and projects.

At the national and international levels, government and non-government organizations, women's groups, and international agencies have critical roles to play. Three points should be made concerning their approaches to women's participation. Women's participation should be part of an integrated approach in the management and support of sustainable water activities; women's issues are an integral part of community and national development concerns; and the emphasis on women's participation does not imply that activities should be carried out by women only. It stresses rather the need for both men and women to address the issue.

The 1990s and beyond call for a holistic approach towards the development and management of water resources, which is a prerequisite for the effective sustainable development of nations. It implies the development of human societies and economies and the protection of natural ecosystems on which the survival of humanity ultimately depends. This includes the need to look not only at the water cycle, but also at inter-sectoral needs, ecological issues, the alleviation of poverty and disease, sustainable rural and urban development, and protection against natural disasters.

Source: Bulajich, 1992, pp. 2–4

Trade globally, pollute locally

Chapter 5

by John Blunden

5.1 Introduction

Chapter 4 has suggested that there is no inevitable global limit to the provision of industrial raw materials, energy or water. Moreover, it has also suggested that unevenness of supply and demand has produced serious local problems even for renewables. This chapter takes a further look at the system of provision of industrial raw materials to see whether comparable local problems exist and, if they do, whether they add up to a problem for sustainability.

regulation

This question is particularly crucial since industrial raw materials are found, extracted and supplied by truly global industries. If it is generating serious local problems, this would pose major issues of *regulation*, and hence of international politics.

The chapter has two major sections, the first analysing the supply system for metals and the second based on a single reading, detailing the environmental impacts. You should read both sections with particular emphasis on conflicts of interest between localities, nation-states and multinational companies.

5.2 Supply and demand for industrial minerals

We begin by considering the global/local relationships which pertain in the extraction of industrial mineral resources. In considering the factors which determine where these are exploited and why, we shall be able to comprehend the increasing globalization of the organization of the industry so that its effects, whether social, economic or environmental, can rarely be considered 'off the map'. Although the processing of minerals can add to global environmental problems, most of their exploitation has a local/regional dimension in terms both of social and economic development and of the environment. However, these local/regional impacts are uneven, with extraction and environmental damage increasingly associated with developing countries, and the benefits of consumption – if that is, indeed, what they are – largely confined at present to the developed world. Indeed, close parallels with our discussion of water can be recognized here insofar as social processes work themselves out unevenly over space, affecting the developed and the developing worlds in different ways. As you move towards the end of this chapter, you will come to appreciate that the so-called benefits of minerals exploitation fall within the arena of contested activity. Moreover, both Reading A from John Young's *Mining the Earth* and the narrative of the chapter increasingly emphasize the idea of more than one value position regarding the exploitation and use of minerals.

5.2.1 Some regional variations

Minerals of the kind that in the industrialized world are considered to be of value, and appear in one form or another in the common artifacts that are used in everyday life, are to be found in many rock formations. But they are not always there in the quantities that might make it possible to consider them as mineral reserves. However, mineralization – and the many different ways in which this comes about as a result of geological and chemical processes within the earth or of the forces of erosion and deposition working upon the surface of it – often concentrates deposits of metallic ores, or many of the other minerals we use, in some places more than in others. To the non-geologist it would seem to be the result of some gigantic geochemical lottery. The outcome is that such minerals are only likely to achieve the status of a reserve where this concentration has occurred. Thus minerals, for this reason alone, are exploited with great spatial variation which can, in turn, give rise to uneven development.

Whilst geological circumstance provides the basic determinant of minerals development, other factors must come into play to establish the viability of minerals as reserves rather than just as a part of the earth's stock. You will remember from the introduction to the previous chapter that we argued that a number of factors, including price (representing the state of supply and demand) and technology, could be important in the international market in determining reserves generally. Thus during periods of buoyant demand and rising prices, stocks of a resource which were on the margins of viability might be drawn into the production arena as reserves.

Where minerals are concerned, however, high levels of investment are now required to investigate and prove reserves, and then to exploit them – especially the metallic ores. In these circumstances, the transnational mining corporations, as the only groups able to function at such global levels, want to be assured of the security of their investment throughout the life of the mine, thus introducing a further factor into the decision-making process regarding location. This will involve their being not merely certain of the integrity of the governments of territories in which workable reserves exist, but assured about such matters as the ownership of mineral rights, the organizational framework within which exploitation can occur, and the nature of the taxation regimes to be imposed upon them as operational minerals companies.

Activity 1 What other constraints might you expect to be imposed on mining activity which can affect location? A brief re-examination of Chapter 4, sections 4.3.1 and 4.3.2 will be helpful here.

You will certainly have noted that environmental considerations can be important. Sometimes potential host governments may impose a set of rules aimed at merely containing the impact of the minerals extraction or processing. However, the protection of wilderness and nature reserves against mining development may come into play. For example, new extractive activities will almost certainly be forbidden in areas such as national parks. Indeed, in a celebrated case in the 1970s in the UK, the transnational company Rio Tinto Zinc was unable to obtain consent to work an economically viable copper ore in Snowdonia National Park, in spite of what the company saw as a formidable array of environmental safeguards. But more than this, the rights of indigenous populations, once largely disregarded, are increasingly to be seen as an important consideration in the decision-making process.

Just what these issues have meant in determining *where* mineral reserves are developed is discussed in sections 5.2.2 and 5.2.3 below.

5.2.2 Organizational and political factors

To understand the interplay between the key players in mining development – the mining companies, the national and sometimes provincial governments of the host territories in which the reserve is located, as well as the local communities that may be affected – it is necessary to know a little more about the protagonists and their power bases. First the mining companies themselves.

At the end of the Second World War, minerals exploitation was carried out largely by relatively small nationally based companies, as it had been for much of the twentieth century up until then. There were some exceptions, particularly in the field of oil production. By the 1970s, however, mineral production was being undertaken mainly by large international concerns – transnational corporations – frequently controlled from the USA but operating on a global basis, often in developing countries. In spite of attempts by the oil-producing countries, mostly located in the Middle East, to control output of oil, and therefore world prices, by acting in concert through their cartel known as OPEC (the Organization of Petroleum Exporting Countries), seven major transnational oil corporations nevertheless dominated the world petroleum industry. Indeed, these corporations managed to break the power of OPEC through successful exploration programmes in non-member countries. The production of aluminium, the commonest and one of the most versatile of all metals and won from the mineral bauxite, was largely determined by six major producers. Three corporations controlled uranium production, essential to the needs of the nuclear power industry, whilst the extraction of the world's nickel supply, a key mineral in a range of metallic ores which impart strength and hardness to steel, was in the hands of two corporations. Most of these transnational organizations not merely acquired the mineral rights and the extraction and processing facilities, but also controlled smelting operations and those of primary fabrication. This *oligopoly control* of output exercised over complex operations in many countries, when allied to highly skilled market intelligence, ensured for these corporations a considerable command of the international market for their products.

oligopoly control

As for the governments in those territories where important reserves had been identified, from the end of the 1960s and into the decade which followed, many of them were developing countries emerging from colonial status. Their achievement of territorial independence fostered a strong desire on the part of many of them to manage their own mineral resources which were often their only hope of earning foreign currency and providing higher levels of domestic employment. This was part of what has been described by Radetski as the 'economic nationalism of the late 1960s and 1970s' (Radetski, 1992, p. 2). Indeed, the governments of these countries believed that the profits that had previously flowed to foreign owners as a result of minerals exploitation would now be captured for their own benefit, and such beliefs, when acted upon, inevitably led to a considerable reduction in the foreign control of mining enterprises. Sometimes the host governments achieved their objectives through expropriation of part or all of the share capital of the mining enterprise. In 1969, the Zambian government, for example, took a controlling interest of just over half the share capital in the two giant transnational copper mining groups, Anglo-American Corporation and Roan Selection Trust – holdings which were later increased by the host nation to

60 per cent. In Botswana the government compulsorily acquired share capital in the copper mines run by an international consortium of foreign companies lead by AMAX, whilst in Zaire the state took over the interests of Union Miniere.

In Latin America, a metalliferous province the potential of which was only just beginning to be realized in the 1970s, development by the transnational corporations largely came to an end. In Brazil, one of the countries most hostile to foreign transnational mining enterprise, only Rio Tinto Zinc remained as a partner with the state to exploit its considerable potential as a major world producer of tin, bauxite and iron ore. As recently as 1988 the restrictions on overseas corporations were further reinforced, preventing their participation in mining in that country unless they were prepared to allow the state to have a majority holding.

These actions by host governments caused the transnational corporations that had been partly or wholly dispossessed of their investments in Latin American or African countries seriously to re-examine their future patterns of mining investment, especially at a time when mining development costs were escalating at rates well in advance of inflation. This situation was largely due to the fact that the richer mineral lodes that had been characteristic of some of the world's mines up to the middle of the twentieth century were, by now, worked out, and it was increasingly necessary at that time to work grades of ore that were much lower in metal content. By the early 1970s a copper mine with its attendant smelting and refining plant could involve an outlay of US $1 billion. Figures of this order of magnitude of necessity have to be raised on the world's financial markets by international corporations with the appropriate expertise, a task generally beyond that of the government of a developing country alone. But where it did seem possible for such a government to enter into a development partnership with a transnational corporation, difficulties could still arise.

For example, in the realization of a copper-mining project at Caujone in Peru, the Peruvian government required that the mining consortium join in a profit-sharing arrangement with it. When, after several years of discussion, negotiations were completed in 1969, it took a further six years to finance the project, involving the World Bank and 29 other commercial banks, before the construction phase, scheduled to last seven years, could begin. Eventually, US $726 million was spent on the development of the project. However, shortly after the mine began operating, the government of Peru unexpectedly imposed a series of export taxes on the output of the mine for domestic reasons, which considerably lowered the return on investment capital.

In many other developing countries the governments found that, for quite legitimate reasons outside their own control, they could only finance their spending through additional taxes on the mining industry. The transnationals in question, unable to resist such measures and with no direct means of redress, found it increasingly difficult to raise the money required for new mining developments in such countries and turned their attention to the safer long-exploited metalliferous provinces of the developed world. Thus, by the mid-1970s, an authoritative report (World Bank, 1977) showed that expenditure on exploration in the developing countries had fallen to 15 per cent of total outlay on such activities, whilst 80 per cent was spent in Australia, Canada, South Africa and the USA, where investment was deemed safe and taxation regimes much more predictable.

Had mining investment in the developed countries continued at such a rate, it could be argued that the inevitable absence of such capital from similar appropriate enterprises in the developing countries represents an opportunity missed for them in terms of the generation of economic growth. Moreover, it is the developing countries that are now known to have the largest share of the reserves of the key industrial minerals. For example, in Latin America alone, Brazil is now known to hold one-third of the world's iron ore reserves, Chile a quarter of the copper and Peru between 10 and 20 per cent of the total reserves of copper and silver and 7 per cent of all zinc.

However, since the beginning of the 1990s there has been a marked change in the attitude of transnationals towards their investment plans. This has been brought about by a combination of factors. First of all, corporations operating in many of the developed countries now perceive themselves as beleaguered by increasing government restrictions, regulation and taxation. In the Canadian provinces in particular, interventionist administrations now exist whose practices have resulted in a lengthening list of government levies, including those relating to licensing fees, fuel taxes and the funding of worker's compensation claims, all of which are unrelated to company profits. As one large minerals corporation operating in British Colombia sees it: 'the resource industries of the province have been milked and drained to support costly government programmes and a standard of living which makes our operation uncompetitive in international markets' (Metals and Mining Research Services, 1992).

Additionally, and especially in North America, mining transnationals see themselves as being under attack by a well-funded environmental lobby which is sapping public support for exploration and mining. The intrusion into wilderness areas by minerals extraction activities, the subsequent damage to ecosystems caused by them, including primary processing and, in some instances, smelting, and the problem of site rehabilitation when the mines close, have all been subject to the antipathy of such groups, even where the proposed developer has carefully carried out an environmental impact assessment and tried to minimize the effects of mining. One corporation, Placer Dome, operating in this same province, estimates that the project approval process for new mines, which is mainly concerned with environmental matters, lasts an average of more than two years without any assurance that they will be eventually allowed to go ahead (personal communication, 1992).

At the same time there has been a sea change in the attitude of many developing nations. In the Latin American countries, for example, under pressure from the International Monetary Fund which sees mining development as an important tool in promoting economic growth, they have had to become, in general terms, committed to financial and trade liberalization, increased foreign investment, and a reduction in their indebtedness to the developed nations. According to Gooding (1992), the governments of these countries are now adopting attitudes which are a far cry from those of two decades ago and have begun to compete for foreign investment and technical expertise. They are now offering transnational mining corporations *preferential tax treatment,* which can include repatriation of capital and profits with limited restrictions and total taxes on profits at 30 per cent or less. This compares with the marginal corporation tax rate in Canada of 49 per cent. They are also offering security of mining tenure;

preferential tax treatment

government joint ventures, the nature of which is guaranteed through third parties; and prompt government response to, and assistance with, development plans and proposals.

Even though environmental and safety standards applied in Latin America are comparable with those of the developed countries, the processes of seeking permission to mine are very much quicker when considered against those of, for example, the USA and Canada. For instance, Cambior, a Canadian company, is bringing (at the time of writing) a large gold mine, producing 250,000 ounces a year, into production in less than two years in Chile. Moreover, the costs of the environmental review and of obtaining the consents to work at between 6 and 10 per cent of total development outlay in the USA or Canada, are, in general, one-third less.

Activity 2 Briefly tabulate in two columns the factors pushing transnational mining companies out of the metalliferous provinces of the developed world and pulling them towards developing countries.

Given the 'push' factors favouring new minerals development outside the developed nations and the 'pull' factors now at work in Latin America, it is perhaps not surprising that transnational corporations are currently seeking to invest in copper, gold and zinc ventures in Chile; silver and copper in Mexico; tin, gold, copper and zinc in Bolivia; copper, zinc and gold in Peru; copper in Argentina; copper and gold in Panama; nickel and gold in Venezuela; and gold in Costa Rica. One bonus for the incoming mining corporation is that, while known reserves of metallic ores are considerable, much of Latin America has still not been fully surveyed using modern techniques, so the potential for finding a very large mineral deposit is much greater than it would be in North America.

5.2.3 Developmental outcomes

It is, then, apparent from sections 5.2.1 and 5.2.2 above that, whilst geological circumstance provides the basic opportunity for minerals exploitation, other factors, including those of an economic and of a political nature, determine where development may take place. From a global viewpoint, decisions about such minerals enterprise represent a formidable contribution to world trade. At the height of the economic boom of the early 1980s and before the first of the two recent major international recessions of 1982–1983 and the early 1990s, the contribution made by minerals stood at US $318 billion. From a low point of US $243 billion in 1986, the figure began to recover, returning to early 1980 levels by the end of the decade, although it fell back again in the early 1990s. However, by 1993 the share of minerals export earnings by developing countries had reached about one-third of the total. This figure is set to rise even further as a result of the new developments already outlined above. For example, in Chile, a country long dependent on its mining activities and one of the first to have to pursue a policy of attracting back the mining transnationals with a liberal foreign investment regime and mining code, minerals already account for half its exports and 15 per cent of its gross domestic product (GDP). This sector is expected to grow by a further 50 per cent by the year 2000.

For many developing countries with mineral reserves, it is not only the export value of minerals that has economic significance. Although in the developed world mining is not a large employer of labour when compared with other sectors of the economy, the situation can be quite different in developing countries where mining can offer one of the few opportunities for both formal and full-time employment. However, there may well be other secondary opportunities which derive from the mining operation itself since it may require massive additional *infrastructural development.* McDivett and Manners (1974) cite the now classic example of the construction of an iron mine in Liberia at Lanico. Even in the early 1970s this involved an outlay of US $100 million. Of this total sum, half was spent on harbour and railway construction, and 10 per cent on electricity supply, as well as the building of a township and a hospital. Only 32 per cent of the total investment went into the mine itself.

infrastructural development

As a more formalized way of attempting to quantify the likely stimulus to regional economies of the opening up of new mining operations, the concept of the *multiplier model* has been developed. Although the practical application of multiplier models involves many complex problems, the basic concept is simple and deterministic. Let us suppose, for instance, that there is an initial injection of incomes into an area through a mining project requiring several hundred persons, mainly local manual workers and perhaps 20 to 30 other people for clerical tasks. At least some of the money generated by this enterprise would be spent locally and would find its way into the pockets of local business people who, in turn, would spend some of their extra income locally with other enterprises, and so on. The total amount of local income generated will be greater when more is spent locally and less is allowed to leak away through the purchase of goods and services from outside the area. Thus it is possible to estimate in any situation the hypothetical size of the local multiplier effect, given assumptions about local spending and after making allowance for rates of taxation and savings. In the UK, studies using this methodology have concentrated on examining the development impact of offshore oil on north-east Scotland, and, more particularly, on the Shetland Islands; the development of a large copper mine in north-west Wales; and the development of a tungsten working in the Plymouth area of Devon in the south-west of England. Box 5.1 considers the nature of these models and their application in the Devon development.

multiplier models

Downstream linkages of the kind referred to in Box 5.1, in which the further processing of metallic ores occurs, can be of considerable importance in terms of economic development. These can involve further improving the metal content of the ore by using concentration processes. Locally, such operations are usually more common than smelting and can more than double the value per tonne of ore produced, as well as offering further employment opportunities. However, while it is apparent that, in the past, additional processing played a disproportionately small role in developing countries – only one-third of the operations in these countries had such facilities between the mid-1950s and 1985 – this is now changing. Developing countries are beginning to be major players in *minerals processing* and ship at least part of their ores as concentrates or refined metals. For example, 75 per cent of copper ore is now exported as metal. In addition, the transfer of 'know-how', particularly on the engineering side of mining activities, can be

minerals processing

Box 5.1 Tungsten mining – an agent for stimulating a regional economy

Four types of relationships were investigated when a multiplier model was used to examine the likely impact of a new tungsten mine on greater Plymouth, the area having been delineated at the beginning of the study. These were (i) purchases of inputs by the mine (transport, plant hire, construction and engineering services, etc.); (ii) purchases by the labour force employed (for everyday needs, recreation, etc.); (iii) induced investment in the infrastructure which may occur as a result of employee demands for extra services (schools, hospitals, social services, etc.); and (iv) 'downstream' linkages (local processing activities and smelting; manufacturing activity based on mine output, etc.), which can involve a further build-up of employment.

From evidence of local consumer spending, collected in the study area, and a comparison with other studies concerning the impact of similar developments both in the south-west region and from elsewhere in the UK, a multiplier of 1.7 was suggested as appropriate for the mine.

Thus for the 340 workers involved in the operation (including 46 clerical and supervisory staff), with an annual payroll estimated at £2.8 million for the first year of operation, it was anticipated that the multiplier would transform this sum into a total local impact of £4.8 million per annum. However, the fact that a third of the workforce was recruited from outside the area and would 'export' some of their earnings, and that assumptions regarding rates of taxation, etc., needed to be taken into consideration, a modest downward adjustment of this figure was required.

As to non-labour expenditure by the operation, since not all goods and services could be supplied from within the area (for example, major items such as electricity and fuel), a considerable leakage of expenditure (more than 80 per cent) was allowed for before arriving at a figure to which the multiplier was applied. This gave a total of just under £1 million for the first year of operation.

No allowance was made for infrastructural investment because investigations showed provision in the area had substantially moved ahead of other similar areas in the south-west region over the previous decade. 'Downstream' linkages also failed to materialize because of the limited size of annual mine production and the specialized nature of its output.

Nevertheless, the study concluded that, at the time the multiplier model was applied (1982), the mine was likely to generate just under an additional £6 million in the study area in its first year of production. The mine was predicted to have an active life of 20 years, assuming an average annual output of 4,500 tonnes of tungsten concentrate. Inputs to the model were subsequently adjusted to permit forecasts of the impact of the mine in later years.

Source: based on Blunden, 1985, pp. 234–8

especially important to other aspects of the economies of developing countries. Indeed, the development of sophisticated environmental control systems for mining operations is now proving helpful in enabling developing countries to cope effectively with more general pollution problems, particularly those of water, as was recognized at the second World Industry Conference on Environmental Management held in Rotterdam in the winter of 1991.

Activity 3 Spooner in *Mining and Regional Development* (1981) has identified what he calls 'the characteristics of minerals resource exploitation as a means of stimulating and maintaining regional economic development'.

All four of these characteristics can be found in the case study in Box 5.1. Try to identify them.

Summary of section 5.2

To complete this first section on industrial minerals as stock resources, let us summarize our discussion so far:

- The exploitation of industrial minerals is no longer the concern of local or regional entrepreneurial initiatives, but the domain of transnational companies or corporations operating at a global level, using capital raised in the international money markets and serving markets primarily in the developed countries.
- The World Bank and other development institutions have pursued a strategy which emphasizes the role of mining as a tool of economic growth.
- Although geology plays a fundamental part in determining the uneven spatial distribution of industrial minerals, the designation of such resources as reserves heightens the idea of their uneven development. This is because other dynamics, frequently technological but also political in character, are also brought into play to determine where minerals are exploited, with the current emphasis particularly upon the winning of metallic ores from the developing nations.

5.3 Mining as a contested activity

5.3.1 Minerals overdependence?

Although minerals development might be represented in terms of economic benefit, there are alternative views. First of all, there are critics of the role of mining as a development tool. Indeed, some commentators point to what they see as evidence of severe harm done to developing countries, suggesting that they have been left ravaged and indebted by mining development strategies. They would maintain that these countries have been encouraged to accept a situation in which there is too high a dependence on one mineral as a single key export. Table 5.1 illustrates such a degree of dependency very vividly. The problem arises because the minerals exports of these countries are traded in an international market which is notoriously volatile, and where a prolonged period of depressed prices can leave in its wake both economic and ecological damage.

Table 5.1 Shares of minerals in value of total exports, selected countries, recent years[a]

Country	Mineral(s)[b]	Share (per cent)
Botswana	Diamonds, copper, nickel	87
Zambia	Copper	86
Zaire	Copper, diamonds	71
Surinam	Bauxite/aluminium, aluminium	69
Papua New Guinea	Copper	62
Liberia	Iron ore, diamonds	60
Jamaica	Bauxite/alumina	58
Central African Republic	Diamonds	46
Mauritania	Iron ore	41
Chile	Copper	41
Peru	Copper, zinc, iron ore, lead, silver	39
Bolivia	Zinc, tin, silver, antimony, tungsten	35
Dominican Republic	Ferronickel	33
Guyana	Bauxite	31
South Africa	Gold	29

[a] Figures are for the most recent year available (1990 or 1991); the earliest (Zaire) is for 1986.

[b] Minerals are listed in order of total export value.

Source: Young, 1992, p. 32, Table 7

5.3.2 Environmental damage

Dependency is, therefore, not all. Mining activity, which is now more frequently undertaken using open pit techniques and heavy earth-moving machinery because of their suitability for the extraction of low-grade ores and their cost-effectiveness, can have very profound impacts on both land and landscapes. More than the vast holes in the ground which are the inevitable outcome, waste rock or *overburden* remains to sterilize considerable areas in the immediate vicinity of the mines, whilst millions of tonnes of fine rock material, resulting from the processing of the ore, are deposited in vast tailings ponds. The smelting of the refined metallic material, if it has originated from a sulphide ore, results in the discharge of considerable quantities of sulphur dioxide into the atmosphere. Deposited sometimes many hundreds of kilometres beyond the sources of discharge, in the form of *acid rain*, profound damage is done to vegetation and to aquatic biota, thus adversely affecting the ecology of wide areas.

overburden

acid rain

The incidence of mining and minerals-processing damage is inevitably increasingly to be encountered in developing countries. However, the need for such mining and minerals-processing activities, particularly on the scale at which they are currently occurring, is a matter which is contested by a number of environmentalists and environmental interest groups. Indeed, this is the subject of a recent paper, *Mining the Earth*, from one of the foremost of these, the Worldwatch Institute in Washington, DC, in the USA (Young, 1992).

Activity 4 Turn now to Reading A 'Mining the earth', by John E. Young, which is taken from this Worldwatch Institute paper. You will find this at the end of the chapter. As you read through the article make notes on how you would answer the following questions:

- What kind of reasoned defence might a mining corporation make against the charges made in Young's article?
- To what extent do you think it would be possible to regulate this so-called 'environmental pariah' more strictly?
- Why is it that the developed world might be deemed to bear a heavy responsibility for the problems of a country such as Zambia?
- What environmental benefits can emerge as a result of recycling?
- What does the article have to say about government support for mining?

5.3.3 Community concerns

So far emphasis has been placed on both misgivings about the role of minerals as a development tool, and of minerals as a general agent of environmental damage, a view well articulated by John Young in his article. However, the impact of mining may also be considered a contested activity where governments and transnational mining corporations may be at one over the mutual benefits of mining development, but local communities are not.

local amenity groups

Sometimes mining operations are contested by *local amenity groups*, particularly in developed countries. For example, the evidence cited in Box 5.1 regarding the multiplier benefits of a tungsten mine, was but one of the arguments – in this case in favour of this particular development – which were aired at a public inquiry held to investigate the proposed mine to be located on the edge of a national park in the south-west of England. Others, including well-organized local amenity groups opposed to it on the grounds of damage to the local landscape and its ecology in a sensitive area adjacent to Dartmoor, drew attention to its disbenefits. Ultimately, the company, in order to achieve its objectives, conceded an expensive post-mining rehabilitation scheme for the area, which also included a tract of adjacent countryside already damaged by china clay extraction (Blunden, 1985).

However, as far as minerals are concerned it is not difficult to identify other forms of conflicts of interest where indigenous peoples, hostile to mining development, either own the land where exploitation is proposed, or believe it to be rightfully theirs, or share an ethos which believes such activities to be alien to their culture. For example, whatever the economic advantages of oil-based development were considered to be in the Shetland Islands – the operational phase of the development directly offered over 900 new jobs alone – the opening up of the North Sea oilfields of Brent and Ninian to the north-east in the mid-1970s, potentially feeding shore-based facilities on the islands, envinced much local opposition. Indeed, the islanders showed themselves in surveys taken at the time to enjoy substantial non-monetary advantages derived from work in their traditional industries. Not surprisingly, they determined that if the infrastructural requirements for bringing oil ashore proved irresistible against wider national interests, as much as possible would be done to reduce the impact on the local culture. Moreover, an articulate, well-briefed local authority, backed with the necessary financial resources, fought hard and successfully to ensure that the earnings on every barrel of oil brought ashore were top-sliced by a special tax linked to world oil prices and company profits, to provide a fund to support viable local lifestyles for the indigenous population once the oil had gone. As Lewis and McNicoll (1978) have argued, if oil development has inevitably transformed the economy of the islanders, they are well placed to receive most of its benefits long after the oil runs out.

Within the developing world, too, there are examples of local communities having been equally empowered in their dealings with mining interests. In the 1980s, Bougainville Island in Papua New Guinea supported the largest open pit copper mine in the world. However, the land-use rights of the mine remained with the local people, who, even if they were subsistence farmers, were bound to have a major influence on the mining project on their land. The significance of this influence grew throughout the decade as the landowners became increasingly dissatisfied with their share of the proceeds from Papua New Guinea's single largest export and tax earner, when measured against its adverse impact on the environment and on their way of life. By 1988, their well-organized agitation became an irresistible imperative for the government, which responded by offering them a 20 per cent share of the royalties on the output of the mine as well as a stake in its equity. However, other villagers in the area, who felt that they were unlikely to have any of these additional benefits passed on to them, destroyed the mine operator's property, thus bringing about the closure of the mine. This has remained the situation, despite the major downturn in foreign earnings accruing to the national government (Kennedy, 1990).

community empowerment

For the most part, however, the interface between mining development, largely supported at either federal or state levels, and the interests of indigenous peoples has invariably been far from positive when seen from the point of view of largely powerless local communities. Australia offers a number of examples of this situation, especially in relation to the indigenous Aboriginal population. For instance, in the Western Australian goldfields Aboriginal land rights have not been recognized, nor has the state government levied royalties on production which might have benefited local communities, as in the Shetland case. Conditions of poverty, poor health and poor educational provision have prevailed, the benefits of locally produced wealth having largely bypassed local economies in favour of the interests of the mining corporations and the state government. Indeed, until recently, concern was being expressed about the continuing survival of Aboriginal communities at all.

In such circumstances, it is, perhaps, not surprising that whilst applications to develop minerals in Australia have, over the last two decades, required a thorough-going preliminary examination of their likely environmental impact, only recently have such *Environmental Impact Statements* been extended to address the social effects of mining on Aboriginal communities. However, some mining developments, started prior to an appreciation of the need to take cognizance of local social and cultural considerations, have undergone a considerable change in attitude more recently. At the East Kimberley diamond mines in the early 1980s controversy surrounded the exploration

Environmental Impact Statements

and development phase of the operation because of damage to sites sacred to the Aborigines. In the early 1990s, however, the mining corporation concluded an agreement with some Aborigines which includes a commitment to mitigate some of the social impacts of mining and to support the work of the East Kimberley Impact Assessment Project. Its task, detailed in a working paper from the Australian National University's Centre for Resource and Environmental Studies (CRES, 1985), is to find ways to harmonize resource development with the interests of the Aborigines.

Such changing attitudes by mining corporations in their approach to Aboriginal communities have been best documented with respect to the Weipa mine in the Cape York region of North Queensland, as Box 5.2 indicates. This has been operated since the mid-1950s by the transnational Comalco.

Box 5.2 Weipa: industrialization and indigenous rights

When bauxite mining was proposed by Comalco, the Federal Government recognized what it saw as the economic imperative of developing an aluminium industry for the whole of Australia, whilst the government of the state of Queensland saw it primarily as a means of expanding its own revenues. The needs of the Aborigines, on whose land the development was to take place, were disregarded through the expedient of unilaterally revoking its reserve status granted in the nineteenth century when vast tracts of Cape York peninsula were dedicated 'for the use and benefit of Aboriginal inhabitants of the state' (Howitt, 1992, p. 225).

Although the state and Comalco expressed the expectation of a high level of Aboriginal employment in Comalco, this was not realized. The cultural unacceptability of the work apart, low pay, which was well beneath that of other workers with different ethnic backgrounds, the in-migration of workers from other areas, poor housing and other community problems alienated the native population.

Nevertheless, in 1978 Comalco recognized its role in exploiting a non-renewable resource and its duty to have a long-term development strategy for the local people. Since then it has become a major source of education and training facilities and services for the community, largely articulated through the Weipa Aborigines Society, on which both local people and company management are represented. In addition, Comalco has since achieved higher levels of Aboriginal employment than other mining companies. It has also sought to fund new forms of development not alien to Aboriginal culture and in harmony with the local ecology, including a fish farm experiment. This is designed to help diversify work opportunities in the area.

In the early 1990s Comalco decided to build an aluminium refinery at Weipa, the largest in Australia. While this is still at an early stage, the degree to which Aboriginal concerns have already shaped the proposal reflects the extent to which Aboriginal people have succeeded in securing recognition. Comalco has already agreed not to develop its preferred site following total community opposition on environmental and social grounds. Moreover, the procedure for constructing the refinery has been modified to reduce the on-site workforce in order to minimize the social disruption associated with such projects.

Source: based on Howitt, 1992, pp. 223–35

Perhaps the crucial difference at Weipa, when considered in relation to many other mining developments in Australia which have involved Aborigines, is the local community's recognition that Comalco was not simplistically 'the enemy', but could in fact be a potential ally and source of support in some important areas. These have included controversial campaigns against oppressive state policies and legislation that have long remained antagonistic to the Aboriginal communities. Indeed, the relationship has provided a setting in which the pursuit of Aboriginal social, economic, political and cultural goals have not had to be put on hold awaiting the gradual erosion of entrenched political and other vested interests in the Queensland legislature under the influence of changing societal attitudes.

Summary of section 5.3

- Mining can have serious environmental impacts at a number of stages – removal of overburden, extraction of ore, concentration and smelting.
- Movement of pollutants in air and water can spread far beyond mine sites and the latter may continue indefinitely, even after mining has ceased.
- Few countries regulate the impacts of mining effectively and even rich countries have only just begun to tackle sites damaged by mining in the past.
- Negative impacts are likely to grow most rapidly in the less developed world as their use of minerals increases and more developed countries impose stricter regulation within their own borders.
- Future policy might use taxation as a means of ensuring that mining pays the full environmental costs of production: the effect will be to raise prices of virgin materials and to encourage recycling.
- Mining often has negative impacts on the local people, though there are now some signs that a partner relationship can replace past exploitation, but this usually relies on states giving land rights to indigenous peoples.

5.4 Conclusion

The two perspectives on the contemporary system of exploitation of mineral resources combine to describe an industry which is increasingly globalized in its organization and which has so many negative local impacts that they add up to a major global problem.

Part of the explanation of unevenness of mineral extraction lies in the geological unevenness of the best concentrations of ores. However, the political and economic processes which translate natural materials into exploitable reserves add their own elements of unevenness. The very high level of domination of the minerals trade by a small number of multinationals based in the developed world emphasizes the fact that the unevenness is caused by inequalities in power, both internationally and between national governments and resource-rich localities. The apparent attractiveness of mineral extraction as an export earner may persuade governments in the less developed world to encourage multinationals with tax incentives and environmental controls which are often unenforced, particularly as the attractiveness of minerals is enhanced by the difficulties of earning adequate amounts through export of agricultural and industrial products.

The imposition of many environmental and social costs on localities in which mining occurs raises the question of how localities can defend their interests (see, for example, **Massey and Jess, 1995***). The analysis in this chapter suggests that current use of mineral resources is certainly not equitable and probably not sustainable, as demand grows and leaner ores are exploited. While some progress can be made through more effective regulation so that mining pays the real costs to environment and society, in Reading A Young suggests that the only long-term remedy is a move towards lifestyles which do not require increasing amounts of mining and materials processing.

References

BLUNDEN, J.R. (1985) *Mineral Resources and Their Management,* London, Longman.

CRES (CENTRE FOR RESOURCE AND ENVIRONMENTAL STUDIES) (1985) *East Kimberley Impact Assessment Project: Project Description and Feasibility Study,* East Kimberley Working Paper No. 1, Canberra, Australian National University.

GOODING, K. (1992) 'Latin American mining', *Financial Times Survey,* 17 September.

HOWITT, R. (1992) 'Weipa: industrialisation and indigenous rights in a remote Australian mining area', *Geography,* Vol. 77, No. 3, pp. 223–35.

KENNEDY, D.M. (1990) 'Papua New Guinea', *Mining Annual Review,* p. A81.

LEWIS, T.M. and MCNICOLL, J.H. (1978) *North Sea Oil and Scotland's Economic Prospects,* London, Croom Helm.

MASSEY, D. and JESS, P. (1995) *A Place in the World? Places, Culture and Globalization,* Oxford, Oxford University Press in association with The Open University.

MCDIVETT, J.F. and MANNERS, G. (1974) *Minerals and Men,* Baltimore, MD, Johns Hopkins University Press.

METALS AND MINING RESEARCH SERVICES (1992) *An Environmental Audit of the Base Metal Industries,* Bath, Metals and Mining Research Services.

RADETSKI, M. (1992) 'The decline and rise of the multinational corporation in the metal mining industry', *Resources Policy,* Vol. 18, No. 1, pp. 2–8.

SPOONER, D. (1981) *Mining and Regional Development,* Oxford, Oxford University Press.

WORLD BANK (1977) *Annual Report,* Washington, DC, World Bank.

YOUNG, J.E. (1992) *Mining the Earth,* Washington, DC, Worldwatch Institute.

* A reference in emboldened type denotes another volume in the series.

Reading A: *John E. Young, 'Mining the earth'*

Introduction

Substances extracted from the earth – stone, iron, bronze – have been so critical to human development that historians name the ages of our past after them. But while scholars have carefully tracked human use of minerals, they have never accounted for the vast environmental damage incurred in mineral production.

Few people would guess that a copper mining operation has removed a piece of Utah seven times the weight of all the material dug for the Panama Canal. Few would dream that mines and smelters take up to a tenth of all the energy used each year, or that the waste left by mining measures in the billions of tons – dwarfing the world's total accumulation of more familiar kinds of waste, such as municipal garbage. Indeed, more material is now stripped from the earth by mining than by all the natural erosion of the earth's rivers.

Scouring the planet for its minerals has damaged large areas of land, often in remote, ecologically pristine areas. Mining projects now threaten four of every 10 national parks in tropical countries [Sheean, 1992]. The smelting of ores pumps millions of tons of sulphur dioxide and other pollutants into the atmosphere each year. Smelter pollution has created biological wastelands as large as 10,000 hectares, and accounts for a significant portion of the world's acid rain. The mineral industry's profligate use of energy makes it a substantial contributor to climate change as well as to more localized environmental problems.

Yet in most discussions of threats to the global environment, mining is conspicuous only by its absence. The damage from mineral extraction is usually considered a local problem, and accepted – or imposed on local inhabitants – as an inevitable cost of economic development. It is also rarely tracked. For instance, the US mining industry – though it is clearly among the largest polluters – is not required to report its toxic emissions to state and federal regulators, as are most manufacturing industries. Cognizant that a country's overall prosperity usually correlates closely with its per capita use of mineral products, industrial nations have focused instead on the question of mineral supplies.

In the United States, for example, periodic waves of concern over future mineral supplies have led to the appointment of at least a half-dozen blue-ribbon panels on the subject since the twenties. Experts have assiduously questioned whether the country is going to have enough copper, tin, uranium and other 'non-renewable resources'. In 1978, a US congressional committee requested a study whose title expressed the central question of virtually all these inquiries: Are we running out? [US Library of Congress, 1978.]

Recent trends in price and availability suggest that for most minerals we are a long way from running out. Regular improvements in exploitive technology have allowed the production of growing amounts at declining prices, despite the exhaustion of many of the world's richest ores. For many minerals, much of the world has yet to be thoroughly explored.

In retrospect, however, the question of scarcity may never have been the most important one. Far more urgent is: can the world afford the human and ecological price of satisfying its present appetite for minerals? If the answer is that it cannot, the challenge will be to find ways to continue developing and improving the quality of human life without constant growth in mineral extraction.

In turn, the question of what the world can afford depends on a true accounting of the costs of taking materials from the earth. Today's low mineral prices reflect only the immediate economics of extraction and distribution; they fail to consider the full costs of denuded forests, eroded land, dammed or polluted rivers, and the uprooting or decimation of indigenous peoples unlucky enough to live atop mineral deposits.

The environmental impacts of mineral extraction are particularly severe in developing countries, which produce a large portion of the world's mineral supplies but use a relatively small share. These nations also harbour some of the globe's greatest remaining concentrations of biological diversity. Mineral projects are among the largest causes of disturbance in such areas.

But while much of the damage is concentrated in the developing world, responsibility for most of it ultimately lies with those who use the most minerals – the fourth of humanity who live in industrial nations, enjoying material comforts others only dream of. The rich nations thus bear a special responsibility to help clean up the messes created to satisfy their needs, and to ensure that new damage is kept to a minimum.

In the long run, the most effective strategy for minimizing new damage is not merely to make mineral extraction cleaner, but to reduce the rich nations' needs for virgin (non-recycled) minerals. Hope for success lies in the economic maturity of today's most prosperous countries. Large quantities of minerals were required to build up their infrastructures – to make the concrete, steel, brick and other materials needed for buildings and transportation systems. But once a society's basic structures are built, the quantities of additional materials it uses need not determine its quality of life. After a certain point, people's welfare may depend more on the calibre of a relatively small number of silicon microchips than on the quantities of copper, steel, or aluminium they use.

The sooner the whole world reaches such a point, the better. At the end of the minerals- and energy-intensive development path taken by today's industrial nations lies ecological ruin. Mining enough to supply a world expected to double in population during the next half century, with everyone using minerals at rates that now prevail in rich countries, would have staggering environmental consequences. Only by adopting a new development strategy – one that focuses on improving human welfare in ways that minimize the need for new supplies of minerals – can such consequences be averted.

[...]

Laying waste

Mining is the original dirty industry. As the German scholar Georgius Agricola put it in his 1550 treatise on mining: 'The fields are devastated by mining operations ... the woods and groves are cut down, for there is need of an endless amount of wood for timbers, machines, and the smelting of metals. And when the woods and groves are felled, then are exterminated the beasts and birds. ... Further, when the ores are washed, the water which has been used poisons the brooks and streams, and either destroys the fish or drives them away' [Agricola, 1950].

Four centuries later, mining's environmental effects remain much the same, but on a vastly greater scale. Modern machinery can do in hours what it took men and draft animals years to do in Agricola's time. Larger equipment reflects the growing scale of the industry. A typical truck used in hard-rock mining in 1960 weighed 20–40 tons, for example; in 1970 it weighed in at 80–200 tons. The size of the shovels used to move ore increased from 2 to 18 cubic metres over the same period [Bosson and Varon, 1977]. Such technological advances allowed world mineral production to grow rapidly – and proportionately increased the harm to the environment.

Mining and smelting have created large environmental disaster areas in many nations (see Table A.1.) In the United States, which has a long history of mining, at least 48 of the 1,189 sites on the Superfund hazardous-waste cleanup list are former mineral operations. The largest Superfund site stretches across the state of Montana, along a 220-kilometre stretch of Silver Bow Creek and the Clark Fork River. Water and sediments in the river and a downstream reservoir are contaminated with arsenic, lead, zinc, cadmium and other metals, which have also spread to nearby drinking-water

Table A.1 *Environmental impacts of selected mineral projects*

Location/mineral	Observation
Ilo-Locumbo area, Peru copper mining and smelting	The Ilo smelter emits 600,000 tons of sulphur compounds each year; nearly 40 million cubic metres per year of tailings containing copper, zinc, lead, aluminium and traces of cyanides are dumped into the sea each year, affecting marine life in a 20,000-hectare area; nearly 800,000 tons of slag are also dumped each year
Nauru, South Pacific phosphate mining	When mining is completed – in 5–15 years – four-fifths of the 2,100-hectare South Pacific island will be uninhabitable
Pará state, Brazil Carajás iron ore project	The project's wood requirements (for smelting of iron ore) will require the cutting of enough native wood to deforest 50,000 hectares of tropical forest each year during the mine's expected 250-year life
Russia Severonikel smelters	Two nickel smelters in the extreme north-west corner of the republic, near the Norwegian and Finnish borders, pump 300,000 tons of sulphur dioxide into the atmosphere each year, along with lesser amounts of heavy metals. Over 200,000 hectares of local forests are dying, and the emissions appear to be affecting the health of local residents
Sabah Province, Malaysia Mamut Copper Mine	Local rivers are contaminated with high levels of chromium, copper, iron, lead, manganese and nickel. Samples of local fish have been found unfit for human consumption, and rice grown in the area is contaminated
Amazon Basin, Brazil gold mining	Hundreds of thousands of miners have flooded the area in search of gold, clogging rivers with sediment and releasing an estimated 100 tons of mercury into the ecosystem each year. Fish in some rivers contain high levels of mercury

Sources: Palacios (1989); Nauru from *Asiaweek* (1991), Weston (1991); Fearnside (1989); Severonikel from Berlin (1991); APPEN (1990); Cleary (1990); Superfund sites from Steve Hoffman, US Environmental Protection Agency (EPA), Washington, DC, private communication, 5 November 1991; Environmental Protection Agency (EPA) and Montana Department of Health and Environmental Sciences (MDHES) (1988); Nielsen and Farling (1991)

aquifers. Soils throughout the local valley are contaminated with smelter emissions.

The Clark Fork Basin was the site of more than 100 years of mining and smelting, including what was at one time the largest open pit in the world, the Berkeley Pit copper mine. The pit and a network of underground mine workings contain more than 40 billion litres of acid mine water that rises a little higher each year, threatening local aquifers and already-tainted streams with contamination. The Clark Fork Coalition, a local environmental group, estimates that cleaning up the pit and other sites in the area could cost over $1 billion. A proposed large new copper mine in the

Cabinet Mountains area of north-west Montana now endangers another section of the Clark Fork's drainage [Environmental Protection Agency and MDHES, 1988; Nielson and Farling, 1991].[1]

The environmental damage done in producing a particular mineral is determined by such factors as the ecological character of the mining site, the quantity of material moved, the depth of the deposit, the chemical composition of the ore and the surrounding rocks and soils, and the nature of the processes used to extract purified minerals from ore (see Table A.2). Damage varies dramatically with the type of mineral being mined. For example, stone ranks first in production, but its extraction probably causes less overall harm than that of several metals. Since stone and other construction materials are usually taken from shallow or naturally exposed deposits and used with little or no processing, the environmental impacts are mostly limited to land disturbance at the quarry or gravel pit, and relatively few wastes are generated.

At the other end of the damage spectrum, metals are produced through a long chain of processes, each of which involves pollution and the generation of waste. Copper production, for instance, typically involves five stages. First, soil and rock (called overburden) that lie above the ore must be removed. The ore is then mined, after which it is crushed and run through a concentrator, which physically removes impurities. The concentrated ore is reduced to crude metal at high temperatures in a smelter, and the metal is later purified, through remelting, in a refinery.

Table A.2 Environmental impacts of minerals extraction

Activity	Potential impacts
Excavation and ore removal	o Destruction of plant and animal habitat, human settlements, and other surface features (surface mining) o Land subsidence (underground mining) o Increased erosion; silting of lakes and streams o Waste generation (overburden) o Acid drainage (if ore or overburden contain sulphur compounds) and metal contamination of lakes, streams and groundwater
Ore concentration	o Waste generation (tailings) o Organic chemical contamination (tailings often contain residues of chemicals used in concentrators) o Acid drainage (if ore contains sulphur compounds) and metal contamination of lakes, streams and groundwater
Smelting/refining	o Air pollution (substances emitted can include sulphur dioxide, arsenic, lead, cadmium and other toxic substances) o Waste generation (slag) o Impacts of producing energy (most of the energy used in extracting minerals goes into smelting and refining)

Source: Worldwatch, compiled from various sources

Most of today's mines are surface excavations rather than underground complexes of tunnels and shafts, so the miner's first task is to remove whatever lies over a mineral deposit – be it a mountain, a forest, a farmer's field, or a town. For any given mineral, surface mining produces more waste than working underground. In 1989, US surface mines produced eight times as much waste per ton of ore as underground mines did. That same year, overburden accounted for more than a third of the 3.4 billion tons of material handled at non-fuel mines. Such material, while it may be chemically inert, can clog streams and cloud the air over large areas. If the overburden contains sulphur compounds – common in rock containing metal ores – it can react with rainwater to form sulphuric acid, which then may contaminate local soils and watercourses [US Bureau of Mines, 1990].

Similar but more severe effects often stem from extraction of the ore itself and from the disposal of tailings, the residue from ore concentration. Up to 90 per cent of metal ore ends up as tailings, which are commonly dumped in large piles or ponds near the mine. The finely ground material makes contaminants that were formerly bound up in solid rock (such as arsenic, cadmium, copper, lead and zinc) accessible to water. Acid drainage, which exacerbates metal contamination, is often a problem, since sulphur makes up more than a third of the commonly mined ores of many metals, including copper, gold, lead, mercury, nickel and zinc. Tailings also usually contain residues of organic chemicals – such as toluene, a solvent damaging to human skin and to the respiratory, circulatory and nervous systems – that are used in ore concentrators. Ponds full of tailings cover at least 3,500 hectares in the Clark Fork area and 2,100 hectares at the Bingham Canyon copper mine in Utah.[2]

A particularly dramatic example of the impact of tailings disposal is the Panguna copper mine on Bougainville, an island in Papua New Guinea that since mid-1989 has been controlled by secessionist rebels. Before it was closed, the mining operation dumped 600 million tons of metal-contaminated tailings – 130,000 tons each day – into the Kawerong River. The wastes cover 1,800 hectares in the Kawerong/Jaba river system, including a 700-hectare delta at its mouth, 30 kilometres from the mine. Environmental writer Don Hinrichsen described the Jaba River as 'so full of sediments from the Bougainville Copper Mine that its slate-grey waters are completely dead. ... Wading into the river to take samples is like inching through moving mud'. Local anger at the destruction of the area by mining was a major cause of the civil war [Howard, 1991; Hyndman, 1991; Scott, 1989; *South*, 1991; Moore and Luoma, 1991; Hinrichsen, 1990].

Smelting, the next stage of the extraction process, can produce enormous quantities of air pollutants. Worldwide, smelting of copper and other non-ferrous (non-iron) metals releases an estimated 6 million tons of sulphur dioxide into the atmosphere each year – 8 per cent of total emissions of the sulphur compound that is a primary cause of acid rain. Non-ferrous smelters can also pump out large quantities of arsenic, lead, cadmium and other heavy metals. If they lack pollution control equipment, aluminium smelters emit tons of fluoride, which can concentrate in vegetation and kill not only the plants but, in some cases, animals that eat them [Möller, 1984; effects of fluoride emissions: Ehrlich *et al.*, 1977].

Uncontrolled smelters have produced some of the world's best-known environmental disaster areas – 'dead zones' where little or no vegetation survives. Such an area around the Sudbury, Ontario, nickel smelter in Canada measures 10,400 hectares; acid fallout from the smelter has destroyed fish populations in lakes 65 kilometres away. Between 1896 and 1936, a smelter at Trail, British Columbia killed virtually all conifers within 19 kilometres and retarded tree growth up to 63 kilometres away. In the United States, a dead zone surrounding the Copper Hill smelter in Tennessee covers 7,000 hectares. In the United Kingdom, 400,000 hectares of agricultural land have been lost to metal smelting since Roman times; and in

The world's largest aluminium smelter, Bratsk, Siberia; powered by hydroelectricity

Japan, about 6,700 hectares of cropland are too contaminated for rice production.[3]

New dead zones, such as the area surrounding the Severonikel nickel smelter in Russia, are still being created. Smelters in industrial countries are now often required by law to have pollution control equipment, but few in developing countries or the formerly socialist nations have any such controls. For each kilogram of copper produced, 12.5 times more sulphur dioxide is released to the air from Chilean smelters than from those in the United States [Berlin, 1991; United Nations, Economic Commission for Latin America and the Caribbean, 1991].

The grade of an ore – its metal content in percentage terms – is a critical factor in determining the overall impact of metal mining. The average grade of copper ores, for example, is lower than that for any of the other major metals. Four centuries ago, copper ores typically contained about 8 per cent metal [Bosson and Varon, 1977]; the average grade of ore mined now is under 1 per cent. One consequence of the drop in grade is that more than eight times as much ore now must be processed to obtain the same amount of copper. An estimated 990 million tons of ore were mined to produce about 9 million tons of copper in 1991 (see Table A.3).

Even this figure understates the total amount of material moved, since it does not include overburden. The scale of the industry is apparent, however, in the size of the holes it creates. Some 3.3 billion tons of material – seven times the amount moved for the Panama Canal – have been taken from Utah's Bingham Canyon copper mine. Now 774 metres deep, this mine is the largest human excavation in the world [see Plate 11] [Goudie, 1990]. It gained another distinction in 1987 when its operator, Kennecott Copper, inadvertently reported its toxic chemical releases to the Environmental Protection Agency's [EPA] Toxics Release Inventory (a national toxic-chemicals reporting system from which the mining industry is exempt) [Horowitz, 1990]. Out of the 18,000 industrial facilities reporting, the Bingham Canyon mine ranked fourth in total toxic releases and first in metals. The company discontinued reporting the following year, but the scale of its releases spurred legislative efforts in 1991 and 1992 to include the mining industry in the inventory in the future.

Gold mining also requires the processing of large amounts of material, since the metal occurs in concentrations best measured in parts per million. An estimated 620 million tons of waste are produced in gold mining each year – even more than is produced in iron mining, which yields 26,000 times as much metal by weight. The operators of the Goldstrike mine in Nevada – the largest in the United States – each day move 325,000 tons of ore and waste to

Table A.3 Estimated ore production, average grade, and waste generation, major minerals, 1991

Mineral	Ore (million tons)	Average grade (per cent)	Waste (million tons)
Copper	1,000	0.91	990
Gold	620	0.00033	620
Iron	906	40.0	540
Phosphate	160	9.3	140
Potash	160	17.0	140
Lead	135	2.5	130
Aluminium/bauxite	109	23.0	84
Nickel	38	2.5	37
Tin	21	1.0	21
Manganese	22	30.0	16
Tungsten	15	0.25	15
Chromium/chromite	13	30.0	9
Total	3,200		2,700

Waste figures do not include overburden. Totals do not add due to rounding.

Sources: Worldwatch Institute, based on production estimates in US Bureau of Mines, 1992, and grade estimates in Rogich, 1992

produce under 50 kilograms of gold [Gooding, 1991]. In Brazil's Amazon Basin, thousands of small-scale gold miners are using a technique called hydraulic mining to extract as much as 120 tons of gold per year [Cleary, 1990]. This involves blasting gold-bearing hillsides with high-pressure streams of water, and then guiding the water and sediment through sluices that separate tiny amounts of gold, which is heavier, from tons of non-valuable material, which then pollutes local rivers. The technique is so environmentally destructive that its use was halted over 100 years ago in California, where it did widespread damage during the state's legendary gold rush [Smith, 1987].

Since 1979, when the price of gold soared to an all-time high of $850 per ounce, a gold rush has swept the world [Cleary, 1990]. Waves of gold seekers have invaded remote areas in Brazil, other Amazonian countries, Indonesia, the Philippines and Zimbabwe [Indonesia: Gurov, 1990; Zimbabwe: from Paul Jourdan, Institute of Mining Research, Harare, Zimbabwe, private communication, 12 April 1991; other nations: Westlake and Stainer, 1989]. Dramatic environmental damage has resulted. Hydraulic mining has silted rivers and lakes, and the use of mercury – an extremely toxic metal that accumulates in the food chain and causes neurological problems and birth defects – to capture gold from sediment has contaminated wide areas. Miners release an estimated 100 tons of mercury into the Amazon ecosystem each year [Cleary, 1990]. An estimated 32 tons are released each year in the watershed of the Madeira River (a major Amazon tributary) alone [Nriagu *et al.*, 1992]. Mercury levels in most carnivorous and some omnivorous fish in the Madeira exceed the maximum safe levels for human consumption set by many nations.

In North America, heap leaching, a new technology that allows gold extraction from very low-grade ores, is now in wide use. Miners spray cyanide solution, which dissolves gold, on piles of crushed ore or old tailings. After repeated circulation through the ore, the liquid is collected and gold is extracted from it. Both cyanide-solution collection reservoirs and the contaminated tailings left behind after leaching pose hazards to wildlife and groundwater. In October 1990, for instance, 38 million litres of cyanide solution spilled from a reservoir at the Brewer Gold Mine, near Jefferson, South Carolina, into a tributary of the Lynches River. The spill, caused by a dam break after a heavy rain, killed as many as

10,000 fish [*Clementine*, 1990]. Thousands of birds also die each year when they mistake cyanide impoundments for lakes [Warhurst, 1991].

Fossil-fuel-powered machinery has allowed mining to expand to such a degree that its effects now rival the natural processes of erosion. An estimated 24 billion tons of non-fuel minerals are taken from the earth each year, of which about 2.7 billion tons are waste (not including overburden). Taking overburden into account, the total amount of material moved is probably at least 28 billion tons – about 1.7 times the estimated amount of sediment carried each year by the world's rivers.[4]

An estimated half-million hectares of land – including mines, waste disposal sites, and areas of subsidence over underground mines – are directly disturbed by non-fuel mining each year.[5] Most of this land will bear the scars indefinitely. Historian Elizabeth Dore, describing the effects of 500 years of mining on the Bolivian landscape, writes: 'Silver and tin are gone; in their place rise mountains of rock, slag, and tailings. ... Saturated with mercury, arsenic, and sulfuric acid, the iridescence of these rubbish heaps provides a psychedelic reminder of the past'. The damage is not limited to the mine site. As Dore puts it, mining initiates 'a chain of soil, water, and air contamination' that can alter the ecosystems of large areas [Dore, 1991].

Moving billions of tons of material and crushing and melting rock requires large amounts of energy, and supplying it can cause major damage to local ecosystems. Ever since Agricola's time, for example, wood-fired smelters have threatened nearby forests. In southern England, the Sussex iron industry was effectively wiped out when it destroyed the local woods that provided its fuel supply [Down and Stocks, 1977]. In the late nineteenth century, more than 2 million cords of wood were used as smelter fuel in Nevada's Comstock Lode [Smith, 1987] – described by one observer as 'the tomb of the forests of the Sierras' [William Wright in 1877, in Smith, 1987].

Today, demand for energy to extract and process minerals is playing a major role in the deforestation and inundation of large parts of the Amazon Basin. A huge iron ore mining and smelting project at Carajás, in the Brazilian state of Pará, threatens a large area of tropical forest. The project's 20 planned pig-iron smelters will need an estimated 2.4 million tons of charcoal each year, which if produced from native trees will require an estimated 50,000 hectares of forest to be logged annually. According to ecologist Philip Fearnside, high costs make it unlikely that plantations will supply much of the wood, and the state enterprise that owns the project has thus far done little to develop plantation production. The mine is expected to operate for 250 years [Fearnside, 1989].

The iron ore facilities are only one piece of Brazil's colossal Grande Carajás Project, a vast state-run development scheme that also includes bauxite, copper, chromium, nickel, tungsten, tin and gold mines; mineral processing plants; hydroelectric dams; deep-water ports; and other enterprises [Dore, 1991]. Aluminium smelters, including the 330,000-ton-per-year Albrás plant (another element of the Carajás project) now take most of the electricity output of the enormous – and enormously destructive – Tucuruí hydroelectric station [Bomsel *et al.*, 1990]. Albrás, a major justification for the dam's construction, receives power at one-fourth the cost of generation (and one-third the average cost of Brazilian electricity).[6] Aluminium smelting took 12 per cent of Brazilian electricity in 1988, and the industry's power requirements more than doubled between 1982 and 1988 [Geller, 1991].

Aluminium production is particularly energy-intensive. Unlike most other metals, which can be obtained by simply heating the ore, aluminium forms such tight chemical bonds that it can only be economically extracted through a process involving the direct application of electrical current. Modern aluminium smelters require 13–18 kilowatt-hours of electricity to produce a kilogram of metal [OTA, 1990]. The world aluminium industry uses an estimated 290 billion kilowatt-hours of electricity each year – more than is used for all purposes on the entire African continent [US Department of Energy, 1992]. Additional energy is used in mining bauxite, and in processing it into the alumina that is smelted [Brown and McKern, 1987]. All told, aluminium production requires an estimated 3.8 billion gigajoules (GJ) of energy each year – around 1 per cent of world energy use.[7] Much of it is purchased at unusually low rates – subsidized by governments at heavy human and environmental expense [Peck, 1988; Graham, 1982].[8]

Though figures are sparse, the mineral industry as a whole is clearly among the world's largest users of energy, and thus a major contributor to the impacts of energy use, including climate change. While aluminium is the most energy-intensive of the metals, steel and copper are also large energy users. Steelmaking, in fact – because of its sheer volume – is probably the largest energy user of all mineral industries; in the United States, which produces only 10 per cent of the world's supply, steelmaking required 2.2 billion GJ in 1988. Worldwide, copper production takes about 1 GJ. All told, the minerals industry probably accounts for 5 to 10 per cent of world energy use.[9]

The efficiency of energy use in smelting and refining metals has improved over time. Today's US aluminium smelters, for example, use between half and two-thirds as much electricity as those built in the late forties. Some new copper smelting technologies use only about 60 per cent as much energy as traditional methods. But while smelting has improved, long-term trends in ore grades and accessibility of deposits tend to increase the energy used per unit of metal mined. Declining ore grades increase energy needs, because more ore must be mined, greater quantities of waste material must be handled, and more effort is required to concentrate and smelt the ore. And as more remote, deeper deposits are mined to replace those more easily reached, more energy is required in order to dig bigger holes and transport the ore longer distances [OTA, 1990; Brown and McKern, 1987].

[...]

Cleaning up

Since the time of Agricola, the destruction from mineral production has been justified in the name of human progress. The sixteenth-century scholar was quite conscious of mining's effects on the environment, yet argued that, without metals, 'men would pass a horrible and wretched existence in the midst of wild beasts'. In an absolute sense, Agricola was right: civilizations have always depended heavily on minerals for survival – and still do [Agricola, 1950].

But as Lewis Mumford put it in his classic *Technics and Civilization*, 'One must admit the devastation of mining, even if one is prepared to justify the end. ... What was only an incidental and local damage in [Agricola's] time became a widespread characteristic of Western Civilization just as soon as it started in the eighteenth century to rest directly upon the mine and its products'. Mining's effects on the earth, far from being merely local as often depicted, are now on the same scale as such hugely destructive natural forces as erosion [Mumford, 1963].

Cleaning up the mineral industry and its legacies will not be easy. Perhaps the hardest task will be to clean up abandoned mineral projects, because doing so often requires moving, treating and containing extraordinarily large amounts of material spread over large areas. For example, about 7 million tons of tailings are present at the Eagle Mine Superfund site in Gilman, Colorado [Environmental Protection Agency, 1987], and more than 200 million cubic metres of materials are stored in the 3,500 hectares of tailings ponds in the Clark Fork area of Montana [Moore and Luoma, 1991]. The latter contain an estimated 200 tons apiece of cadmium and silver, 9,000 tons of arsenic, 20,000 tons of lead, 90,000 tons of copper and 50,000 tons of zinc.

The consequences of not cleaning up such abandoned operations can be severe and long-lasting. The very reason these sites will be difficult to clean up is the most compelling reason to do so, since such huge volumes of material can cause abandoned mines to continue paying negative dividend – in such forms as sediment-choked streams, acid drainage and metal contamination – for centuries [Moore and Luoma, 1991].

Aside from technical challenges, however, the chief problem in cleaning up old mines is that the mine operators are gone – and with them, the money for cleanup. Hence, governments often end up with the bill, which can be huge. The price tag for the Clark Fork cleanup, for example, is estimated to be $1 billion. Furthermore, the number of sites to be evaluated for cleanup is quite large: EPA estimates that between 800 and 1,500 mining sites need assessment, and that 70–100 will require remedial action. Nations with similar mining histories are likely to have similar cleanup needs.

The United States has chosen to fund mine cleanups through its Superfund programme which covers all types of abandoned industrial facilities. The programme is primarily funded by a tax on chemical feed-stocks. Progress on mineral sites has been sluggish, however. The special scale and characteristics of such sites may merit establishment of a separate programme for former mineral facilities. One possible approach would be to fund mine cleanups through taxes on virgin mineral consumption. A similar programme already exists for abandoned coal mines.[10]

Such taxes would serve another purpose as well, since they would also help create incentives for more efficient use of minerals, reducing the need for new mines. Some of these funds could be directed to third world countries for cleanup of their abandoned mines. Without outside assistance, it is unlikely that many developing countries will be able to afford cleanups of their old mineral sites, which in most cases were developed for export to richer nations.

New technologies may help reduce the costs of cleanup. The high metal content of some tailings – which can pose the threat of contamination – can be turned to advantage in mine cleanups if methods are available for extracting the remaining metals. Thus, reprocessing old tailings can sometimes not only help reduce environmental hazards at a site, but also yield a saleable product to help pay for cleanup.

Such methods are not only useful at abandoned sites, but also may help in reducing pollution at operating mines. Biological leaching – in which bacteria are used to extract metal from ore – is a promising new method. At the Los Bronces copper mine in Chile, a biological extraction project now being put into place is designed to avoid pollution in the Mantaro River, the source of Santiago's drinking water, by extracting copper from water repeatedly circulated through waste, overburden and marginal ore dumps. The project's designers believe it will eventually recover more than a half-million tons of pure copper. An even more ambitious project at La Escondida, another giant Chilean copper mine, will recover pure metal from ore without smelting. Instead, copper will be extracted from concentrates by an ammonia solution, and then precipitated by electrolysis. In addition to avoiding the pollution and expense of a smelter, the project will have the advantage of producing pure copper rather than less-valuable concentrates [Warhurst, 1992; Crawford, 1992].

A variety of other practices can help reduce environmental impacts at every stage of the mineral extraction process. While destruction of surface features – be they forests or villages – is usually inevitable with surface mining, a variety of techniques can help cut down air, water and soil pollution, and sometimes can return mined land to stable (if not its original) condition. In the initial excavation and mining process, careful storage of topsoil can ensure its availability for reclamation after mining is finished. If soil and rock are stored in well-designed impoundments, runoff and sedimentation problems can be kept to a minimum. Similarly, more careful storage and disposal of tailings can minimize the opportunity for them to contaminate their surroundings. Air pollution controls can substantially reduce emissions from smelters. Advanced methods – especially biological techniques – for extracting metals from ore could also offer substantial energy savings if they replace thermal (smelting) or mechanical methods.

Such careful attention by mine operators to minimizing environmental damage is unlikely, however, unless governments have the resources – and political will – to require it. While the mineral sector is currently subject to a broad range of environmental rules in most industrial and some developing countries, legal loopholes, lack of government funds and weak enforcement are still allowing the creation of new environmental disaster areas.

For example, in the United states, which is widely regarded as a leader in such regulation, smelter emissions are regulated under the Clean Air Act and mining-caused water pollution under the Clean Water Act. Unfortunately, federal regulation of mining itself remains quite weak. EPA has done little to regulate disposal of mining wastes, despite their status as the single largest category of waste produced. In the 1980 Bevill Amendment to the Resource Conservation and Recovery Act of 1976, Congress exempted most mining wastes from regulation as hazardous waste, pending an EPA determination of their status. EPA has since decided to retain the exemption for most types of mineral industry waste, though final rules are still in process. In general, however, the states play a more important role in mining regulation, and the level of attention and enforcement varies dramatically [OTA, 1992].

Nonetheless, industrial nations have available to them the funding and government personnel to put in place and more effectively enforce environmental laws for mineral producers – if the political will exists to do so. In the United States – by far the largest mineral producer among the industrial market nations – lawmakers now have a chance to strengthen environmental provisions in several laws affecting the impacts of mining, including the Resource Conservation and Recovery Act, the Clean Water Act and the General Mining Act. Amendments to the Emergency Planning and Community Right-to-Know Act now pending would require the mining industry to report its toxic emissions to state and federal regulators.[11]

For developing countries, the challenge is much greater. While many of them have broad environmental protection laws, specific regulation of the minerals industries is rare. Where environmental laws do exist, funding and staff for enforcement are usually scarce. Chile, for example, has comprehensive and stringent environmental rules for mining, but they are virtually unenforced. The Chilean government has been particularly loath to force state-owned mineral operations to comply with laws [Warhurst, 1991]. Other countries with large state mining companies face similar conflicts of national interest between state mining companies and regulators, with local people and the environment most often the losers.

At times, the prospect of major revenue from projects leads government officials to simply ignore environmental rules or studies. At the Ok Tedi copper and gold mine in Papua New Guinea, the government allowed the project's operator – an international consortium of private firms – to dump up to 150,000 tons of tailings a day into the nearby Fly River rather than contain them at the mine site, despite studies showing the potential for major damage to the river system [Howard, 1991].

The pressure to neglect environmental concerns in favour of continued mineral output will continue to be strong unless broad changes occur in development policies and international debt service requirements. A legislative foundation already exists in some developing countries, however, for improved regulation of mineral industries. Several major mineral producers, including Chile, Brazil and Peru, have recently started looking at mineral production's impacts on the environment and are attempting to improve the regulation of such activities [Warhurst, 1991; *Journal of Commerce*, 1992].

Additional pressure for environmental improvements could be created through substantial international assistance for environmental regulation and enforcement, as well as through the attachment of environmental conditions to mineral development funding. A portion of virgin mineral taxes levied in industrial countries could be allocated to improving the capacity of mineral-exporting developing countries to regulate their industries. Loans from development banks and their affiliates could include substantial components earmarked specifically for environmental protection, as non-governmental organizations in both industrialized and developing countries have urged in recent years. Some progress is already being made in this direction. The International Finance Corporation (the World Bank affiliate that lends to private-sector projects) has begun insisting on environmental impact assessments for mineral projects it funds. Not until environmental concerns play a major role in the decisions on whether and how to fund projects, however, will the situation improve dramatically.

In the short run, better regulation of the environmental impacts of mining is the most obvious way to reduce the damage done in supplying the world with minerals. There is considerable room for improvement of current mining practices – in increased attention to environmental safeguards, more sensitivity to local people and their concerns, and better planning for the indirect impacts of mineral development. More attention and new approaches to reducing the environmental impacts of currently operating mineral facilities could help to prevent the creation of more abandoned sites in the future. Operations can be made even more benign if they are designed with environmental concerns in mind from the outset. Additional research, such as that of the Mining and Environment Research Network – a far-flung group of independent researchers investigating the impacts of mineral extraction in developing countries – should ease the task of cleaning up the mineral industries.[12]

In the long run, however, the benefits to be gained through mining regulation, while critically important, are not enough. Even well-managed mines are often enormously destructive. Careful reclamation may reduce erosion and pollution problems at mine sites, but ecological complexity and high costs usually preclude restoring the land to its previous condition. High energy use in mining and smelting makes reuse and recycling of metal-containing products almost always preferable to virgin production. To dramatically reduce the impacts of the minerals industry, attention must be paid not only to the extraction process, but to how mineral products are used.

Digging out

The ultimate solution to the problem of mining's environmental destruction will require profound changes in both minerals use and in the global economy. No country has yet developed and put into place comprehensive policies on the use of minerals and other raw materials. The assumption that prosperity is synonymous with the quantities of minerals taken from the earth has shaped the industrial development strategies of both capitalist and socialist nations. But that assumption is open to question. The environmental damage from non-stop growth in mineral production will eventually outweigh the benefits of increased materials supplies – if it does not already.

The way out of the trap lies in a simple distinction: it is the extraction and processing of minerals, not their use, that poses the greatest threat. The *de facto* materials policies of industrial nations have always been to champion the production of virgin minerals. Although such an approach has effectively promoted mining, it has also helped make minerals artificially cheap. This has led to widespread waste of mineral products, and has diverted funds that might have been used more productively to serve other needs.

A far less destructive policy would be to maximize conservation of mineral stocks already circulating in the global economy, thereby reducing both the demand for new materials and the environmental damage done to produce them. The world's industrial nations, the leading users of minerals, offer the most obvious opportunities for cutting demand. Minerals use in those nations is still rising, but increases have been slower in the last two decades than before [Bomsel *et al.*, 1990; Larson *et al.*, 1987; Drucker, 1986; Ayres, 1989; Herman *et al.*, 1989; US Bureau of Mines, 1991]. A growing body of evidence suggests that per capita needs for virgin minerals have already peaked there, and that major shifts are underway in the mix of minerals needed.

National governments could accelerate the transition to more materials-efficient economies through basic changes in policies that govern the exploitation and use of raw materials. Tax policy offers the most obvious tool with which to start. Taxing, rather than subsidizing, production of virgin minerals would create stronger incentives to use them more efficiently. It could also provide governments with a way of paying for mine cleanups, as well as augmenting general revenues.

Many technical possibilities exist for using minerals more efficiently. The most obvious is recycling, as there is ample room to increase recycling rates for many metals. A 1992 US Bureau of Mines study found that 10.6 million tons of iron and steel, 800,000 tons of zinc and 250,000 tons of copper are discarded in US solid waste each year [Rogich, 1992]. Though they recycle 45 per cent of the aluminium they use, US residents still throw away so much of the metal each year – 2.3 million tons – that the energy saved by recycling it could meet the annual electricity needs of a city the size of Chicago.[13]

Beyond recycling, however, even more opportunity lies in making mineral-containing products more durable and repairable. More than a decade ago, a study by the US Office of Technology Assessment concluded that reuse, repair and remanufacturing of metal-containing products were the most promising methods of conserving metals. Governments could promote such practices by requiring manufacturers to offer longer warranties or to take products back at the end of their useful lives. Deposit/refund systems for items as diverse as beverage containers and automobiles, can encourage consumers to return products for reuse instead of throwing them away [OTA, 1979].

A particularly promising initiative has been undertaken by several European auto manufacturers, including BMW, Mercedes-Benz, Peugeot, Renault, Volkswagen, Audi and Volvo, to make their vehicles entirely and easily recyclable. Engineers at the firms are designing cars with an eye toward easy disassembly, reuse and recycling of various parts, and are attempting to minimize the use of non-recyclable or hazardous materials. The approach could easily be adopted for other products [*Financial Times*, 1991; Siuru, 1991; *Multinational Environmental Outlook*, 1991; *International Environment Reporter*, 1991; Miller, 1991; *New York Times*, 1991; Marshall, 1991].

Another option is to substitute more benign materials for those whose production is judged to be the most environmentally damaging. Such judgments are inherently difficult, since comparison of the environmental impacts of different materials is an inexact science. But some minerals stand out from the crowd. Production of copper, for example, is exceptionally destructive. The use of optical fibres made of glass, in place of copper wires used in communications, is an encouraging example of a shift to a less-damaging substitute. Fibre-optics also offer a much greater information-carrying capacity than copper wire. Similarly, the large amounts of energy required for aluminium production make it a logical candidate for replacement with other materials in applications where its light weight does not compensate by saving even greater amounts of energy than its production requires. The energy taxes now proposed as measures to reduce carbon emissions would speed shifts to less energy-intensive materials.

More difficult than shifting industrial nations to a minerals-efficient economy will be the search for a path to a sustainable future for developing countries. For those now heavily dependent on mineral exports, a rapid decline in demand could have dire consequences. Development planners need to recognize that mineral projects have generally failed to deliver long-term national economic success, and that current trends in minerals markets and use make mining an unpromising sector for future investment. More attention – and funding – needs to be devoted to diversifying the economies of mineral-producing nations and regions. Industrial nations will need to consider devoting a substantial share of any taxes levied on virgin minerals to development assistance to producing nations. One prerequisite for successfully rehabilitating those economies is relief from the crushing burden of international debt.

The greatest challenge, however, lies in the search for a third world development strategy that allows poor countries to improve human welfare dramatically without using and discarding hundreds of kilograms of metals and other minerals per person each year. Roughly three-fourths of the world's people now live in countries that generally have yet to build the railways, roads, bridges, buildings, and other basic infrastructure so essential to rich nations' economies. These countries will inevitably need more minerals as their development proceeds.

But basic choices in development plans – in the location, scale, density and character of urban development, for example, or in the types of transportation systems emphasized – could have dramatic impacts on the amount and mix of materials needed, and on the environmental costs of producing them. It is critical that development planners take minerals use into account in the same way that they consider energy use, water use and the production of food. The urgency of this issue cannot be overestimated: if more countries follow the development path of the rich industrial nations, world demand for minerals will continue to rise for many years. Given current geographical trends in production, it is the third world itself that will suffer most of the environmental damage from meeting the demand.

In the past decade, concerns about the declining quality of the environment have brought anguished re-examinations of virtually all the major economic activities on which civilization depends. Nearly all – from energy production to transportation, from forestry to agriculture – have been called upon to reduce their toll on the natural systems that underpin the global economy. Despite this, mineral production – perhaps because of its remoteness or because of its growing concentration in countries that depend on it so heavily – has been relatively free of scrutiny. Yet this sector, on which so many others depend, is one of the largest users of energy, despoilers of air, water and land, and producers of industrial waste, and therefore one of the most desperately in need of reform.

With analysis of the environmental impacts of other industries has come a growing realization that prevention is better than cure – that the greatest environmental benefits are usually yielded by basic alterations in production processes and consumption patterns, rather than through 'pollution control'. Mining's devastating environmental effects make the ultimate case for a strategy emphasizing pollution prevention, and not just control. As an inherently destructive industry that supplies raw materials throughout the economy, its impacts are best reduced by basic changes in *other* industries, and in the societies that eventually use mineral products.

While mining companies and the governments of nations heavily dependent on mining exports may feel the costs of such a strategy are unacceptable, we have now passed the point where we can continue to live with anything less than a full accounting for today's policies. Reducing the global appetite for minerals, and the environmental toll of the huge industries that satisfy it, will not be easy. But allowing the devastation to grow unchecked could prove to be an even more difficult – and costly – course for humanity.

Notes

1 Peter Nielsen, Executive Director, Clark Fork Coalition, Missoula, MT, private communication, 16 October 1991; Cabinet Mountains from 'Shame on Montana' (videotape), World Wide Film Expedition, Missoula, MT, 1991.
2 Share of metal ore discarded as tailings, metal contaminants in tailings and tailings pond examples from Moore and Luoma (1991); sulphur content of metal ores from Kelly (1988); organic contaminants from Down and Stocks (1977);

toluene use in concentrators from Horowitz (1990); health effects of toluene from Jorgenson (1989).

3 Dead zones from Moore and Luoma (1991); Sudbury fish kills and Trail smelter pollution from Down and Stocks (1977).

4 Total world mineral extraction and waste generated are Worldwatch Institute estimates, based on production data in US Bureau of Mines (1992), and average ore grades in Rogich (1992); the estimated sediment load in the world's rivers is 16.5 billion tons/year, according to J.D. Milliman and R.H. Meade, cited in Skinner (1989).

5 Area mined is a Worldwatch estimate, derived by multiplying world production data by 1980 US land-use/production ratios; world production from US Bureau of Mines (1992); ratios derived from Johnson and Paone (1982).

6 Albrás production capacity from *Gazeta Mercantil*, 1991; Tucuruí electricity used by Albrás from Liliana Acero, Centro de Investigacion y Promocion Educativa y Social, Buenos Aires, private communication, 7 May 1992; for information on the destructive nature of the Tucuruí dam, see Cummings (1990).

7 Total energy consumption in aluminium production is a Worldwatch estimate derived from these sources and from 1991 production estimate in US Bureau of Mines (1992); electricity use in smelting converted into GJ assuming a 64/36 per cent mix of fossil and non-fossil electricity generation (while a large portion of aluminium smelting is powered by hydroelectric sources, much of that energy could be directed into other uses now dependent on fossil energy sources).

8 For additional information on the environmental impacts of the aluminium industry, see Young (1992a).

9 Energy use in copper production is a Worldwatch estimate derived from 1990 production estimate in US Bureau of Mines (1992), and from estimates of energy requirements per ton of copper in Brown and McKern (1987) and in Gelb and Pliskin (1979); figure includes some energy used in fabrication as well as crude production, but does not include that used in mining and ore concentration; minerals industry share of world energy use is a Worldwatch estimate based on information in sources above.

10 The Surface Mining Control and Reclamation Act of 1977 established an Abandoned Mine Reclamation Fund, financed by a tax on coal production; Freedman (1987).

11 The Senate Environment and Public Works Committee voted to approve amendments to the right-to-know law on 30 April 1992; the proposal still awaits action on the Senate floor and in the House of Representatives; 'Key vote advances right-to-know-more', *Working Notes on Community Right-to-Know*, Working Group on Community Right-to-Know, Washington, DC, May 1992.

12 The Mining and Environment Research Network consists of researchers in more than 20 countries and is co-ordinated by Professor Alyson Warhurst of the Science Policy Research Unit, University of Sussex, Brighton, UK.

13 A ton of recycled aluminium takes about 15,700 kilowatt-hours less electricity to produce than a ton of primary aluminium; thus about 3.6 billion kilowatt-hours – about 1.3 per cent of annual US electricty use – could have been saved by recycling 2.3 million tons of aluminium; the population of Chicago is about 1.3 per cent of that of the United States.

References

AGRICOLA, G. (1950) *De Re Metallica*, New York, Dover Publications.

APPEN (1990) 'Sabah mining pollution – part one: villagers demand US $6 million compensation', *APPEN (Asia-Pacific Peoples Environment Network) Features*, Penang, Malaysia.

ASIAWEEK (1991) 'Who will clean up paradise?', *Asiaweek*, 4 January.

AYRES, R.U. (1989) 'Industrial metabolism', in Ausubel, J.H. and Sladovich, H.E. (eds) *Technology and*

Environment, Washington, DC, National Academy Press.

BERLIN, V.E. (1991) 'Conservation efforts in Kola Peninsula's Lapland Preserve detailed', *JPRS Reports*, 17 June.

BOMSEL, O. *ET AL.* (1990) *Mining and Metallurgy Investment in the Third World: the End of Large Projects?*, Paris, Organization for Economic Co-operation and Development (OECD).

BOSSON, R. and VARON, B. (1977) *The Mining Industry and the Developing Countries*, New York, Oxford University Press.

BROWN, M. and McKERN, B. (1987) *Aluminium, Copper and Steel in Developing Countries*, Paris, Organization for Economic Co-operation and Development (OECD).

CLEARY, D. (1990) *Anatomy of the Amazon Gold Rush*, Iowa City, IA, University of Iowa Press.

CLEMENTINE (1990) 'Heavy rains burst South Carolina dam: major cyanide spill', *Clementine*, Mineral Policy Center, Washington, DC, Winter.

CRAWFORD, L. (1992) 'Chile's giant copper mine to boost production', *Financial Times*, 8 February.

CUMMINGS, B.J. (1990) *Dam the Rivers, Damn the People*, London, Earthscan.

DORE, E. (1991) 'Open wounds', NACLA *Report on the Americas*, September.

DOWN, C.G. and STOCKS, J. (1977) *Environmental Impact of Mining*, New York, John Wiley and Sons.

DRUCKER, P.F. (1986) 'The changed world economy', *Foreign Affairs*, Spring.

EHRLICH, P.R. *ET AL.* (1977) *Ecoscience: Population, Resources, Environment*, San Fransisco, CA, W.H. Freeman.

ENVIRONMENTAL PROTECTION AGENCY (EPA) and MONTANA DEPARTMENT OF HEALTH AND ENVIRONMENTAL SCIENCES (MDHES) (1988) *Clark Fork Superfund Master Plan*, Helena, MT.

ENVIRONMENTAL PROTECTION AGENCY, OFFICE OF EXTERNAL AFFAIRS, REGION VIII (1987) *Mining Wastes in the West: Risks and Remedies – Overview*, Denver, CO.

FEARNSIDE, P.M. (1989) 'The charcoal of Carajás: a threat to the forests of Brazil's eastern Amazon region', *Ambio*, Vol. 18, No. 2.

FINANCIAL TIMES (1991) 'Old cars get a new lease on life', *Financial Times*, 3 September.

FREEDMAN, W. (1987) *Federal Statutes on Environment Protection: Regulation in the Public Interest*, New York, Quorum Books.

GAZETA MERCANTIL (1991) 'A month to assess Albrás damage', *Gazeta Mercantil*, 18 March.

GELB, B.A. and PLISKIN, J. (1979) *Energy Use in Mining: Patterns and Prospects*, Cambridge, MA, Ballinger Publishing.

GELLER, H. (1991) *Efficient Electricity Use: a Development Strategy for Brazil*, Washington, DC, American Council for an Energy-Efficient Economy.

GOODING, K. (1991) 'American Barrick's glittering run of luck continues', *Financial Times*, 4 October.

GOUDIE, A. (1990) *The Human Impact on the Natural Environment*, Cambridge, MA, MIT Press.

GRAHAM, R. (1982) *The Aluminium Industry and the Third World*, London, Zed Books.

GUROV, A. (1990) 'Gold rush in Kalimantan', *Asia and Africa Today*, No. 2.

HERMAN, R. *ET AL.* (1989) 'Dematerialization', in Ausubel, J.H. and Sladovich, H.E. (eds) *Technology and Environment*, Washington, DC, National Academy Press.

HINRICHSEN, D. (1990) *Our Common Seas*, London, Earthscan.

HOROWITZ, D.M. (1990) 'Mining and right-to-know', *Clementine*, Winter.

HOWARD, M.C. (1991) *Mining, Politics, and Development in the South Pacific*, Boulder, CO, Westview Press.

HYNDMAN, D. (1991) 'Digging the mines in Melanesia', *Cultural Survival Quarterly*, Vol. 15, No. 2.

INTERNATIONAL ENVIRONMENT REPORTER (1991) 'Volvo announces plans to recycle cars as part of environmental impact scheme', *International Environment Reporter*, 11 September.

JOHNSON, W. and PAONE, J. (1982) 'Land utilization and reclamation in the mining industry, 1930–80', Bureau of Mines Information Circular 8862, Washington, DC.

JORGENSON, E.P. (ED.) (1989) *The Poisoned Well: New Strategies for Groundwater Protection*, Washington, DC, Island Press.

JOURNAL OF COMMERCE (1992) 'Chile mining, environment leaders to talk', *Journal of Commerce*, 6 May.

KELLY, M. (1988) *Mining and the Freshwater Environment*, London, Elsevier Science Publishers.

LARSON, E.D. *ET AL.* (1987) 'Materials, affluence, and industrial energy use', *Annual Review of Energy, Vol. 12*, Palo Alto, CA.

MARSHALL, S. (1991) 'Green scrapyards', *Financial Times*, 23 March.

MILLER, K. (1991) 'On the road again and again: auto makers try to build recyclable car', *Wall Street Journal*, 30 April.

MÖLLER, D. (1984) 'Estimation of the global man-made sulphur emission', *Atmospheric Environment*, Vol. 18, No. 1.

MOORE, J.N. and LUOMA, S.N. (1991) 'Large-scale environmental impacts: mining's hazardous waste', *Clementine*, Washington, DC, Mineral Policy Center, Spring.

MULTINATIONAL ENVIRONMENTAL OUTLOOK (1991) 'Peugeot developing facility to recycle junk automobiles', *Multinational Environmental Outlook*, 5 March.

MUMFORD, L. (1963) *Technics and Civilization*, New York, Harcourt Brace Jovanovich.

NEW YORK TIMES (1991) 'Daimler has 10% of recycler', *New York Times*, 29 March.

NIELSEN, P. and FARLING, B. (1991) 'Hazardous wastes endanger water, wildlife, land: mining catastrophe in Clark Fork', *Clementine*, Washington, DC, Mineral Policy Center, Autumn.

NRIAGU, J.A. *et al.* (1992) 'Mercury pollution in Brazil', *Nature*, 2 April.

OTA (US CONGRESS, OFFICE OF TECHNOLOGY ASSESSMENT) (1979) *Technical Options for Conservation of Metals: Case Studies of Selected Metals and Products*, Washington, DC, Government Printing Office (GPO).

OTA (US CONGRESS, OFFICE OF TECHNOLOGY ASSESSMENT) (1990) *Nonferrous Metals: Industry Structure – Background Paper*, Washington, DC, Government Printing Office (GPO).

OTA (US CONGRESS, OFFICE OF TECHNOLOGY ASSESSMENT) (1992) *Managing Industrial Solid Wastes From Manufacturing, Mining, Oil and Gas Production, and Utility Combustion*, Washington, DC.

PALACIOS, J.D. (1989) 'Environmental destruction in southern Peru', *Earth Island Journal*, Summer.

PECK, M.J. (ED.) (1988) *The World Aluminium Industry in a Changing Energy Era*, Washington, DC, Resources for the Future.

ROGICH, D.G. (1992) 'Trends in material use: implications for sustainable development', unpublished paper, US Bureau of Mines, Division of Mineral Commodities, April.

SCOTT, D.C. (1989) 'Rebels keep Papua New Guinea mine closed', *Christian Science Montior*, 7 July.

SHEEAN, O. (1992) 'Fool's gold in Ecuador', *World Wildlife Fund News*, January/February.

SIURU, B. (1991) 'Car recycling in Germany', *Resource Recycling*, February.

SKINNER, B.J. (1989) 'Resources in the 21st century: can supplies meet needs?', *Episodes*, December.

SMITH, D.A. (1987) *Mining America: The Industry and the Environment, 1800–1980*, Lawrence, KS, University Press of Kansas.

SOUTH (1991) 'A mine of controversy', *South*, June/July.

UNITED NATIONS, ECONOMIC COMMISSION FOR LATIN AMERICA AND THE CARIBBEAN (1991) *Sustainable Development: Changing Production Patterns,*

Social Equity and the Environment, Santiago, Chile.

US BUREAU OF MINES (1990) *1989 Minerals Yearbook*, Washington, DC, Government Printing Office (GPO).

US BUREAU OF MINES (1991) *The New Materials Society, Volume 3: Materials Shifts in the New Society*, Washington, DC.

US BUREAU OF MINES (1992) *Mineral Commodity Summaries 1992*, Washington, DC.

US DEPARTMENT OF ENERGY (DOE) (1992) Energy Information Administration: *International Energy Annual 1990*, Washington, DC.

US LIBRARY OF CONGRESS (1978) Congressional Research Service: *Are We Running Out? A Perspective on Resource Scarcity*, Washington, DC, Government Printing Office (GPO).

WARHURST, A. (1991) 'Environmental degradation from mining and mineral processing in developing countries: corporate responses and national policies', draft discussion document for meeting of the Mining and Environment Research Network, Steyning, UK, 10–13 April.

WARHURST, A. (1992) 'Environmental management in mining and mineral processing in developing countries', *Natural Resources Forum*, February.

WESTLAKE, M. and STAINER, R. (1989) 'Rising gold fever', *South*, March.

WESTON, M. (NAURU GOVERNMENT OFFICE) (1991) Letter to the Editor, *Economist*, 23 February.

YOUNG, J.E. (1992a) 'Aluminium's real tab', *World Watch*, March/April.

Source: Young, 1992, pp. 5–7, 16–26, 35–50, 52–53

Uneven development and sustainability

Chapter 6

by Philip Sarre

6.1 Introduction

Under the overall question 'is the world overcrowded?' this book set out to focus on the narrower question of the sustainability of current lifestyles. This was further focused on the implications for sustainability of rapid population growth and of growing consumption of resources.

The purpose of this chapter is to summarize the answers to the questions posed, on the basis of the arguments of earlier chapters. Like many important academic and political questions, these are not susceptible to straightforward 'yes' or 'no' answers. Assessments of overcrowding and of sustainability depend on values and on expectations about the way that related factors will change in the future. As a result, there are different emphases and interpretations. The approach to be taken here is first to sum up as directly as possible the answers to the questions given in the chapters and then to analyse some of the key issues which the chapters identify, explicitly or implicitly, as influencing those answers. The main emphasis will be on two issues which were identified in the book introduction as of particular interest to geographers – uneven development and globalization. Throughout the chapter it is necessary to keep in mind that the questions we are discussing ultimately raise the whole issue of the relationship of society to nature – not only what this relationship is now and how it will change, but how it ought to change. This is a very major issue which, in the same way in which it engages the professional attention of geographers and the passionate commitment of environmentalists, ought to be a concern of everyone.

Activity 1 Make brief notes of your own answers to the questions 'is the world overcrowded?' and 'are current lifestyles sustainable?', drawing on Chapters 1 to 5. What role does uneven development have in relation to problems of overcrowding and sustainability? How do you think relations between society and environment should be changed to overcome present and probable problems?

6.2 The story so far

6.2.1 Wilderness

The discussion of the issue of wilderness suggests that there is a clear answer to the question of overcrowding: the growth of competition for space and resources in all parts of the world – except the poles – shows that we are in some ways pressing against natural limits. At one level, this is just the conclusion of an outward spread of industrializing societies from Europe and then across the Americas, Africa and Australasia. However, this overcrowding is not just a straightforward issue of pioneers against the wilderness: what seems to be occurring is that the globalization of the economy is now seeking resources in even the most remote locations, and repeatedly finding that those remote locations have indigenous human inhabitants. Because those inhabitants are hunter-gatherers, they may leave little obvious trace of their presence in the landscape and lack any legal title, but they do have the distinction of having carried human settlement into almost the whole world. This previous mode of production was global in the sense of being practised almost everywhere, though it was not globally connected. The pre-existence of these indigenous cultures questions the argument of romantics, scientists and environmentalists in more developed societies that wilderness should be preserved, since there has been no untouched wilderness for millennia. This, in effect, strengthens the dominant tendency: to insist that society should push back the frontier and exploit natural resources, even if this involves dispossessing indigenous groups.

This process of dispossession throws a particular light on the question of the sustainability of current lifestyles: it shows that the answer depends on whose lifestyles we are discussing. Although hunter-gatherers have changed environments significantly, the fact that they have been resident in many areas for tens of thousands of years shows that their lifestyles have been sustainable in the ecological sense. Loss of land, habitat change and competition for fish and game have, however, now made their situation unsustainable in many cases. Even where active extermination has not been practised, the outcome for indigenous peoples has usually been a decline in their population numbers as a result of contact with new diseases, alcohol and drugs, plus confinement to reservations which have been too small to cater for their needs. For many, incorporation into the new society has meant a new identity as cheap casual labour or welfare recipient. In recent decades, particularly in Australia, Canada and the USA, some restoration of land rights has been occurring, most notably in the Alaska Native Claims Settlement Act, but even where land rights have been conceded, native groups find themselves under strong pressure to allow resource exploitation by outside interests in return for short-term cash benefits. They have been incorporated into an unevenly developed world as the least prepared, if not the poorest, of the poor. In some parts of the world, they find themselves dealing directly with mining or timber companies, but in Latin America and South East Asia they often find themselves under pressure from the poorest groups in the wider society, such as displaced peasants and mineral prospectors.

One counter strand to the general rule of dispossession and disdain is that native peoples have been involved in international debates about environmental issues, notably at the 1992 Rio Earth Summit. The ways of life of some have been taken up by environmentalist groups as examples of sustainable management, whereas others have organized themselves and used new technologies to establish contact with other groups. The Inuit of Greenland have become part of the independent government, and North American groups have begun to be very assertive both extra-legally (notably when the Mohawks took up arms against construction of a golf course) and legally (the Cree campaigning against Quebec seceding from Canada, for instance). There may be tensions in these new roles – environmentalists, for example, may advocate indigenous peoples acting as managers of wilderness areas when some, at least, wish to abandon their traditional culture and 'develop'. Traditional cultures wishing to hunt seals or whales may find themselves under attack by environmentalists and animal rights advocates. Nevertheless, indigenous hunter-gatherers are already part of a global debate about the relations between environment and society. If, as Plumwood asserts (Plumwood, 1993), European society is built on a Platonic 'master narrative' of the subjugation of nature, non-European cultures and women, then the voices of hunter-gatherers, as well as those of old Asian agricultural societies, may be an essential part of the globalization of environmental responsibility.

The armed resistance of the Mohawks against development was a dramatic demonstration of the conflict between native peoples' attitudes to land and capitalist commodification of resources

As regards wilderness itself, the philosophical rethinking of society–nature relations is highly relevant. If the dualism is abandoned and the fact that humans are part of nature is recognized, the existence of low-impact groups need not prevent us from recognizing the existence of 'wild nature', if not the wilderness conceived of as untouched by humans. Given the tendency of visitors to perceive wilderness even in managed areas like British or American national parks, there is room for many people to have 'wilderness experiences'. Similarly, biodiversity can be maintained by saving relatively small areas as reserves. Perhaps the crucial question is whether there is a strong enough demand from voters for contact with wild nature and for the preservation of biodiversity. These are issues beyond sustainability: it is possible to conceive of a well-managed but totally humanized world which would be ecologically and socially sustainable. All it would lack would be a knowledge of the existence of wild places. This would be a new era because every previous human society has been conscious of wild nature beyond its boundaries. In this sense, the imminent extension of the humanized world to cover all the continents except Antarctica ushers in a new challenge to society. However, in the face of immediate pressures like population growth and resource exploitation, it is a challenge which can all too easily go unnoticed until it is too late.

6.2.2 Population in Africa

Chapter 2 challenged one of the assumptions of media debates about population: that Africa is overpopulated. Although the image of overpopulation is apparently supported by the rapid rate of population growth, the prevalence of famine in the Sahel and environmental problems like deforestation and soil erosion, these do not amount to a conclusive demonstration of overcrowding. Other factors challenge the image. First, the density of Africa's population is lower than that of the other continents. Second, there is little evidence that population pressure is a prime cause of environmental damage: damage occurs in areas of sparse population; it often has natural causes, notably drought; some cases exist where relatively dense and growing populations achieve better environmental management; the low standards of living of many Africans mean that they make modest overall demands on resources. As a result of these complex interactions, different analysts adopt very different interpretations, ranging from optimistic to pessimistic. The one thing that is very clear is that there are no simple generalizations which apply everywhere in Africa: relations between population and environment are influenced by uneven development and unequal interdependence.

Uneven development is apparent in the strong contrasts between different parts of the continent. In part, these result from environmental differences – climate ranging from hot desert to tropical forest, with strong variations in altitude, notably in the highlands of Kenya and Tanzania. African societies are also variable, influenced by tribal divisions as well as external interests, but in relation to population a key factor is the degree of dominance of lineage systems as forms of social organization. Urban influences especially can bring incentives to intensify production and hence environmental management, as shown by the examples of Kano and the Machakos District near Nairobi.

These internal variations have been exacerbated by Africa's experience of globalization. Nineteenth-century colonization brought substantial numbers of Europeans to Africa as farmers. They concentrated in more productive areas, especially in the East African highlands, and built up export-oriented agriculture. In areas less agreeable to European settlers, merchants assembled cash crops such as cocoa and dates and encouraged Africans to move away from subsistence agriculture. Plantations were established by outside interests. Even after political independence, European interests continued to shape African agriculture – for example, through French promotions of groundnuts in semi-arid areas of West Africa such as Niger. The intensification of production left farmers and pastoralists more vulnerable to drought, with devastating consequences in the 1970s.

Rapid decolonization also left a legacy of political instability which erupted as civil wars in a number of African countries and attracted outside powers to support particular factions and prolong the fighting, to the detriment of agriculture, economic security and environmental protection. Most recently, imposition of 'structural adjustment packages' on many African countries by the International Monetary Fund has required cuts in health and education spending in an attempt to foster enterprise, even though debt repayments were transferring large amounts of capital from Africa to developed countries. As a result, in the 1980s Africa suffered the Malthusian fate of population growing faster than the economy, and hence declining per capita income.

Even though the root cause of environmental degradation is difficult to disentangle from the complex set of influences at work, one clear message does emerge in relation to population. Since Africans have large numbers of children – first to carry out essential domestic work and ultimately to provide for security in their old age – the best incentive for family planning is the improvement of living standards and security. This would entail some form of *globalization of welfare*, a form of globalization scarcely contemplated in the developed world. The alternative may well be increased migration from poor countries to rich ones – another form of globalization not appreciated by rich countries.

globalization of welfare

Finally, to sum up the implications of the chapter for sustainability, two key points need to be made. First, if sustainability is conventionally defined as 'to meet the needs of the present without compromising the ability of future generations to meet their own needs' (World Commission on Environment and Development, 1987, p. 8), Africa is an extreme case of non-sustainable lifestyles. Current needs are not being met and the trajectory is towards worse prospects for the next generation. Second, it must be emphasized that this is largely an anthropocentric definition of sustainability, based on social equity and not in terms of ecological sustainability. In ecological terms the jury is still out: some argue that Africa is being desertified or at least degraded irretrievably, while other scholars argue that the continent has a long history of rapid change followed by rapid recovery. It may well be that the forms of environmental damage experienced in Africa are inherently more reversible than the kind of damage experienced in industrialized countries, where toxic and radioactive materials threaten ecosystems and health over many generations and rapid use of energy and raw materials suggests diminished availability of both in the future.

6.2.3 Population in Europe

The analysis of Europe's population in Chapter 3 confirms the complexity of the factors influencing population change, and presents very different issues in relation to overcrowding and sustainability. Many of the differences result from Europe's very different role in the system of global interdependence over the last few centuries.

Europeans do seem to feel overcrowded, in two very different but linked ways. First, high population densities and levels of urbanization lead many people, especially in countries like England and the Netherlands, to feel that they live in a crowded and humanized world with relatively difficult access to natural areas. Many have the resources to overcome this through recreational visits and holidays, but in so doing they displace the problem of crowding to recreational areas and tourist destinations. The second feeling of crowding is more political and underpins attitudes to immigration, especially

The 'export' of 'surplus' people to the USA and to the other continents was both a key factor in stabilizing European population growth and a contributor to the globalization of the economy

immigration from the less developed world. It is commonplace to justify controls on immigration in terms of overcrowding, and especially of housing shortages and limited availability of jobs. Fear of immigration seems to combine an insecurity about Europe's economic prospects in an increasingly competitive world with apprehension about the fertility of foreigners. Both are exacerbated by racist attitudes which date back to the age of European domination of the world.

The chapter shows that Europe's current population regime reflects a history of economic growth through industrialization, with population pressure reduced by emigration associated with the colonial period. As in the case of Africa, variation between countries and regions shows that cultural variations have a significant effect. France has long reminded demographers that factors such as forms of land inheritance and the education of women can influence fertility to some extent independently of industrialization and economic growth. France's stable population in the 1920s and 1930s anticipated the most unheralded 'problem' of European population today – the rapid collapse of fertility in southern Europe to well below replacement levels. This suggests the possibility of labour shortages in the medium-term future and a more extreme form of the pension problems expected everywhere as the labour force shrinks and the retired population grows. The extremely low levels of fertility in Italy and Spain seem to result from late and rapid industrialization beginning to falter in the face of Asian competition and leaving workers with hard choices about jobs versus family in societies with low levels of welfare support but high standards of family life. More generous welfare systems in northern Europe seem to have encouraged some recovery of fertility towards replacement levels – certainly a stronger trend than that observed where pro-natalist policies have focused on a narrow range of incentives. A broad emphasis on citizenship and welfare seems more effective in promoting replacement level fertility than specific incentives within an insecure society.

One of the most obvious features of European attitudes to population and development is that there has been a shift from a commitment to free trade and free movement of people in the pre-1914 period to an extremely defensive attitude to trade and immigration from outside the European Union today. This clearly reflects both apprehension about growing export penetration by Asian industry and about the numbers of potential immigrants from less developed countries, especially from Eastern Europe and North Africa. This apprehension is strongest in the countries with a long history of immigration, and less so in Italy and Spain, which have only recently become countries of immigration. However, those are precisely the countries where population decline is a cause for concern. Some European countries have a good record of settling refugees, which is a small step towards *globalization of citizenship* rights. However, growing numbers of refugees have provoked stricter rules, so denial of citizenship rights to immigrants seems likely to remain the general rule.

globalization of citizenship

Given these anxieties about economic competitiveness, it is paradoxical that Europe's population regime is actually unsustainable because of its high level of resource use and its dependence on imported materials, from high protein animal feed to petroleum and metals. Undoubtedly, the long history of industrialization has left Europe with a legacy of contaminated land, polluted water and acid rain. Indeed, the scale of pollution has increased from local to regional and, in the case of air pollution, to global. What is not clear is whether resource exhaustion will join air pollution as a truly global problem.

6.2.4 Sustainable resources?

The argument of Chapter 4 is relatively optimistic about the prospects for resource sustainability in developed areas, including Europe. Although resource use has mushroomed since 1945, the rate of growth has been slower in the first industrialized countries. Their high incomes and high levels of technological development put them in a good position to apply the 'technological fixes' discussed in the chapter. Increased efficiency in resource use and exploration, recycling and substitution are all likely to be implemented quickly in Europe, especially as most EU countries are well advanced in environmental consciousness and regulation. Even in the case of water, a crucial and non-substitutable resource, Europe is advantaged by relatively high rainfall and cool climates. The problems seem most likely to concern pollution of waterways which flow between countries, and exploitation of international waterways for hydro power.

Unfortunately, the readings for the chapter strongly suggest that less developed countries will face much more severe problems in relation to resources. Lack of capital seems likely to force these countries to rely on technologies which are cheap but not economically or environmentally efficient. So fossil fuels are likely to supply a large part of their future increase in energy consumption, and increased pollution is very likely. Similarly, many of the less developed countries are in semi-arid areas where conflicts over water supply are likely to grow as populations increase and development is pursued. In short, uneven development will limit the implementation of existing technological fixes.

6.2.5 Mining: global organization, local impact

Uneven development, in the form of unequal interdependence, is absolutely central to the argument of Chapter 5. Extraction, processing and trade in energy minerals and metals is one of the industries in which concentration of ownership is most pronounced, with almost all significant corporations headquartered in the developed world.

The location decisions of the major corporations are conditioned by geology: resources can only be extracted where they exist. This is most apparent in the case of petroleum, where all major known reservoirs are being exploited, even where they occur in countries which are politically hostile to the developed countries. However, the balance of economic power between supply and demand varies in relation to the demand for oil. In the 1970s, when demand was high, OPEC (the Organization of Petroleum Exporting Countries) was able to dominate. In the 1980s and 1990s, demand was lower and new suppliers outside the Arab heartland of OPEC were reluctant to subordinate their own interests to those of OPEC on aggregate. As a result, the oil companies were able to regain the power to keep raw material prices down and to direct exploration and contracts to areas where governments are more co-operative. If predictions of future exhaustion of oil reserves prove accurate, the balance of power may shift back towards countries with reserves, but this prospect may well encourage substitution of alternatives, such as bio-fuels, in the first world. Security of energy supply is a powerful incentive for countries and companies to override short-term economic considerations – if they are wealthy enough to afford more expensive substitutes.

In the case of metals, where the potential for substitution of materials and locations is greater, the multinationals have had somewhat different priorities. In the 1960s, after decolonization had made many less developed countries less stable or more radical, exploration and supply were concentrated in North America, South Africa and Australia – countries where the massive investments necessary for a modern mine were seen as safe. In more recent years, as those countries began to regulate environmental impacts more effectively, exploration and extraction have shifted back towards the less developed world, where shorter time periods for decisions and less effective regulation make mining operations more economically attractive. The result is that the catalogue of local impacts identified by Young in Reading A are increasingly occurring in poor countries, and doing so on a progressively larger scale as world demand grows. The fact that some countries are willing to accept environmental damage in return for jobs and export earnings reflects their disadvantaged position in other forms of economic activity. The past record of falling mineral prices suggests that this is an economic strategy with poor prospects – unless the pessimists are correct in forecasting a future upturn in prices in the face of shortages of key minerals.

The growth in scale of local impacts of mining and mineral processing, including its contributions to acid rain and energy consumption, suggests that it is at least locally unsustainable. Arguably, the growing extent and severity of impacts make it globally unsustainable. Like economic development, the impacts are uneven over space, but this is no less a form of globalization, because the interconnections that generate the activities causing the impacts are themselves global.

One odd aspect of the mineral production system is that it is not even economically optimal: the use of virgin materials has been both directly subsidized by some governments through tax exemptions and indirectly subsidized by exemption from paying the costs of pollution. Withdrawal of these subsidies would make the use of recycled materials more cost-effective and stimulate improved technologies to overcome problems of reclaiming materials from complex products. The move to more realistic price and tax structures would seem to be beneficial to the environment and to the quality of life of most people. It would require a move from a mass production mentality to a more flexible system with more small plants using reclaimed materials – in other words, a reduction in the concentration of power in the global system of mineral supply.

Summary of section 6.2

- The near elimination of wilderness by human use of the whole earth requires a reappraisal of the relations between society and environment.
- Generalizations about population, technology, lifestyle and environment are made difficult by the unevenness of development.
- Overcrowding is not an absolute state but is relative to expectations, lifestyle and technology, as well as to environment.
- Resources are finite but abundant (and many can be substituted as technology improves): the problem at present is not exhaustion but wasteful and damaging use.
- Solving increasingly global environmental problems will require globalization of human welfare, security and citizenship, as well as globalization of environmental responsibility.

6.3 Sustainable development?

6.3.1 Uneven development and sustainability

It is clear from earlier chapters that debates about population and resources, about overcrowding and sustainability, are intrinsically difficult because they involve complex explanations of past and present events, predictions of future change and value judgements about what is desirable. As a result, there is a wide range of interpretations, some optimistic and some pessimistic. The discussions in the chapters narrow the range of plausible views but do not eliminate uncertainty. They also make clear that uneven development is both a key cause of complexity and a serious obstacle to establishment of a global consensus about policy.

At the time of writing, a very similar message is beginning to emerge from a major international project organized from Clark University in the USA (Kasperson and Kasperson, 1994). This project involved 60 academics in nine teams dealing with nine areas suffering from a range of environmental problems. The areas were: the North Sea, the Aral Sea, Mexico City, Amazonia, the High Plains of the USA, the Kenya dry hills, Nepal, East Sundaland, and the loess plateau in China.

The aims of the project were to document change over 50 years and to assess the roles of population growth, technological change, affluence, poverty, trade and politics in bringing about threats to sustainability. The project recognized a range of departures from sustainability, from impoverishment through endangerment to criticality. Criticality was defined as precluding current use systems and/or precluding human well being: in other words, it was an anthropocentric and not an ecocentric definition.

Two broad sets of findings are relevant to our concern with uneven development. First, the project teams found different arrays of driving forces in the different regions. In all cases they faced a problem in recognizing the effects of natural variability interacting with human factors. Population growth was seen to be an important factor in the less developed regions. Technology proved to be multi-faceted: in some ways worsening problems, in other ways helping to adapt to change or improving the situation. Affluence was only felt to be an important cause of environmental problems in the North Sea and Mexico City, while poverty was judged to be a minor cause, significant in only three cases. The dominant cause in most cases was held to be political economy – the interaction between economic factors and state policy. Unevenness of development does not just mean the interaction of complex mixtures of causes which happen to occur in, or affect, particular areas: those causes also generate distinctive political regimes which apply very different policies in different places. Differences between places help to generate different policy responses which often generate further differences.

The second set of findings concerned the way in which different places are linked together. Every region in the study showed increasing dependency on distant world markets. In many of them a shift towards capitalism was judged to accelerate damage because new forms of resource allocation were adopted, with loss of traditional safety nets for people and environment. Many areas showed increasing differences in vulnerability between social

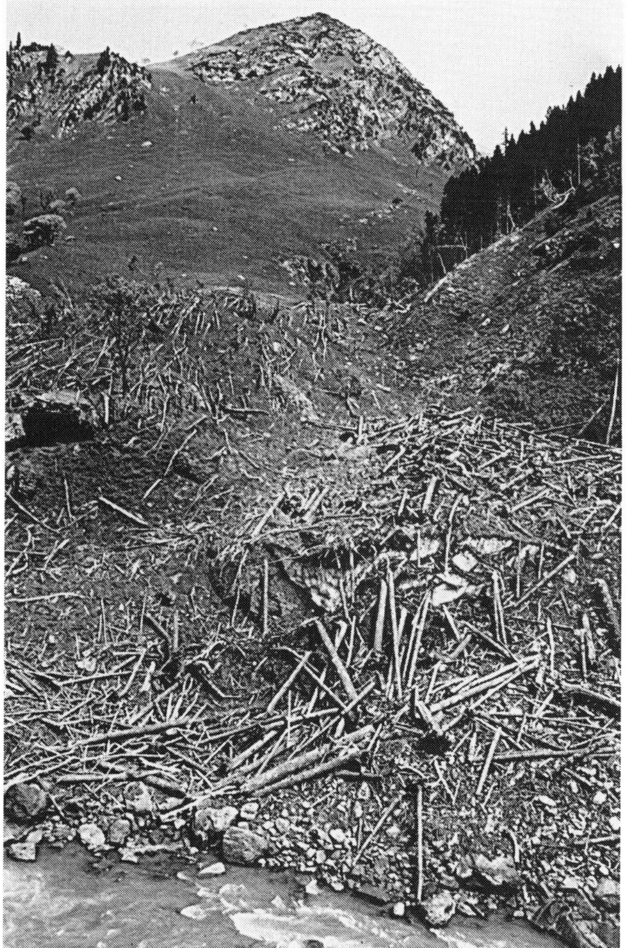

Although there are disagreements about the overall extent of deforestation in the Himalayas, clear felling of steep slopes causes soil erosion which both reduces local fertility and clogs streams and rivers, reducing water quality and increasing the severity of flooding

groups within the region, though some were less polarized than others. Distant political control could allow local corruption and increase polarization.

The study has two sets of implications, one explanatory and the other relevant to policy. First, the authors argued that aggregate claims about trends or causes are highly suspect until related to events at regional and local scales. Second, the important role of local political economy means that there is scope for the reduction of environmental problems and of social polarization at local and national scales. However, the pervasive effects of the globally integrated economy mean that the problems cannot be overcome without an international effort to regulate the economy in the interests of sustainability. Overall, the study reinforces the arguments of this book: uneven development is a key part of the explanation of environmental and social problems. Solving those problems requires local and international action. Such action requires new forms of globalization.

Activity 2 Think about the environmental problems of an area (or areas) known to you. What roles do technology, population growth, affluence or poverty play as causes of those problems? Do you agree that local or national policy affects the outcomes significantly, or is the international economy the dominant influence?

6.3.2 Sustainable development

At the time of writing, the most sustained effort to promote global change to solve environmental problems has come from the United Nations Commission on Environment and Development, which generated the Brundtland Report in 1987 and the Rio Earth Summit in June 1992.

The Brundtland Report was the culmination of a three-year process of consultation, analysis and negotiation (World Commission on Environment and Development, 1987). It tackled two massive problems – development and environment – and claimed to offer solutions to both. It demands attention because it has been so influential, but since its publication the problems have worsened and the response is inadequate, so the attention must be critical, seeking the limitations as well as the achievements.

A central part of the Report's argument concerns the problems generated by the 'breakdown of compartments' between nations and between sectors, and the creation of an interlocking global crisis of population, environment, development and energy. Rather than being daunted by the complexity of the crisis, the Commission, as noted earlier, proposed an integrated response, a new form of development, to be known as 'sustainable development', which would *'meet the needs of the present without compromising the ability of future generations to meet their own needs'* (p. 8, emphasis added).

The focus on needs has the strength that it puts the interests of the less developed world high in its priorities. This is clearly positive in principle, though it could be seen as a weakness in practical terms, since this issue had been emphasized a decade earlier by the Brandt Report (Brandt Commission, 1980) but little had been done to change the economic and political processes which lead to uneven development. However, there is also a serious weakness in principle: the needs being addressed are human needs and, in spite of the 'one earth' in the title of the extract from the Brundtland Report included in the Introduction to this book, and the recognition of 'spaceship earth', nature is regarded only as a resource base to be used wisely. The stress on greater equity makes the Report humanist in the best sense, but the neglect of nature's interests makes it anthropocentric in the derogatory sense.

The main body of the Report consists of six chapters dealing with specific topics, followed by three dealing with key areas of policy. The six topics were population and human resources, food security, species and ecosystem conservation, energy, industry, and urbanization. The analysis suggested that the six topics were interlinked and that all posed issues in the policy areas of managing the commons, peace and security and institutional change. The argument of the Report makes a provocative comparison with that of this book, reaching similar conclusions about population, ecosystems conservation and resources, and adding food security and urbanization to the problems considered here.

The Report's concluding chapter is a call for massive institutional and legal change to promote sustainable development policies on all scales from local to global. The problem here is motivation: such changes could be made if all governments accepted the need, but how can the governments and citizens of the first world be persuaded to prioritize the present needs of the less developed world and the needs of future generations over the established goals of higher consumption in the immediate future? The nearest approach offered by the Commission is in the chapter on 'Peace and Security', in which conflicts over migration and resources are predicted to rise if sustainable development is not achieved. However, the potential problems for the first world are not stressed as much as the potential danger from nuclear war or the opportunity cost of global spending on arms and military forces. Having pointed out that 'environmental stress is both a cause and an effect of political tension and military conflict' (p. 290), the large majority of such conflicts cited (including a short paragraph on water) are in the less developed world, leaving the impression that the developed countries need not be involved. Perhaps most puzzling is that the chapter on 'managing the commons' covers the oceans, Antarctica and space, but omits the most obvious common resource – the atmosphere.

It may be that the Brundtland Commission was unfortunate in that most of its work was done in the mid-1980s, before ozone depletion and global warming had emerged as major issues. However, both issues are mentioned sporadically in the Report without ever coming centre stage, so it is more likely that the pervasive anthropocentrism of the Report marginalized these geographical problems. Paradoxically, in subsequent years public concerns about 'the ozone hole' and the effects of global warming were much more extreme than concern about the current devastation of lives and environments by poverty. Indeed, in the late 1980s, action against ozone depletion was more rapid and determined than action over any other environmental problem before or since. When the Brundtland process reached its climax at the Earth Summit in Rio in 1992, most of its concerns were expressed only as desirable directions for policy in Agenda 21. Two issues had generated sufficient support to be put forward as conventions for signature – the Climate Convention tackling emission of greenhouse gases, and the Biodiversity Convention, focusing on conserving genetic resources. Problems which involve the destabilization of environmental systems seem to stimulate greater concern than human problems even when their consequences for humans are uncertain. Most environmentalists are very critical of what was achieved at Rio, but more was done for certain global environmental issues than for development.

Today, the Brundtland Report and the concept of sustainable development are seen in profoundly ambiguous ways. On the one hand, they have stimulated the most significant political response ever achieved by environmental issues. Most world governments have adopted the concepts and are involved in ongoing policy debates. On the other hand, environmentalists criticize the anthropocentrism and the degree of vagueness which makes it possible to espouse the concept without any significant change to priorities or practices. Most politicians assume that the degree of change advocated by the Report is far too radical to be contemplated, let alone implemented. Environmentalists are beginning to argue that not only is sustainability virtually impossible to define in practical terms, but that it is also, in principle, the *minimum* environmental standard anyone could possibly advocate. To opt explicitly for non-sustainable treatment of resources and environments is to suggest handing on a poisoned chalice to future generations. No-one would advocate this as desirable, though this is precisely what global society is doing in practice.

In these circumstances there seem to be two options for environmentalists: put more energy and ingenuity into promoting sustainable development or find new ways of reconciling environmental goals with those of citizens in all parts of the world.

Summary of section 6.3

- Different local social and environmental outcomes are influenced by local politics as well as by global economics and politics.
- The concept of sustainable development has had positive but limited effects on national and international policy.
- The ideal of sustainability is minimal and has not proved an effective counter to the desire for material consumption.
- Adequate political response seems to require more global social and environmental views to counter the power of the globalized economy.

Activity 3 Find out what has been done by your national government to promote sustainable development since this book went to press. How does its performance compare with that of other developed countries?

6.4 Think global

Earlier chapters of this book have identified a range of different ways in which humans relate to nature. These can be summarized under five headings:

Membership

Humans are a part of nature, evolved from apes and dependent on nature for air, water and food. They face dangers from carnivores, disease organisms and hazards such as volcanoes and floods.

Reproduction

As part of nature, humans depend on biological mechanisms for reproduction of the species, and their reproductive behaviour has major impacts on society and on nature. Reproduction of society involves gender relations, groups and institutions and issues of security and access to resources.

Transformation

Humans, individually and especially organizationally, have transformed nature, through labour and technology. They have produced new kinds of environment (called by some 'second nature') and products for use and/or exchange. This transformation is nearly global and ranges from positive (gardens) to negative (toxic dumps). However, the transformation falls short of total dominance, and wildness and natural hazard persist.

Consumption

Through most of human history, people have had direct contact with nature every day. In an industrialized and urbanized society there remains a strong tendency to consume nature in the form of national parks and holiday resorts as well as through books, television and films. The demand is for the sublime as well as the pleasant, and, for some, retains an element of testing oneself against nature.

Conception

Chapter 1 showed that all societies have explicit or implicit conceptions of nature which may describe how it works, explain how it came to its present form and imply how it may be used. These conceptions may include supernatural, social and scientific elements and are often expressed through myth or metaphor.

The main thrust of the argument of this book, the Brundtland Report and most environmentalist writing is that these five kinds of relationship are out of balance. Society's technical ability to transform nature on a global scale is much greater than our ability to decide what transformations are desirable or to allocate the benefits and detriments equitably. The result is that rapid change both threatens the environmental quality and the welfare of many of the planet's inhabitants and poses potential threats to long-term survival. Hence the strong doubts that current practices are sustainable into the future. A crucial factor in promoting the imbalance is uneven development: disparities between areas, and especially between most and least developed areas, play a crucial role in generating the problems and local responses which add up to a global crisis. They also make it harder to recognize and accept the need for radical changes in the interests of global society and its whole environment. The post-Brundtland process and negotiations over global warming (Greene, 1989) confirm that national and local differences of interest are the major obstacle to building a global consensus over what should be done.

This problem is widely recognized and there are many proposals to resolve it. These range from calls for renewed efforts through the United Nations system, more collaboration between environmentalist non-governmental organizations worldwide, a reactivation of democratic socialism (Pepper, 1993), new forms of eco-anarchism (Bookchin, 1990; Eckersley, 1992) to espousal of an environmental dictatorship. Ultimately, political and economic mechanisms will have to be put in place, but they cannot be implemented without a major shift in the priorities of governments and citizens. That seems to depend on a shift to a way of conceptualizing the environment which is truly global and which stresses common interests and a re-evaluation of environmental and material benefits. The need for such a unifying conceptualization was tacitly recognized by the Brundtland Commission when they started their Report by referring to the view of the earth from space. However, it was not followed through, as was noted earlier.

Gaia hypothesis

The main exponent of a truly, and literally, *global representation* of the environment is James Lovelock (1979), through his *Gaia hypothesis.* Lovelock is an unconventional scientist who was inspired to formulate the idea of Gaia while working for NASA (the US National Aeronautics and Space Administration) as a contributor to the design of a lander to seek for traces of life on Mars. He realized that the obvious indicator of life on earth is the composition of the atmosphere: a large proportion of a highly reactive gas like oxygen is dependent on living things to offset the chemical processes which would naturally capture the oxygen. From this insight, he has developed not only a holistic view of geology and biology (much of which is acceptable to other scientists), but also the suggestion that the whole system is a self-regulating being, named Gaia after the Greek earth-goddess. Lovelock denies any supernatural or teleological intent and insists that he regards Gaia as a testable hypothesis, but his view has been more attractive to New Age environmentalists than to most scientists. However, as more work is done on global warming and other problems of global linkages, more scientists are beginning to accept the existence of natural feedback processes which act to regulate the state of the atmosphere and biosphere. Arguably, it is a most significant development in the geographical imagination as it integrates all environmental processes and situates human society.

Lovelock's account of the development of Gaia brings together geology, evolution and the atmosphere. It explains how living things transformed the early atmosphere, which consisted mostly of carbon dioxide, to the current nitrogen/oxygen mix. Oxygen was first released by blue-green bacteria until levels were high enough to allow green plants, at first in water and later on land, to survive and start fixing the carbon. Over billions of years fossil fuels and limestones locked up almost all of the atmospheric carbon, a process recently reversed by the burning of fossil fuels. The evolution of living things progressively changed the atmosphere, and with it the temperature of the earth's surface, in ways which allowed more complex life forms and ecosystems to evolve. Lovelock shows that this process effectively insulated living things on the surface from cosmic rays and fluctuations of solar radiation through the structure and properties of the atmosphere. A similar argument relates the earth, Mars and Venus to the Goldilocks story: from rather similar planetary beginnings, the earth has ended up 'just right' for life, Mars too cold and Venus too hot. It could be that it is just tautology: we would not be here to marvel at the improbability of the earth's evolution if we had not been part of it. However, for many people Gaia offers a creation story which both excites a sense of wonder and seems scientifically credible.

By itself, it does not offer detailed policy prescriptions, but it does supply a powerful imperative. Lovelock's hypothesis is planet centred and not human centred, and, rather than suggesting that the human species is the culmination of evolution, it tends to suggest that we may be an aberration. From a planetary point of view we are seen more as a runaway disease, growing rapidly at the expense of other organisms and disrupting essential natural cycles. Fortunately for the planet, Lovelock recognizes great resilience in the face of such disruptive events (including recovery from the huge comet impact which is now seen as the likely cause of the extinction of the dinosaurs). Unfortunately for the human species, he suggests that if we are too disruptive this resilience will allow the planet to eliminate us. Overcrowding will lead to catastrophic effects on humanity, though Gaia will survive.

The Gaia hypothesis does not represent the environment as fragile, as do many environmentalists, but as an extremely powerful self-regulating system. As an example, Lovelock is far less impressed by nuclear hazards than most environmentalists – pointing out that earlier in history the earth had much larger amounts of naturally radioactive materials and a much higher incidence of cosmic radiation. He is confident that the earth can survive a high level of nuclear contamination – but notes that human beings are much less resilient. The hazards he takes most seriously are the 'three Cs': cattle, cars and chain saws, because they seem locally harmless, but add up to a massive assault on biodiversity, resources and atmosphere at a global scale.

The Gaia hypothesis dramatically contradicts the Modernist idea of science as analytical, atomist, reductionist and mechanical. It builds on earlier thinkers like James Hutton (who argued that processes such as mountain building and erosion were necessary to produce the soils on which plants and animals depend) and Charles Darwin (who explained how natural selection could have guided evolution towards the complex organisms and ecosystems of the natural world) to portray a world which is highly integrated, changing over time, and adaptable to changing circumstances. It is a world which should excite admiration and even awe when we learn not to take it for granted. One possible disadvantage is that cynics might conclude that if Gaia is so resilient we need not worry about sustainability. However, the point is that we are dependent for survival on the water and carbon cycles and could disrupt them sufficiently to wipe out our own agricultural systems, and hence ourselves, without doing irreparable damage to the planet. Indeed, one of the paradoxical lessons of the Gaia hypothesis is that reversibility acts on a longer timescale for the planet than for the human species. Changes may be irreversible within a human timescale but reversible over millions of years. We need to be more careful about irreversible changes from our own point of view, and from those of other contemporary species, than from Gaia's point of view. However, to reduce biodiversity and/or ecosystem variety dramatically may reduce Gaia's resilience against major shocks. As I write this months after a major comet hit Jupiter, the reduction of the earth's ability to recover from a similar impact does not seem wise.

globalization of environmental responsibility

More positive is the view that, whereas the Brundtland Report extended the principle of stewardship only to future generations of humans, Gaia makes the human species (as 'nature grown self-conscious') stewards for the whole process of planetary resilience and species evolution. To spread human domination to the whole of the earth's surface and to exercise it only for the sake of our consumption would be extremely short-sighted. After all, *Homo sapiens* has lived for 100,000 years on an earth 4.6 billion years old, and industrial society has existed for only a few hundred years. We should be thinking of handing the earth in good order to our descendants, not in the shape of our children but in terms of our evolutionary successors. To do that, we need to give a high priority to the maintenance of biodiversity, not frozen in gene banks or zoos, but living free in varied ecosystems. We need also to avoid pushing atmospheric changes so far that they cause a rapid shift to a new climatic system in which many species are threatened.

Activity 4 What are the implications of the Gaia hypothesis as a global representation of environment for policy in relation to population and resources? How could more environmentally responsible policies be implemented?

Thinking globally and ecocentrically leads to recognition of a need for diversity of habitats and ecosystems. This seems to allow for many ecosystems with strong human influence, but to require some which are returned to the wild. This would suggest lower populations and resource use than are required by anthropocentric analyses. Such a world would seem unlikely to emerge from the current processes of economic or cultural globalization, except possibly as a result of Balkanization. It also seems unlikely that it could be achieved by central planning or international treaties. It requires an enhanced local contribution to decision making over development issues.

Strong local inputs are evident in many parts of the world, but have come to have a negative reputation, summed up in the acronym NIMBY (not-in-my-backyard) and the implication of dog-in-the-manger behaviour. More positive views exist: Bookchin (1990) has long advocated a form of social ecology where decisions are made by independent municipalities in which direct citizen participation replaces hierarchical government. Eckersley (1992) has questioned local autonomy on the grounds that it may give a platform to extremist political views and support xenophobic attitudes. A possible middle way might be to work towards a local veto against proposals with negative environmental impacts. Proponents of 'locally unwanted land uses' (LULUS) would have to assure local residents of precautions being taken and perhaps offer compensation if damage could not be prevented. Such a local veto would be a start towards making LULUS globally unwanted. They may seem a far cry from current reality, but are partly anticipated by the planning system, which gives local authorities some powers to resist proposals. They are also anticipated by a long history of direct action by citizens and institutions of particular places.

Rechem plant, New Inn, near Pontypool in South Wales. Although the incineration of PCBs is officially accepted as a desirable way of destroying otherwise long-lasting pollutants, the fears of local residents make this a 'locally unwanted land use' or LULU

Summary of section 6.4

- Lovelock's Gaia hypothesis offers a global and evolutionary view which puts human society into its natural context.
- It displaces humans from the summit of creation and demands respect for natural systems and for other species.

6.5 Conclusion

Fears of overcrowding and non-sustainability rest ultimately on the notion that there are natural limits to human activities. Many of the pessimistic 'population bomb' and 'resource exhaustion' publications suggest that those limits will be met abruptly and in the near future. The analyses of wilderness, population and resources in this book challenge those simplistic assumptions. Human impacts on wilderness, overpopulation and resource constraints all appear to be more complex issues where limits are relative to expectations, lifestyles, technology and development. As a result, they can be represented in very different ways by people with different interests and used to support very different policy proposals. The conclusion of this book is that the natural resource base is abundant but finite, so that the demands we make upon it must ultimately be restrained in order that the future human population can enjoy quality lifestyles.

The concept of sustainability also remains somewhat elusive: it certainly involves an element of equity between areas and groups as well as a concern with the conservation of the natural resource base. Some critics argue that it uses environmental fears to bolster a social democratic agenda, but the argument of this book is that, because uneven development is both a cause of environmental damage and an obstacle to more global responses, a reduction of disparities in standards of living is a necessary part of a needed global change in lifestyles away from material consumption and towards quality of environment. Such a move implicitly requires global rather than fragmented national bases for security, welfare and even citizenship. It is a move which seems unlikely while political domination, economic exploitation and demographic fears remain prominent. Perhaps only shared global risk of environmental disaster offers the hope of persuading the human species to act together. Attempts to generate this 'one world' commitment are discussed by **Yearley (1995)***.

The argument of this book is that the elimination of wilderness, and the closure of the 'global frontier', ushers in a new era in which humans cannot rely on untouched nature to provide a safety net if we damage nature's capacity to support us. In the new era, we have to accept long-term responsibility for the maintenance of natural systems so that we can provide the basis for our survival. In doing so, we have no choice but to achieve sustainability or go extinct. Unfortunately, sustainability might be achieved with monotonous and degraded environments and a high degree of social constraint to achieve the levels of efficiency needed to supply large populations. More ecocentric environmentalists, including those inspired by

* A reference in emboldened type denotes a chapter in another volume of the series.

Gaia, demand voluntary limits on population and material consumption to allow for lifestyles which include access to varied environments including large areas where nature can be left unmanaged.

Rather than speculate on what such an uncrowded world might be like, the best conclusion to the book is to state a series of priorities. First, an obvious step is to stop causing irreversible damage to natural systems. Second, policies should be geared towards achieving sustainability, in the sense of bringing population, lifestyle and environment into a sustainable balance. Achievement of those two goals would give time and resources for debates about the kind of world – natural and social – that humans want to inhabit in the future.

To achieve these goals would be to terminate two millennia of dominance by Plato's 'master narrative'. This would entail replacing domination of nature, women and non-Europeans by male 'rationality' and power by the emancipation of those interests. It would require replacing the glorification of death in the service of the state by a celebration of life – human and natural – and an admiration for diversity and creativity over uniformity and obedience. This would require globalizing our geographical imagination to recognize the mutual interdependence of society and nature, not as abstractions, but as they interact in an increasingly uneven and interdependent, as well as overcrowded, world.

References

BOOKCHIN, M. (1990) *The Philosophy of Social Ecology: Essays on Dialectical Naturalism*, Montreal, Black Box.

BRANDT COMMISSION (1980) *North–South: a Programme for Survival*, London, Pan Books.

ECKERSLEY, R. (1992) *Environmentalism and Political Theory: Toward an Ecocentric Approach*, London, UCL Press.

GREENE, O. (1989) 'Tackling global warming', in Smith, P.M. and Warr, K. (eds) *Global Environmental Issues*, London, Hodder and Stoughton in association with The Open University.

KASPERSON, D. and KASPERSON, J. (1994) 'Environmentally threatened regions: a global perspective', paper presented at the International Geographical Union Conference, Prague, August.

LOVELOCK, J.E. (1979) *Gaia: a New Look at Life on Earth*, Oxford, Oxford University Press.

PEPPER, D. (1993) *Ecosocialism: From Deep Ecology to Social Justice*, Andover, Routledge, Chapman and Hall.

PLUMWOOD, V. (1993) *Feminism and the Mastery of Nature*, London, Routledge.

WORLD COMMISSION ON ENVIRONMENT AND DEVELOPMENT (1987) *Our Common Future* (The Brundtland Report), Oxford and New York, Oxford University Press.

YEARLEY, S. (1995) 'The transnational politics of the environment', in Anderson, J., Brook, C. and Cochrane, A. (eds) *A Global World? Reordering Political Space*, Oxford, Oxford University Press in association with The Open University.

Acknowledgements

We have made every attempt to obtain permission to reproduce material in this book. Copyright holders of material which has not been acknowledged should contact the Rights Department at The Open University.

Grateful acknowledgement is made to the following sources for permission to reproduce material in this volume:

Text

Introduction: Hamilton, A. (1994) 'Population is not the agenda', *The Observer,* 11 September 1994; ***Chapter 1:*** *Reading A:* Nuttall, M. (1993) *Environmental Policy and Indigenous Values in the Arctic: Inuit Conservation Strategies* (Paper presented to Values and the Environment Conference, University of Surrey, 23 and 24 September 1993). Copyright © 1993 Mark Nuttall; *Reading B:* Merchant, C. (1990) 'Gender and environmental history', in *The Journal of American History,* Vol. 76, No. 4, March 1990, Organization of American Historians; *Reading C:* Eckersley, R. (1992) *Environmentalism and Political Theory: Toward an Ecocentric Approach,* UCL Press Limited, © 1992 State University of New York; ***Chapter 2:*** Tierney, J. (1990) 'Betting the planet', *The Guardian,* 28 December 1990; *Reading A:* from *The Population Bomb* by Dr Paul R. Erlich. Copyright © 1968, 1971 by Paul R. Erlich. Reprinted by permission of Ballantine Books, a division of Random House Inc.; *Reading B:* Caldwell, J.C. (1982) *Theory of Fertility Decline,* Academic Press; *Reading C:* Mortimore, M. (1993) 'Population growth and land degradation', *GeoJournal,* Vol. 31, No. 1, pp. 15–21. © 1993 (Sep) by Kluwer Academic Publishers. Reprinted by permission of Kluwer Academic Publishers; *Reading D:* Hardin, G. (1968) 'The tragedy of the commons', *Science,* Vol. 162, © American Association for the Advancement of Science; *Reading E:* Tiffen, M. and Mortimore, M. (1992) 'Environment, population growth and productivity in Kenya: a case study of Machakos district', *Development Policy Review,* Vol. 10, No. 4, pp. 376–82, by permission of Sage Publications Ltd; ***Chapter 3:*** Nundy, J. (1994) 'Pasqua in new war on immigrants', *The Independent,* 7 January 1994; *Reading A:* Livi-Bacci, M. (1992) in Ipsen, C. (trans.) *A Concise History of World Population,* Basil Blackwell Ltd; *Reading B:* Flinn, M.W. (1981) *The European Demographic System 1500–1820,* © 1981 by The Johns Hopkins University Press; *Reading D:* Humphries, S. and Gordon, P. (1993) *A Labour of Love: The Experience of Parenthood in Britain 1900–1950,* Sidgwick and Jackson, © Steve Humphries and Pamela Gordon 1993; *Reading E:* Hartmann, B. and Boyce, J.K. (1983) *A Quiet Violence: View From A Bangladesh Village,* Zed Books Ltd, London. Copyright © Betsy Hartmann and James K. Boyce 1983; ***Chapter 4:*** *Reading A:* Duncan, N. (1993) 'Norman Duncan sees a declining future for renewables', *Oxford Energy Forum* August 1993, Oxford Institute of Energy Studies; *Reading B:* Bulajich, B. (1992) 'Women and water', *Waterlines,* Vol. 11, No. 2, October 1992, Intermediate Technology Publications; ***Chapter 5:*** *Reading A:* Young, J. (1992) *Worldwatch Paper 109: Mining the Earth,* Worldwatch Institute.

Figures

Figures 1.1 and 1.2: Kummer, D.M. (1991) *Deforestation in the Postwar Philippines,* University of Chicago Press, © 1991 by The University of Chicago; *Figure 1.3:* adapted from Dalby, R. (1994) *The Alaska Guide,* Fulcrum Publishing. Copyright © 1994 Ron Dalby; *Figure 2.1:* reprinted from *Beyond the Limits.* Copyright © 1992 by Meadows, Meadows and Randers. With permission from Earthscan Publications Ltd and Chelsea Green Publishing Co., White River Junction, Vermont; *Figure 2.2:* Demeny, P. (1984) 'A perspective on long-term population growth', *Population and Development Review,* Vol. 10, No. 1, March 1984, The Population Council; *Figure 3.1:* Hall, R. (1989) *Update World Population Trends,* Cambridge University Press; *Figure 3.2:* Wrigley, E.A. and Schofield, R.S. (1981) *The Population History of England 1541–1871: A Reconstruction,* © E.A. Wrigley and R.S. Schofield 1981. Reproduced by permission of Edward Arnold (Publishers) Ltd; *Figures 3.3, 3.10 and 3.11:* World Bank (1992) *World Development Report 1992,* Development and the Environment, © 1992 by the International Bank for Reconstruction and Development / The World Bank; *Figure 3.4:* World Bank (1984) *World Development Report 1984,* © 1984 by the International Bank for Reconstruction and Development / The World Bank; *Figure 3.5:* Livi-Bacci, M. (1993) in Ipsen, C. (trans.) *A Concise History of World Population,* Basil Blackwell Ltd; *Figure 3.6:* Coale, Ansley J. and Watkins, Susan Cotts, *The Decline of Fertility in Europe.* Copyright © 1986 by Princeton University

Press. Reprinted by permission of Princeton University Press; *Figure 3.8:* OPCS (1993) *General Household Survey 1991*, Series GHS No 22, © Crown Copyright. Reproduced with the permission of the Controller of Her Majesty's Stationery Office; *Figure 3.9:* Hall, R. (1993) 'Europe's changing population', *Geography*, No. 338, Vol. 78 Part 1, January 1993, © Geographical Association; *Figure 4.1:* Blunden, J. (1985) *Mineral Resources and their Management*, Butterworth-Heinemann Ltd, © John Blunden; *Figure 4.2:* reprinted with permission from *Global Change in the Geosphere-Biosphere.* Copyright 1986 by the National Academy of Sciences. Courtesy of the National Academy Press, Washington, DC; *Figure 4.3:* Northumbrian Water Authority, by kind permission; *Figures 4.4 and 4.5: Adelphi Paper 273* (1992) International Institute for Strategic Studies.

Tables

Table 1.1: Oelschlaeger, M. (1991) *The Idea of Wilderness: From Prehistory to the Age of Ecology*, Yale University Press. Copyright © 1991 by Yale University; *Table 3.1:* Sundström, M. and Stafford, F.P. (1992) 'Female labour force participation, fertility and public policy in Sweden', *European Journal of Population*, 8, pp. 199–215, © 1992 – Elsevier Science Publishers BV. All rights reserved; *Table 3.2:* reprinted from *State of the World 1994*, A Worldwatch Institute Report on Progress Toward A Sustainable Society, Project Director: Lestor R. Brown. By permission of W.W. Norton & Company, Inc. Copyright © 1994 by Worldwatch Institute. Also by permission of Earthscan Publications Ltd; *Table 4.1:* Shea, C.P. (1988) 'Renewable energy: today's contribution, tomorrow's promise', *Worldwatch Paper 81*, Worldwatch Institute; *Table 4.3: ALTENER Targets for 2005.* Copyright © The European Community; *Table 4.4: Energy Policy*, September 1992, by permission of the publishers, Butterworth-Heinemann Ltd, ©; *Table 4.6:* Rogers, P. (1993) 'The value of cooperation in resolving international river basin disputes', *Natural Resources Forum*, Vol. 17, No. 2, May 1993, by permission of the publishers, Butterworth-Heinemann Ltd, ©; *p. 207:* Hoagland, P. (1993) 'Manganese nodule price trends: dim prospects for the commercialization of deep seabed mining', *Resources Policy*, Vol. 19, No. 4, December 1993, by permission of the publishers, Butterworth-Heinemann Ltd. ©; *Tables 5.1, A.1, A.2 and A.3:* Young, J. (1992) *Worldwatch Paper 109: Mining the Earth*, Worldwatch Institute.

Photographs

p. 12: reproduced from Wells, E. (1819) *An Historical Geography of the Old and New Testament*, Oxford, Clarendon Press, Plate 1, Volume 1; *p. 13:* The Minneapolis Institute of Arts. Bequest of Mrs Kate Dunwoody. Oil on canvas, 23.75 x 31.5 in.; *p. 14:* publisher: Currier & Ives, 1868. The Museum of the City of New York, 56.300.107. The Harry T. Peters Collection; *p. 16:* reproduced from the original in The Huntington Library, San Marino, California; *p. 61:* both photographs: Chris Steele-Perkins / Magnum; *p. 83:* Kenya view 1937: photo by R.O. Barnes, 1937. From the Kenyan National Archives / courtesy of Dr Michael Mortimore, African Drylands Research. Reproduced by permission of the Overseas Development Institute; *p. 83:* Kenya view 1991: courtesy of Dr Michael Mortimore, African Drylands Research. Reproduced by permission of the Overseas Development Institute; *p. 127:* Nineteenth-century family: Hulton-Deutsch Collection; parents and two children: Brenda Prince / Format; *p. 130:* photo: Lars-Kristian Crone / Copyright Royal Danish Ministry for Foreign Affairs; *p. 134:* Hulton-Deutsch Collection; *p. 140:* J. Allan Cash; *p. 172:* Mike Levers / The Open University; *p. 177:* BT Pictures / A BT photograph; *p. 181:* Jorgen Schytte / Still Pictures; *p. 189:* Nigel Cattlin / Holt Studios International; *p. 201:* Gilles Saussier / Frank Spooner / Gamma; *p. 236:* Environmental Picture Library, © Morgan; *p. 252:* Associated Press; *p. 255:* by courtesy of the Ellis Island Immigration Museum; *p. 260:* Environmental Picture Library; *p. 267:* Environmental Picture Library, © Hoffman.

Colour plate section

Plate 1: NASA / Science Photo Library; *Plate 2:* Galleria Doria Pamphilj [Inventory FC 341] / Foto Alessandro Vasari, Rome; *Plate 3:* The Royal Collection. © Her Majesty Queen Elizabeth II; *Plate 4:* Galleria Doria Pamphilj [Inventory FC 241] / Foto Alessandro Vasari, Rome; *Plate 5:* Mark Edwards / Still Pictures; *Plate 6:* Mark Edwards / Still Pictures; *Plate 7:* Museum of the City of New York, 33.169.1. Gift of Mrs Robert M. Littlejohn; *Plate 8:* Mark Edwards / Still Pictures; *Plate 9:* Daniel Dancer / Still Pictures; *Plate 10:* NASA / Science Photo Library; *Plate 11:* J. Allan Cash.

Index